Principles of Mathematical Modeling

Numerical Insights

The Numerical Insights series aims to show how numerical simulations provide valuable insights into the mechanisms and processes involved in a wide range of disciplines. Such simulations provide a way of assessing theories by comparing simulations with observations. These models are also powerful tools which serve to indicate where both theory and experiment can be improved.

In most cases the books will be accompanied by software on disk demonstrating working examples of the simulations described in the text.

The editors will welcome proposals using modelling, simulation and systems analysis techniques in the following disciplines: physical sciences; engineering; environment; ecology; biosciences; economics.

Volume 1
Numerical Insights into Dynamic Systems: Interactive Dynamic System Simulation with Microsoft® Windows 95 ™ and NT ™
Granino A. Korn

Volume 2
Modelling, Simulation and Control of Non-Linear Dynamical Systems: An Intelligent Approach using Soft Computing and Fractal Theory
Patricia Melin and Oscar Castillo

Volume 3
Principles of Mathematical Modeling: Ideas, Methods, Examples
A.A. Samarskii and A.P. Mikhailov

This book is part of a series. The publisher will accept continuation orders which may be cancelled at any time and which provide for automatic billing and shipping of each title in the series upon publication. Please write for details.

Principles of Mathematical Modeling

Ideas, Methods, Examples

A.A. Samarskii
Founder of the Institute of Mathematical Modeling, Moscow, Russia

and

A.P. Mikhailov
Head of Department, Institute of Mathematical Modeling, Moscow, Russia

TAYLOR & FRANCIS
Founded 1798

London and New York

First published 2002 by Taylor & Francis
11 New Fetter Lane, London EC4P 4EE

Taylor & Francis is an imprint of the Taylor & Francis Group

This book has been produced from camera-ready copy supplied by the authors
Printed and bound in Great Britain by TJ International Ltd, Padstow, Cornwall

Originally published in Russian in 1997 as MATEMATICHESKOE MODELIROVANIE: IDEI. METOID. PRIMERI by Physical and Mathematical Literature Publishing Company, Russian Academy of Sciences, Moscow.

British Library Cataloguing in Publication Data
A catalogue record for this book is available from the British Library

ISBN 0-415-27280-7

Contents

Dedicated to the bright memory of Academician
Andrei Nikolaevich Tikhonov

INTRODUCTION

It is impossible to imagine modern science without the wide application of mathematical modeling. The essence of this methodology is the replacement of an initial object by its "image" – the mathematical model - and the further study of model with the help of computing-logical algorithms. This "third method" of research, construction and design, combines many advantages both theoretical, and experimental. Working not with the object itself, but with its model enables one to investigate cheaply, easily and quickly its properties and behavior in any conceivable situation (this is the advantage of the theory). At the same time, and thanks to the power of modern computing methods, the numerical experiments with models of objects using computations, simulations and imitation allow detailed and deep study of objects which was not possible using pure theoretical approaches (this is the advantage of the experiment). It is not surprising that the methodology of mathematical modeling has developed intensely, covering new spheres – from the development of technological systems and their control up to the analysis of complex economic and social processes.

Elements of mathematical modeling were used from the very beginning of the development of fundamental science, and it is not by chance that certain methods of calculations are named after Newton and Euler, and the word "algorithm" originates from the name of medieval Arabic scholar Al-Khorezmi. The renaissance of this methodology occurred in the late 40s and early 50s, at least for two reasons. The first of them was the appearance of computers, which although modest from a modern point of view, nevertheless enabled the scientists to avoid a huge amount of routine computing work. The second was an unprecedented social problem – the Soviet and American national nuclear arms programs, which could not be realized by traditional methods. Mathematical modeling solved that problem: both nuclear explosions and flights of rockets and satellites were previously tested within the computers using mathematical models, and only later were realized in practice. This success in many respects determined the further development of the methodology, without which no large-scale technological, ecological or economic project is now considered seriously (this is also true in relation to some social and political projects).

Nowadays mathematical modeling is entering the third essentially important phase of its development, being built within the structures of the so-called *information society.* The impressive progress in means of processing, transferring and storing information corresponds to the global tendencies towards complication and overlap in various spheres of human activity. Without possessing information resources it is impossible even to think about a solution to the large and diverse problems posed to the world community. However, the information itself is often not enough for the analysis and forecast, for choice of solutions and control by their performance. Reliable ways of processing raw information into a ready "product" are necessary. The history of mathematical modeling methodology convinces: it can and should be the *intellectual core* of information technologies, of the whole process of creating the information society.

Technical, ecological, economic and other systems investigated by modern science cannot be studied adequately using regular theoretical methods. Direct experimentation in nature is time-consuming, expensive, often even dangerous, or simply impossible, since many systems are unique. The price of mistakes and wrong decisions is unacceptably high. Therefore *mathematical modeling is an inevitable component of scientific and technical progress.*

The very formulation of the problem of mathematical modeling an object leads to a precise plan of actions. It can be conditionally split into three stages: *model – algorithm – code* (see the diagram).

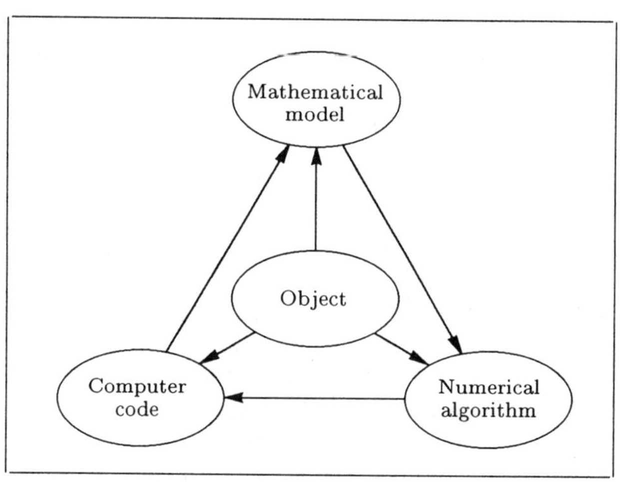

At the first stage, the "equivalent" of the object is chosen (or constructed), reflecting its major properties in a mathematical form – the laws, controlling it, connections peculiar to the components, and so on. The mathematical model (or its fragments) is investigated using theoretical methods, enabling one to obtain important preliminary knowledge about the object.

The second stage is the choice or development of the algorithm for the realization of a model on the computer. The model is represented in a form convenient for the application of numerical methods. The sequence of computing and logic operations are defined, enabling us to find out the sought quantities with required accuracy. The computing algorithms should not distort the basic properties of the model and, hence, of the initial object; they should be economical and convenient for the considered problems and the computers used.

At the third stage the codes are created, "translating" the model and algorithm into a language accessible to the computer. They also have to fulfill the economy and convenience criteria. One can call them the "electronic" equivalents of the investigated object, already suitable for direct tests on "the experimental facility" – the computer.

By creating a *triad* "model – algorithm – code", the researcher gets a universal, flexible and inexpensive tool, which first has to be debugged and tested in computer experiments. When the adequacy of the triad with respect to the initial object is confirmed, detailed and diverse tests are performed revealing the qualitative and quantitative properties and characteristics of the object. The process of modeling is accompanied by the improvement and specification, as far as is possible, of all parts of the triad.

Being a methodology, mathematical modeling is not the substitute for mathematics, physics, biology and other scientific disciplines; it does not compete with them. On the contrary, it is difficult to overestimate its synthesizing role. The creation and application of a triad is impossible without relying on the different methods and approaches – from qualitative analysis of nonlinear models to modern programming languages. It gives additional stimulus to quite different areas of science.

Considering the broader question, recall, that modeling is present in all kinds of human creative activity in the work of researchers and businessmen, politicians and military commanders. Introducing exact knowledge into these spheres helps to restrict intuitive, speculative modeling, and expands the field of application of rational methods. Certainly, mathematical modeling is good only at carrying out well-known professional requirements: the precise formulation of basic concepts and assumptions, a posteriori analysis of the adequacy of used models, guaranteeing the accuracy of computing algorithms and so on. Regarding the modeling of systems involving human factor, one has to add to these requirements the accurate differentiation of mathematical and everyday terms (sounding identical, but having different content), the cautious application of already available mathematical tools to the study of phenomena and processes (the preferable way is "from problem to method", and not vice versa), etc.

While solving the problems of the information society, it might be naive to rely only on the power of computers and other means of computer science. Permanent development of the triad of mathematical modeling and its introduction into the modern information-modeling systems is a *methodological imperative*. Only its performance enables us to obtain adequately high technological, competitive and diverse material and intellectual production.

Many good books are devoted to the various aspects of mathematical modeling. While working on the present book, the authors were aiming to select and describe the approaches to the construction and analysis of mathematical models common for various disciplines and not dependent on particulars. The world surrounding us is uniform, and researchers effectively use this gift of nature also via *the universality* of mathematical models. Certainly, the content of the book is connected in certain way with the authors' personal experience. Nevertheless, adhering the above formulated viewpoint, it was easier to expand the framework of the account and to demonstrate the broad possibilities of mathematical modeling – from mechanics to social sciences. Because of this the authors believe the present volume differs from the books of our colleagues.

As for the style of the book, we tried to avoid bulky and strict procedures (the interested reader can find them in more specialized editions), and paid more attention to the description of corresponding ideas and examples. Therefore the book contains plenty of illustrations and exercises; its simpler material can be used for educational programs on various directions of mathematical modeling.

The constraints of the book have not only allowed us to expand on a number of important topics; it might be necessary to consider in more detail the approaches to the construction of discrete models and numerical methods, questions of the equipment of models and their identification, of their synthesis and decomposition, etc. For the same reason, the bibliography includes only the required minimum – either the works directly reflected in the text, or the key books, where further references can be found.

The selection of material and the manner of its representation corresponds to the concepts of our school of mathematical modeling, which operates at a global level and widely treats the given methodology. Therefore, we cannot mention all our colleagues, who have influenced the content of the book. The authors consider it a pleasant duty to thank N.V.Zmitrenko (doctor of physico-mathematical sciences), who undertook the task of reading the manuscript and helping with a number of valuable remarks; B.N.Chetverushkin, V.F.Tishkin (the doctors of physico-mathematical sciences) and Academician A.A.Petrov for their help with material and productive discussions on the crucial problems of mathematical modeling. The authors are grateful also to V.Ya.Karpov (doctor of physico-mathematical sciences), A.E.Korolev (candidate of physico-mathematical sciences), V.A.Ovsyannikov (candidate of engineering sciences), V.I.Zelepnev (engineer), O.L.Busygin (economist) and to N.I.Colen'ka (financial director) for useful discussions of particular questions. We are grateful also to the staff members of the Institute of Mathematical Modeling of Russian Academy of Sciences N.G.Sirotenko and A.S.Boldarev, who helped prepare the draft of the book.

We hope that our book will be interesting and useful both for beginners and experienced researchers and teachers using the methods of mathematical modeling and computer experiments in their scientific work.

Moscow, June, 1997.

A.A.Samarskii, A.P.Mikhailov

Chapter I

THE ELEMENTARY MATHEMATICAL MODELS AND BASIC CONCEPTS OF MATHEMATICAL MODELING

1 Elementary Mathematical Models

Let us consider some approaches to the construction of elementary mathematical models, illustrating the application of the fundamental laws of nature, variational principles, analogies and hierarchical chains. Despite its simplicity, the material involved will enable us to start the discussion of such concepts as adequacy of models, their "equipment", nonlinearity, numerical realization and a number of other basic questions of mathematical modeling.

1. Fundamental laws of nature. The most widespread method of constructing of models is by applying the fundamental laws of nature to a particular situation. These laws are conventional, repeatedly confirmed by experience, and are the basis of scientific and technical achievements. Therefore their validity is doubtless, which besides everything, provides a powerful psychological support to the researcher. The main questions are:

which laws should be applied in any given case and how should they be applied.

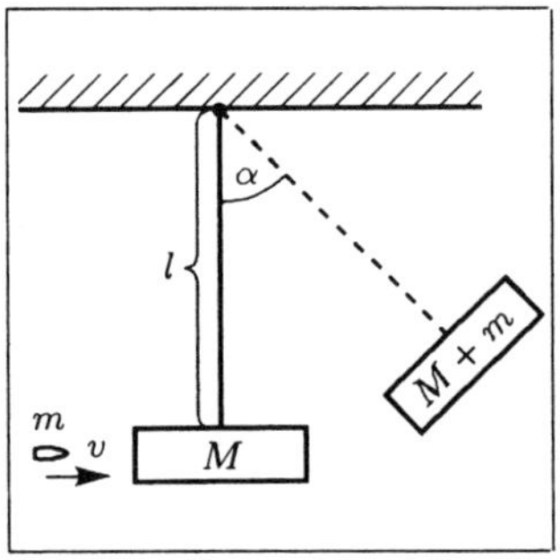

Fig. 1.

a) Conservation of energy. This law has been known for almost two hundred years and occupies perhaps the most honorable place among the great laws of nature. Based on it, an expert in ballistics, aiming to determine quickly the velocity of a revolving bullet without laboratory conditions, can take advantage of a rather simple device such as a pendulum – a load hung on a light, rigid and freely rotating rod (Fig. 1). The bullet embedded in a load, will pass to the system "bullet–load" a kinetic energy, which at the moment of the maximal shift of the rod from the vertical will completely transform into the potential energy of system. These transformations are described by a chain of equalities

$$\frac{mv^2}{2} = (M+m)\frac{V^2}{2} = (M+m)gl(1-\cos\alpha)\,.$$

Here $mv^2/2$ is the kinetic energy of a bullet of mass m and velocity v, M is the mass of a load, V is the velocity of the system "bullet–load" immediately after the collision, g is the free-fall acceleration, l is the length of the rod, α is the angle of the maximal shift. The required velocity is determined by the formula

$$v = \sqrt{\frac{2(M+m)gl(1-\cos\alpha)}{m}}\,, \tag{1}$$

which will be quite exact, if one neglects the losses of energy on the heating up of the bullet and load, the resistance of the air, the speeding up of the rod, etc. These at first sight reasonable assumptions are actually incorrect. The processes occurring at the "merging" of the bullet and the pendulum, are no longer purely mechanical. Therefore when calculating the value v, the law of conservation of mechanical energy is not valid: the total energy, rather than the mechanical one is conserved. It provides only the lower limit for the estimation of the velocity of the bullet (for the correct solution of this simple problem it is necessary to use also the law of momentum conservation – see exercise 1).

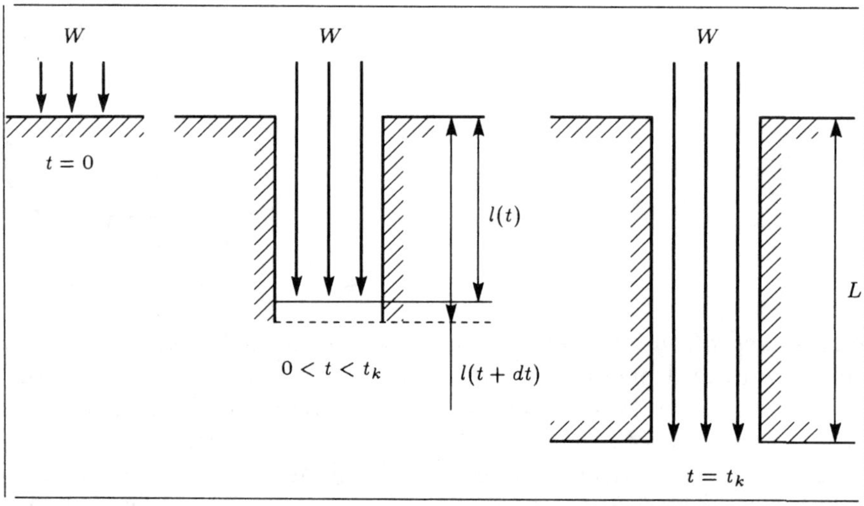

Fig.2. *The initial, intermediate and final stages of drilling metal using a laser.*

Similar considerations can be used by the engineer to estimate the time t_k taken to drill a layer of metal of thickness L using laser of a power W, with radiation perpendicular to the surface of the material (Fig. 2). If the energy of the laser completely spent on the evaporation of the metal of mass is $LS\rho$ (S is the irradiated area, LS is the volume of the metal, ρ is the density of the matter), the law of conservation of energy is expressed by equality

$$E_0 = Wt_k = hLS\rho\,, \tag{2}$$

where h is the energy required for the evaporation of a unit of mass. The size h has compound structure: $h = (T_{\text{mel}} - T)h_1 + h_2 + h_3$, since the matter has to be consistently heated to melting point T_{mel}, and then smelted and transformed into vapor (T is the initial temperature, h_1 is the specific heat, h_2 and h_3 are the specific heat of melting and evaporation, respectively).

The variation of the depth of the hole $l(t)$ in course of time is determined by the detailed balance of energy within the interval of time from t up to $t + dt$. To evaporate during this time a mass:

$$[l(t + dt) - l(t)]\,S\rho = dl\;Sp$$

the energy $dl\,hS\rho$ equal to the energy $W\,dt$ has to be passed to the matter by the laser:

$$dl\, hS\rho = W\, dt\,,$$

Whence the differential equation yields

$$\frac{dl}{dt} = \frac{W}{h\, S\rho}\,.$$

Its integration (in view of the fact that the initial depth of the hole is equal to zero) gives

$$l(t) = \frac{W}{hS\rho}t = \frac{E(t)}{hS\rho}\,, \tag{3}$$

where $E(t)$ is the total energy released by the laser up to the moment of time t. Hence, the depth of the hole is proportional to the spent energy (and value t_k, when $l(t_k) = L$, and coincides with the energy calculated by the formula (2)).

Actually, the process of drilling is much more complicated than the considered scheme: the energy is spent on heating the sample and removing vapors from the hole, which can have a complicated form, etc. Therefore the correctness of the considered mathematical description is much less certain than in the case of the bullet. The question of the correspondence of the object and its model is one of the central ones in mathematical modeling; we shall be dealing with this later.

b) Conservation of matter. This law is used, say, by a schoolboy while solving the problem of filling of a pool with water, inflowing and outflowing from two pipes. Certainly, the area of application of this law is incomparably wider.

Consider, for example, a small amount of radioactive substance (uranium) surrounded by a thick layer of a regular material (lead), – a situation typical either for the storage of decaying materials, or for their use as energy sources (Fig. 3). The word "small" implies the simplifying circumstance, namely that all products of decay not undergoing collisions with atoms of the substance, escape from the area I. In other words, the length of free path of the products of decay λ_{I} in the first substance is significantly larger than the characteristic sizes of the material L_{I}, i.e. $\lambda_{\text{I}} \gg L_{\text{I}}$. The words "a thick layer" implies that in accordance with the purpose of the storage the products of the decay are completely absorbed in area II. It is guaranteed by the fulfilling of the opposite condition $\lambda_{\text{II}} \ll L_{\text{II}}$, where λ_{II} is the free path of products of disintegration within the second substance, L_{II} is its characteristic size.

Thus, everything that escapes from area I is absorbed in the area II, and the total mass of both substances in course of time does not vary. This is the law of conservation of matter applied to the given situation. If at the

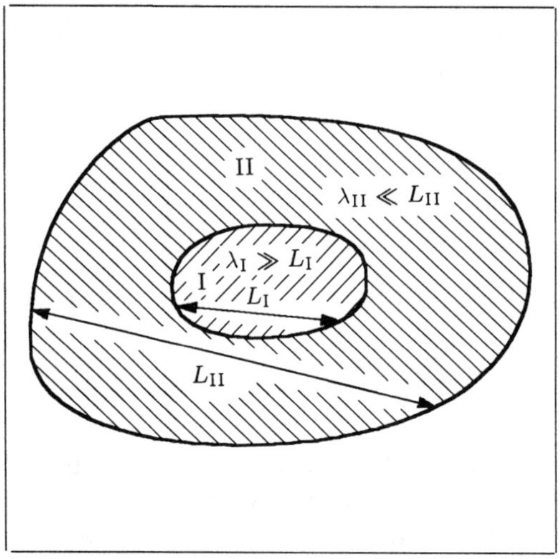

Fig.3.

initial moment in time $t = 0$ masses of the substances were equal $M_{\rm I}(0)$ and $M_{\rm II}(0)$, at any moment in time the following balance is valid

$$M_{\rm I}(0) + M_{\rm II}(0) = M_{\rm I}(t) + M_{\rm II}(t)\,. \tag{4}$$

The only equation (4), obviously, is not enough to determinate the current values of both masses – $M_{\rm I}(t)$ and $M_{\rm II}(t)$. For a complete mathematical formulation it is necessary to involve an additional consideration on the character of disintegration. Namely, that the rate of decay (the number of atoms decaying in a unit of time) is proportional to the total number of atoms in the radioactive substance. For a small time interval dt between the moments t and $t + dt$

$$N_{\rm I}(t+dt) - N_{\rm I}(t) = -\alpha N_{\rm I}(t+\xi dt), \qquad \alpha > 0, \qquad 0 < \xi < 1,$$

atoms will decay. Here, again the law of conservation of matter is used, but applied not to the whole process, but to an interval of time dt. In the last equation describing the balance of atoms, in the right hand side is with the sign minus (substance decreases), and $N_{\rm I}(t + \xi dt)$ describes the average number of atoms during the considered time. Rewrite it in a differential form:

$$\frac{dN_I(t)}{dt} = -\alpha N_I(t)\,.$$

Taking into account that $M_{\rm I}(t) = \mu_{\rm I} N_{\rm I}(t)$, where $\mu_{\rm I}$ is the nuclear weight of substance I, we have

$$\frac{dM_{\rm I}(t)}{dt} = -\alpha M_{\rm I}(t)\,. \tag{5}$$

At spontaneous radioactivity any atom, independent of the condition of the environmental substance, has a certain probability of disintegration. Hence the greater (smaller) the products of disintegration in a unit of time, the greater (smaller) is the amount of the radioactive substance. The coefficient of proportionality $\alpha > 0$ (constant decay) is determined by the particular substance.

The equations (4), (5) together with conditions $\lambda_{\rm I} \gg L_{\rm I}$, $\lambda_{\rm II} \ll L_{\rm II}$ and with the given values α, $M_{\rm I}(0)$, $M_{\rm II}(0)$ represent the mathematical model of the considered object.

Integrating (5), we obtain that the mass of the decayed material decreases by exponential law

$$M_{\rm I}(t) = m_{\rm I}(0)e^{-\alpha t} ,$$

and at $t \to \infty$ the matter in the area I disappears completely.

In so far as the total mass according to (4) remains constant, the amount of the matter in area II increases

$$M_{\rm II}(t) = M_{\rm II}(0) + M_{\rm I}(0) - M_{\rm I}(0)\, e^{-\alpha t} = M_{\rm II}(0) + M_{\rm I}(0)\, (1 - e^{-\alpha t}) ,$$

and at $t \to \infty$ the products of disintegration completely move from area I to II.

c) Conservation of momentum. A motionless boat in a lake in windless weather will begin to move forward if one steps from bow to the stern. This displays the law of conservation of momentum: the total a momentum of a system not undergoing actions of external forces is preserved. The boat reacts to the movement by displacement in the opposite direction.

The principle of jet propulsion is the basis of many remarkable technical devices, for example, a rocket lifting an artificial satellite ("sputnik") to an orbit around the Earth; this needs to develop a speed of approximately 8 km/s. The elementary mathematical model of the movement of a rocket follows from the law of conservation of momentum if one neglects the resistance of the air, the gravitation and other forces, excluding, certainly, the power of the jet engines.

Let the products of combustion of rocket fuel outflow from the rocket nozzle with a speed u (for modern fuels u is equal to 3–5 km/s). For a small interval of time dt between the moments t and $t + dt$, some part of fuel will be burnt out, and the mass of the rocket will be changed on dm. The momentum of the rocket will be changed also, however the total momentum of the system "the rocket plus products of combustion" will remain the same as in the moment t, i.e.

$$m(t)\, v(t) = m(t + dt)\, v(t + dt) - dm[v(t + \xi\, dt) - u] ,$$

where $v(t)$ is the speed of the rocket, $v(t+\xi dt) - u$, $(0 < \xi < 1)$ is the average speed of gas escaping from the muzzle within interval dt (both speeds are relative to the Earth). The first member on the right hand side of this equality is the momentum of the rocket in the moment $t + dt$, the second is the momentum transferred to the escaping gas during dt.

Taking into account, that $m(t + dt) = m(t) + (dm/dt)dt + O(dt^2)$, the law of conservation of momentum can be written as a differential equation

$$M\frac{dv}{dt} = -\frac{dm}{dt}u\,,$$

where $-(dm/dt)u$, obviously nothing other than the drag force of the rocket engines, and which, rewritten in the form

$$\frac{dv}{dt} = -u\,\frac{d\ln m)}{dt}\,,$$

is easily integrated:

$$v(t) = v_0 + u\,\ln\left(\frac{m_0}{m(t)}\right)\,,$$

where v_0, m_0 are respectively the speed and the mass of the rocket at the moment $t = 0$. If $v_0 = 0$, the maximal speed of a rocket achieved with complete use of the fuel, is equal

$$v = u\,\ln\left(\frac{m_0}{m_p + m_s}\right)\,. \tag{6}$$

Here m_p is the useful mass (mass of the satellite), m_s is the structural mass (the mass of actual rocket including the fuel tanks, engines, control systems, etc.).

Tziolkovsky's simple formula of (6) allows us to make a fundamental conclusion about the design of a rocket for space flights. Consider the size $\lambda = m_s/(m_0 - m_p)$, which characterizes at $m_p = 0$ the ratio of structural and initial masses of the rocket. Then for practical values of $\lambda = 0.1$, $u = 3$ km/s at $m_p = 0$ we have

$$v = u\ln(1/\lambda) = 7 \quad \text{km/s}\,.$$

Even in the ideal case where the useful mass is equal to zero, gravity and resistance of air are absent, etc. it follows that a rocket of the considered type is not capable of achieving the first space speed. Thus it is necessary to use multistage rockets – a conclusion drawn by the founders of cosmonautics.

The given example also illustrates a kind of a principle of favored conditions, frequently used in the mathematical modeling of complex objects: if an object in best conditions is not capable of achieving the required characteristics, it is necessary to change the approach to the object or to soften the requirements; if the requirements basically are achievable, the following steps are connected to the study of the influence of additional complicating factors.

2. Variational principles. One more approach to the construction of models, comparable by its breadth and universality with the opportunities provided by the fundamental laws, concerns the application of so-called *variational principles.* They represent rather general statements on the considered object (system, phenomenon), namely, among all possible variants of the behavior (movement, evolution), only those are chosen which satisfy a certain condition. Usually, according to this condition, certain quantities associated with the object achieve extreme values while in transition from one state to another.

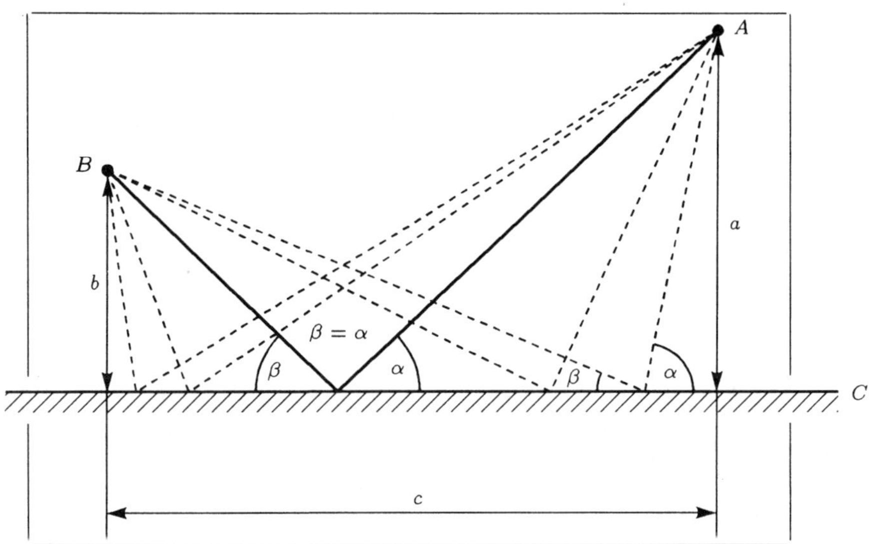

Fig.4. *Various trajectories of movement from a point A to a point B with a contact with the line C. The bold line indicates the fastest way.*

Assume that a car has to move with a constant speed v from point A to point B, touching the line C (Fig. 4). The driver of the car is in a hurry and among various trajectories is choosing the one with the least spent time. Let us represent the spent time as a function of α, of the angle between the line and the trajectory starting from A:

$$t(\alpha) = \frac{a}{v \sin \alpha} + \frac{b}{v \sin \beta(\alpha)} .$$

Here a and b are the lengths of the perpendiculars from the points A and B on the line, $\beta(\alpha)$ is the angle between the line and the trajectory ending on B.

The condition of extremity $t(\alpha)$ with respect to the argument α implies

$$\left.\frac{dt(\alpha)}{d\alpha}\right|_{\alpha=\alpha_{\text{ext}}} = 0\,,$$

or

$$\frac{a\cos\alpha}{\sin^2\alpha} + \frac{b\cos\beta(\alpha)}{\sin^2\beta(\alpha)}\frac{d\beta}{d\alpha} = 0\,. \qquad (7)$$

For any values of α the following equality is valid

$$c = \frac{a}{\tan\alpha} + \frac{b}{\tan\beta(\alpha)}\,,$$

where c is the distance between the projections of points A and B on the line (the same for all trajectories). Differentiating, we obtain the ratio

$$\frac{a}{\sin^2\alpha} + \frac{b}{\sin^2\beta(\alpha)}\frac{d\beta}{d\alpha} = 0\,, \qquad (8)$$

which together with a condition of the minimum (7) means

$$\cos\alpha = \cos\beta(\alpha)\,,$$

i.e. the equality of angles α and β (see exercise 5).

Now it is not difficult to find out the values of $\alpha_{\min}$, $t_{\min}$ through the given a, b, c. However of more importance for us is the condition of minimal spent time, which leads to the choice of the appropriate trajectory by the law "the fall angle is equal to the reflection angle". We see that this is just the law of the reflection of a light beam from a reflecting surface! Is it possible that in a common case the light beams move via trajectories ensuring the fastest transition of a signal from one point to another? Yes, this is exactly the case according to the renown variational principle of Fermat, leading to the basic laws of geometrical optics.

Let us show this via consideration of the refraction of beams on the boundary of two media (Fig. 5). The light moving from the point A via the first medium with a speed v_a refracts and, passing through the boundary, moves within the second medium with a speed v_b up to the point B. If α is the fall angle of the beam, and $\beta(\alpha)$ is the angle of its refraction, the time of passage from A to B is

$$t(\alpha) = \frac{a}{v_a\,\sin\alpha} + \frac{b}{v_b\,\sin\beta(\alpha)}\,.$$

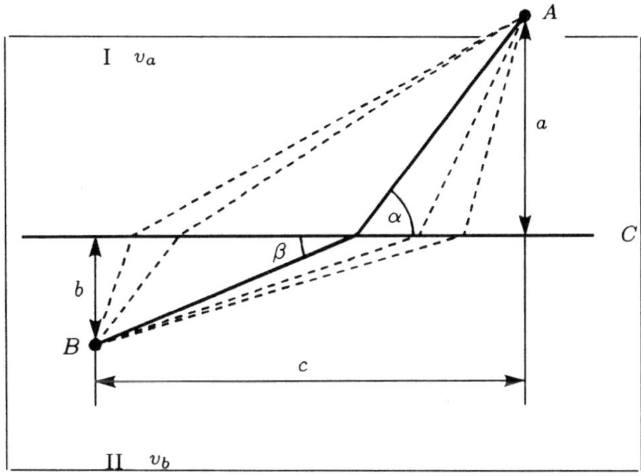

Fig.5. *Possible trajectories of the light beams from point A to point B and refracting on the line C – to the boundary of two media. The bold line indicates the trajectory corresponding to the refraction law* $\cos\alpha/\cos\beta = v_a/v_b$.

The condition of the minimum of $t(\alpha)$ is written as follows (compare with (7))

$$\frac{a\,\cos\alpha}{v_a\,\sin\alpha} + \frac{b\,\cos\alpha}{v_b\,\sin\beta(\alpha)}\,\frac{d\beta}{d\alpha} = 0\,.$$

Differentiating via α, the condition of constancy of c is still expressed by the formula (8). Here the quantities a, b, c have the same meaning, as in the previous case. Excluding from the last formula the derivative $d\beta/d\alpha$, we arrive at the equality

$$\frac{\cos\alpha}{\cos\beta} = \frac{v_a}{v_b}\,, \tag{9}$$

i.e. the known law of refraction of light.

Formulated for a certain class of phenomena, the variational principles allow us to build the appropriate mathematical models uniformly. Their universality is expressed also by the fact that their application enables in certain degree to neglect the specific nature of the process. Thus, the driver of a car following the principle of the minimal time and wishing to get from point A, located on a sandy ground (one speed), to point B, located on a grassy area (another speed), is obliged to move not via the straight line connecting A and B, but via the certain "refracting" line.

3. Use of analogies in the construction of models. In plenty of cases where one is attempting to construct a model of a given object it is either impossible to specify directly the sought fundamental laws or variational principles, or, from the point of view of our present knowledge, there is no confidence in the existence of such laws admitting mathematical formulation. One of the fruitful approaches to such objects is to use analogies with already investigated phenomena. Indeed, what can be common between radioactive decay and the dynamics of populations, in particular, the change in the population of our planet? Even at the elementary level such an analogy is quite visible, as it is clear for one of the simplest models of population – *the Malthus model.* It is based on the simple assumption that the speed of change of the population in time t is proportional to its current number $N(t)$, multiplied on the sum of factors of the birth $\alpha(t) \geq 0$ and the death rate $\beta(t) \leq 0$. As a result one comes to the equation

$$\frac{dN(t)}{dt} = [\alpha(t) - \beta(t)]\, N(t)\,, \tag{10}$$

which is rather similar to the equation of radioactive decay and coinciding with it at $\alpha < \beta$ (if α and β are constants). It is not surprising, since identical assumptions were made for their derivation. The integration of the equation (10) gives

$$N(t) = N(0) \exp \left(\int_{t_0}^{t} [\alpha(t) - \beta(t)]\, dt \right),$$

where $N(0) = N(t = t_0)$ is the initial population.

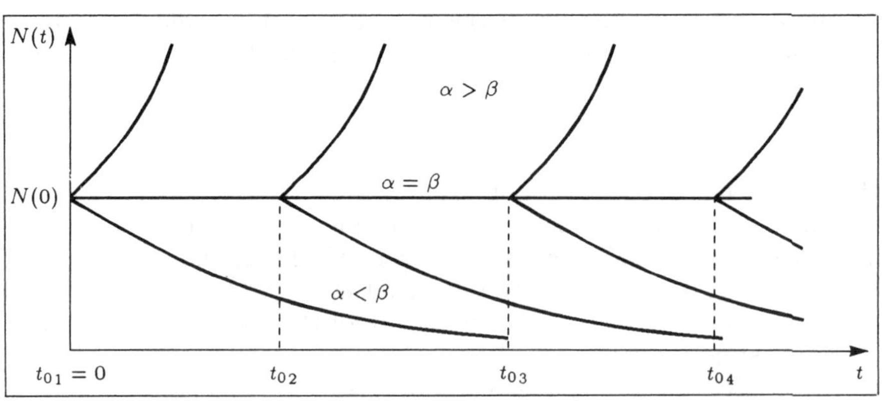

Fig.6. *Change of the population in time in the Malthus model.*

In Fig.6 the diagrams of function $N(t)$ are given at constant values of α and β (to different similar to each other curves correspond different values of time t_0 of the beginning of the process). At $\alpha = \beta$ the population remains constant, i.e. in this case the equilibrium value $N(t) = N(0)$ is the solution of the equation. The balance between the rates of birth and death is unstable in the sense that even the small violation of equality $\alpha = \beta$ results in a large deviation of function $N(t)$ from its equilibrium value $N(0)$. At $\alpha < \beta$ the population decreases and tends to zero at $t \to \infty$, and at $\alpha > \beta$ increases by some exponential law, up to infinity at $t \to \infty$. This last circumstance has served as the basis for fears about the future overpopulation of the Earth with all the consequences following from here.

Both in the given example, and in several other cases considered above, it is possible to specify many obvious restrictions of the applicability of the constructed model. Certainly, the complex process of variation of the population depending also on the conscious intervention of humans themselves, cannot be described by a simple law. Even in an ideal case of the isolated biological population the considered model is not real enough because of the limitation of resources necessary for its existence.

These remarks nevertheless do not refute however, the role of analogies in the construction of mathematical models of very complex phenomena. The application of analogies is based on one of the major properties of models – their universality, i.e. their applicability for objects of essentially different natures. Thus, the assumption that the rate of variation of a quantity is proportional to its value or to its certain function are widely used in various areas.

4. Hierarchical approach to the construction of models. Only in rare cases it is convenient and justified to construct complete mathematical models at once, even of quite simple objects, in view of all the factors essential for their behavior. Therefore it is natural to proceed in accordance to the principle "from the simple to the complex", when the following step is made after the detailed study of models which are not too complex. Then, a chain *(hierarchy)* of more and more complete models is appearing, each of which generalizes the previous ones, including the former as a particular case.

Let us construct such a hierarchical chain on an example of a model of a multistage rocket. As was established at the end of Section 1, a real one-stage rocket is unable to develop the first space speed. The reason is due to the amount of fuel to be used for the speeding up of the unnecessary parts of the structural mass of the rocket. Hence, with a movement of a rocket it is necessary to periodically get rid of a ballast. In terms of practical design it means that the rocket consists of several stages, which are discarded in the process of their use.

Let m_i be the total mass of i-th stage, λm_i be the corresponding structural mass (so that the fuel mass is $(1-\lambda)m_i$), m_p be the mass of the useful loading. The value of λ and speed of the escape of gases u are the same for all stages. Consider for clarity the number of stages $n = 3$. The initial mass of such a rocket is equal

$$m_0 = m_p + m_1 + m_2 + m_3 .$$

Consider the moment when all the fuel of the first stage is spent and the mass of the rocket is equal

$$m_p + \lambda m_1 + m_2 + m_3 .$$

Then by the formula (6) of the initial model, the speed of the rocket equals

$$v_1 = u \ln \left(\frac{m_0}{m_p + \lambda m_1 + m_2 + m_3} \right) .$$

After achieving of the speed v_1 the structural mass λm_1 is removed and the second stage is operated. The mass of the rocket in this moment is equal

$$m_p + m_2 + m_3 .$$

Since this time and up to the moment in the second stage when the fuel is completely spent, nothing prevents us from applying the model already constructed for this case. All the reasoning on the conservation of total momentum and the corresponding calculations remain valid (it is necessary only to take into account that the rocket already has an initial speed v_1). Then in accordance with the formula (6), after the fuel in the second stage is spent, the rocket achieves the speed

$$v_2 = v_1 + u \ln \left(\frac{m_p + m_2 + m_3}{m_p + \lambda m_2 + m_3} \right) .$$

The same considerations are applicable for the third stage of the rocket. After the engines are switched off the speed of the rocket is equal

$$v_3 = v_2 + u \ln \left(\frac{m_p + m_3}{m_p + \lambda m_3} \right) .$$

This chain can be easily continued for any number of stages, with the derivation of corresponding formulae. In the case $n = 3$ for the final speed we have

$$\frac{v_3}{u} = \ln \left\{ \left(\frac{m_0}{m_p + \lambda m_1 + m_2 + m_3} \right) \left(\frac{m_p + m_2 + m_3}{m_p + \lambda m_2 + m_3} \right) \left(\frac{m_p + m_3}{m_p + \lambda m_3} \right) \right\} ,$$

or, inserting $\alpha_1 = \frac{m_0}{m_p+m_2+m_3}$, $\alpha_2 = \frac{m_p+m_2+m_3}{m_p+m_3}$, $\alpha_2 = \frac{m_p+m_3}{m_p}$, we obtain

$$\frac{v_3}{u} = \ln\left\{\left(\frac{\alpha_1}{1+\lambda(\alpha_1-1)}\right)\left(\frac{\alpha_2}{1+\lambda(\alpha_2-1)}\right)\left(\frac{\alpha_3}{1+\lambda(\alpha_3-1)}\right)\right\}.$$

The given expression is symmetrical with respect to the values α_1, α_2, α_3, and it is not difficult to show that its maximum is achieved in the symmetrical case, i.e. at $\alpha_1 = \alpha_2 = \alpha_3 = \alpha$. Thus for $i = 3$

$$\alpha = \frac{1-\lambda}{P-\lambda}, \qquad P = \exp\left(-\frac{v_3}{3u}\right).$$

It is easy to check that the product $\alpha_1\alpha_2\alpha_3 = \alpha$ is equal to the relation m_0/m_p or

$$\alpha^3 = \frac{m_0}{m_p} = \left(\frac{1-\lambda}{P-\lambda}\right)^3.$$

Similarly, for a multistage rocket we have

$$\frac{m_0}{m_p} = \left(\frac{1-\lambda}{P-\lambda}\right)^n, \qquad P = \exp\left(\frac{v_n}{nu}\right), \tag{11}$$

where n is the number of stages.

Consider the formula (11). Let us adopt $v_n = 10,5$ km/s, $\lambda = 0,1$. Then for $n = 2,3,4$ we have $m_0 = 149m_p$, $m_0 = 77m_p$, $m_0 = 65m_p$, respectively. It means that the two-stage rocket is suitable for taking to the orbit of some useful mass (however for one ton of payload it is necessary to have a 149 ton rocket). The transition to the third stage reduces the weight of the rocket by almost half (but certainly complicates its design), while the four-stage rocket does not give any remarkable advantage in comparison with the three-stage one.

The construction of a hierarchical chain has allowed us to arrive at these important conclusions in a simple way. The hierarchy of mathematical models is frequently applied in the opposite direction "from the complex to the simple". In such cases the way down is used, i.e. from general and complex models and simplifying assumptions to a sequence of more simple models (which nevertheless have decreasing area of applicability).

5. On the nonlinearity of mathematical models. The simplicity of the considered above models in many respects is connected with their *linearity*. From the mathematical point of view this important concept means that *the principle of superposition* is valid, i.e. any linear combination of the solutions (for example, their sum) is also a solution of the problem. Using the principle of superposition it is not difficult, finding out the solution in any special case, to construct the solution for a more general situation. Therefore it is possible to judge the qualitative properties of the general case based on the properties of the particular ones – the difference between two solutions is of purely quantitative character. For example, the doubling of the speed of the gases escaping from a rocket doubles the speed of the rocket, the reduction of the angle of fall of a light beam to the reflecting surface leads to the same change of the reflection angle and so on. In other words, in case of linear models the response of object to change of any conditions is proportional to the value of this change.

For *nonlinear phenomena* – with mathematical models not satisfying the principle of superposition, the knowledge of behavior of a part of the object does not guarantee the knowledge of behavior of the whole object, and its response to change of conditions can depend qualitatively on the values of those changes. Thus, the reduction of the fall angle of a light beam to the boundary of two media results in the reduction of the angle of refraction, but only up to a certain limit. If the fall angle becomes less than critical (see formula (9)), a qualitative change occurs – the light fails to penetrate to the second medium, if it is less dense than the first one. Thus, the refraction of light is an example of a nonlinear process.

The majority of real processes and their corresponding mathematical models are nonlinear. Linear models correspond to rather special cases and, as a rule, serve only first as a approximation to reality. For example, the population models at once becomes nonlinear if one takes into account the limitation of resources accessible for population. In that case, it is assumed that:

1. An "equilibrium" population N_p does exist, which is determined by the environment;

2. The speed of population variation is proportional to the number multiplied (as distinct to the Malthus model) on the amount of its deviation from the equilibrium value, i.e.

$$\frac{dN}{dt} = \alpha \left(1 - \frac{N}{N_p}\right) N \,, \qquad \alpha > 0\,. \tag{12}$$

The member $(1-N/N_p)$ in this equation describes the mechanism "of saturation" of number – at $N < N_p (N > N_p)$ the growth rate is positive (negative) and tends to zero, if $N \to N_p$.

Representing equations (12) as

$$\frac{dN}{N_p - N} + \frac{dN}{N} = \alpha\, dt$$

and integrating it, we obtain

$$-\ln(N_p - N) + \ln N = \alpha t + C\,.$$

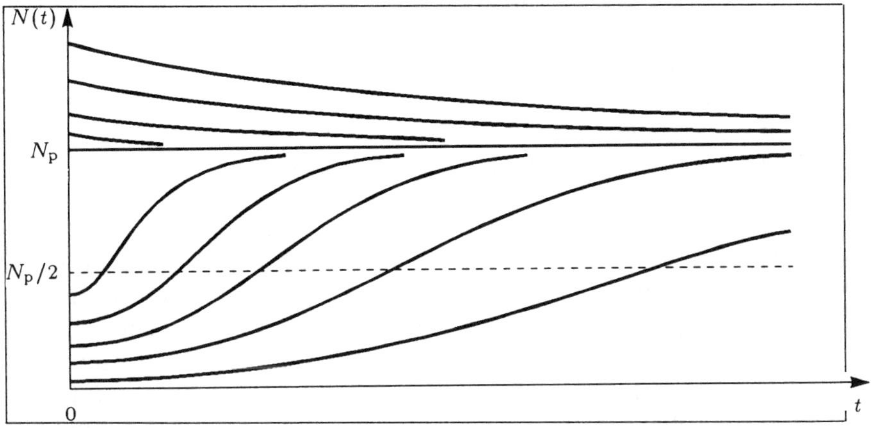

Fig.7. *Logistic curves, corresponding to various values of the initial number* $N(0)$.

The constant of integration is determined from the condition $N(t=0) = N(0)$, i.e. $C = \ln((N_p - N(0))^{-1}N(0))$. As a result, we find

$$N = N_p \frac{N(0)}{N_p - N(0)} e^{\alpha t} - N \frac{N(0)}{N_p - N(0)} e^{\alpha t},$$

or, in a final form

$$N(t) = \frac{N_p N(0) e^{\alpha t}}{N_p - N(0)\left(1 - e^{\alpha t}\right)}.$$

The behavior of the function $N(t)$ is described by a so-called *logistic curve* (Fig. 7). At any $N(0)$ the number tends to equilibrium value N_p more slowly, the closer is $N(t)$ to $N(0)$. Thus the equilibrium, as distinct from the model (10), is stable.

The logistic model reflects more realistically the dynamics of population as compared with the Malthus model, but it becomes nonlinear and consequently more complex. Note that the assumptions on the mechanisms of saturation are used in the construction of many models in various fields.

6. Preliminary conclusions. The process of constructing models can be conditionally split into the following stages.

1. The construction of the model starts from the semantic description of an object or phenomenon. Besides general data on the nature of the object and the purposes of its study, this stage can also contain some assumptions (weightless axis, thick layer of matter, propagation of light beams via straight lines, etc.). We can call this step the formulation of the premodel.

2. The following stage is the end of the idealization of the object. All the factors and effects which are not too crucial for its behavior are abandoned. For example, while considering the balance of the matter (Sect.1b), the defect of mass at the radioactive decay was not taken into account, in view of it smallness. Whenever possible, idealizing assumptions are represented in mathematical form (similar to the condition $\lambda_I \gg L_I$ in Sect.1b), so that their validity can be checked quantitatively.

3. After the performance of the first two stages one can move to the choice or formulation of a law (variational principle, analogy and so forth) governing the object, and its record in the mathematical form. The additional data on the object are used if necessary and are again given in mathematical form (for example, constance of size c for all trajectories of light beams, following from the geometry of the problem; Sect.2). One should take into account that even for simple objects the choice of the appropriate law is by no means trivial (see exercise 1).

4. The formulation of the model is completed by looking at its "equipment". For example, it is necessary to give the data on the initial conditions of the object (speed of a rocket and its weight at the moment $t = 0$) or its other characteristics (quantities l, g in Sect.la); α, λ_I, λ_{II} in Sect.1b); $\alpha(t)$ and $\beta(t)$ in Sect.3), without knowledge of which it is impossible to determine the behavior of the object. And finally, the aim of studying model is formulated (to find out the law of refraction of light, to understand the laws of variation of population, to determine the requirements for the design of a rocket launching a sputnik, etc.).

5. The constructed model is studied by all methods accessible to the researcher, including the mutually checking the various approaches (see, for example, exercises 4, 7). As distinct from elementary cases, considered in Section 1, the majority of models cannot be treated purely theoretically, and consequently one has to widely use computational methods. This circumstance is especially important because the study of nonlinear objects, as with their qualitative behavior, is generally unknown.

6. As a result of studying a model not only is the aim achieved, but also its adequacy – correspondence with the object and formulated assumptions – has to be established. A non-adequate model can give results too far from the true ones (compare the formula (1) and the result of exercise 1), and should be either rejected, or modified correspondingly.

EXERCISES

1. In the first problem of Subsection 1a to obtain the value of V (speed of the system "a bullet-load" immediately after the collision) apply the law of conservation of momentum instead of the law of conservation of energy. Be sure, that for the speed of the bullet v one gets a value smaller by a factor $((M + m)/m)^{1/2}$, than

predicted by formula (1).

2. Let the power of the laser drilling a material (Subsection 1a), depends on time: $W = W(t)$. How will the formula (3) be changed? Will the conclusion that the depth of the hole is proportional to the spent energy remain valid?

3. Find the moment in time when the last atom of a radioactive substance (Subsection 1b) will decay. In the model (5), why the matter is disintegrated completely only at $t \to \infty$?

4. Assume that in Subsection 1c an "ideal" one-stage rocket is considered, which is dropping the unnecessary fraction of its structural mass, (at the moment of complete combustion of the fuel $m_s = 0$). Using the law of conservation of momentum, show that the maximal speed of such a rocket is determined by the formula $v = (1 - \lambda)u \ln(m_0/m_p)$. Compare it with the formula (6). Why can the ideal rocket reach any speed?

5. Check, with the use of (8), that the condition (7) is a condition for minimal value of $t(a)$. From Fig. 5 determine which speed is more – v_a or v_b? Using the formula (9), find out at what angles the light beam does not penetrate from medium a into medium b, i.e. when the effect "of complete internal reflection of light" used in a number of technical devices is realized.

6. Determine the behavior of $\tau(t) = \alpha(t) - \beta(t) > 0$ at large t in Malthus model (10), in order the keep the population finite at $t \to \infty$.

7. In Eq. (11) for the multistage rocket take the limit $n \to \infty$, be convinced that its limiting speed is determined via the formula for an ideal rocket from exercise 4. Why do their results coincide?

8. In logistic model (12) consider small deviations from the equilibrium, i.e. when the solution is $N(t) = N_p + \delta N(t)$, where $|\delta N(t)| \ll Np$. Show that for $\delta N(t)$ in the first approximation the linear Malthus model (10) is valid.

2 Examples of Models Following from the Fundamental Laws of Nature

We shall now consider in more details than in Section 1.1, models following from the laws of Archimedes, Newton, Coulomb and other well known laws. Let us discuss some properties of considered objects.

1. The trajectory of a floating submarine. Let a submarine at the moment of time $t = 0$ situated at depth H from the sea surface and moving with constant horizontal speed v (Fig. 8), receive an order to come up to the surface. If only a short time interval is needed for the tanks of the submarine to be released from the water and be filled by air, so that its average density ρ_1 has become less than density of water ρ_0, it is possible to assume that at the moment $t = 0$ a pushing force is acting on the submarine, greater than the weight of the boat. In accord to the law of Archimedes the pushing force is equal $F = gV\rho_0$, where g is the acceleration of free fall, V is the volume of the boat. The total force acting on the submarine in a vertical direction, i.e. the difference between F and weight of the body $P = gV\rho_i$, and its acceleration in accordance with Newton's second law is equal

$$\rho_1 V \frac{d^2h}{dt^2} = F - P = gV(\rho_0 - \rho_1)\,.$$

Coordinate l, describing the horizontal position of the submarine, varies due to the movement of the body with a constant speed:

$$\frac{dl}{dt} = v\,.$$

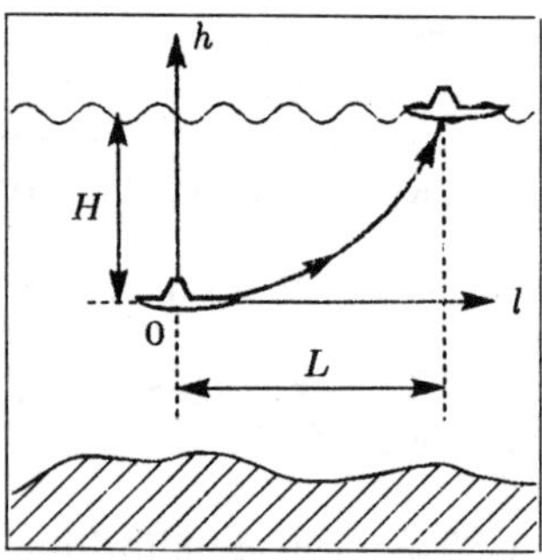

Fig.8.

Solving these equations, we find

$$h(t) = g\frac{\rho_0 - \rho_1}{\rho_1}t^2\,, \qquad l(t) = vt\,, \tag{1}$$

and that the boat will arrive on the surface at the moment $t = t_k$, when

$$h(t_k) = g]\frac{\rho_0 - \rho_1}{\rho_1}t_k^2 = H\,, \qquad t_k = \left(\frac{\rho_1 H}{g(\rho_0 - \rho_1)}\right)^{1/2}.$$

Thus in horizontal direction a submarine will cover the distance

$$L = vt_k = \left(\frac{\rho_1 H}{g(\rho_0 - \rho_1)}\right)^{1/2}.$$

Excluding the time from (1) we shall find the trajectory of motion of the submarine in coordinates $(l,\, h)$

$$h = g\frac{\rho_0 - \rho_1}{\rho_1 v^2}\,l^2\,,$$

which appears to be a parabola with its top at $l = 0$, $h = 0$ (at derivation of (1) the vertical speed of the boat, and the values of l and h were taken to be equal to zero in the moment $t = 0$). We have also assumed that no other vertical forces except F and P, are acting on the submarine. This assumption is correct only at small speeds, when one can neglect the effect of water resistance on the movement of the boat (see exercise 1).

Thus, the direct application of Archimedes' law determining the pushing out force, and Newton's law connecting the force acting on the body and its acceleration, has permitted us easily to find out the trajectory of the submarine. Obviously, any body moving in a plane has a parabolic trajectory when its velocity is constant in one direction and a constant force is acting on the other direction (equations (1) actually give the parametric representation of a parabola). This type of motion includes, for example, the motion of a stone thrown from a height H with horizontal velocity v or electron's motion in an electrical field of a flat capacitor. However in the latter case it is impossible to obtain the trajectory of the body directly from the fundamental laws and a more detailed procedure is required. Let us consider this problem in more detail.

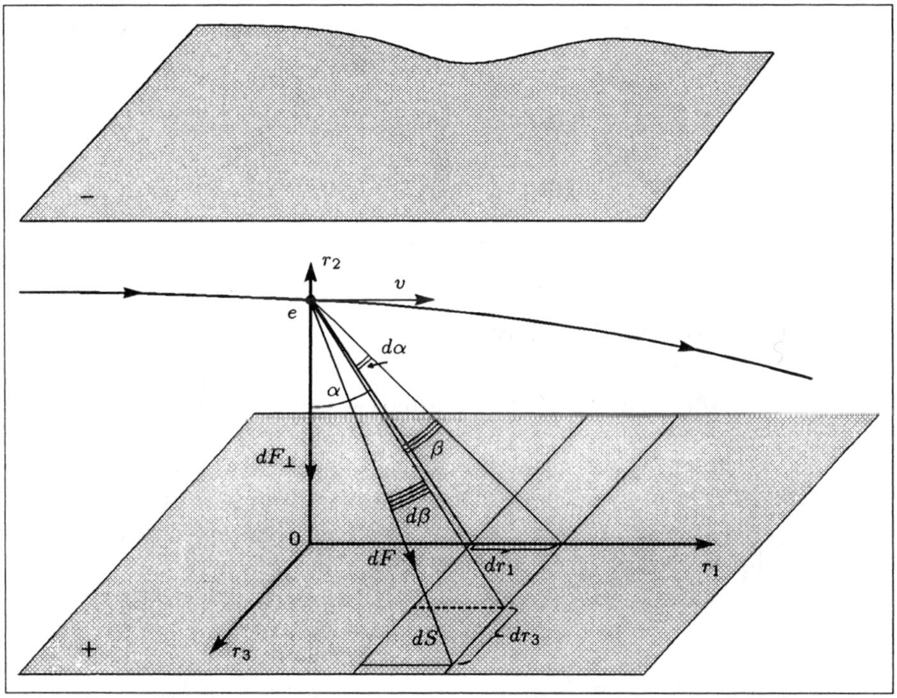

Fig.9.

2. Deviation of a charged particle in an electron-beam tube. We shall assume that the plates of the capacitor of an electron-beam tube (Fig. 9) are infinite surfaces (this assumption is justified when the distance between the plates is much less than their sizes, and the electron is moving far from their edges; see exercise 2). Obviously, the electron will be attracted to the lower plate and repelled from the upper. The force of attraction F of two opposite charges is determined easily from the Coulomb law

$$F = \frac{q_1 q_2}{r^2},$$

where q_1 and q_2 are the magnitudes of the charges, r is the distance between them. The difficulty of this example is that the infinite number of charges are located on the plate, each of them at a certain distance from the moving electron. Therefore, it is first necessary to find out the force induced by each charge, and then to combine all elementary forces to determine the resulting action of plates on the electron.

Divide the whole surface of the lower plate into elementary "belts", given by coordinates r_1, r_2, r_3; $-\infty < r_1, r_3 < \infty$; $r_2 \equiv 0$ (see Fig. 9).

Let us estimate the attraction force of an electron by a charge located on an elementary area $ds = dr_1\, dr_3$ and equal $dq = q_0\, ds$, where q_0 is the surface density of charge on the plate. If the particle is at a distance r_2 from the charged plane, then

$$dr_1 = r_2(\tan(\alpha + d\alpha) - \tan\alpha) = r_2 \frac{d\alpha}{\cos^2\alpha}$$

(the smallness of $d\alpha$ is taken into account here.) To estimate dr_3, we have

$$\frac{r_3 + dr_3}{r_1 + dr_1} = \frac{\tan(\beta + d\beta)}{\sin(\alpha + d\alpha)}, \qquad \frac{r_3}{r_2} = \frac{\tan\beta}{\tan\alpha}.$$

From the last two formulae we obtain

$$dr_3 = (r_1 + dr_1)\, \tan(\beta + d\beta) - r_1\, \tan\beta = \frac{r_1\, d\beta/(\cos^2\beta) + dr_1\, \tan\beta}{\sin\alpha},$$

where, as in the previous case, the smallness of $d\beta$ is taken into account. Multiplying dr_1 on dr_3 and neglecting the term of higher order of smallness, we obtain $ds = r_2 r_1\, d\alpha\, d\beta/(\cos^2\alpha\, \cos^2\beta\, \sin\alpha)$. The attraction force of the electron with the charge q_e of an elementary area ds is

$$dF = \frac{q_e q_0\, r_2\, r_1\, d\alpha\, d\beta}{\bar{r}^2\, \cos^2\alpha\, \cos^2\beta\, \sin\alpha},$$

where $\bar{r}$ is the "average" distance from the electron to the area element, which in view of the small size of $d\alpha$, $d\beta$ is estimated from the formula $\bar{r} = r_2/(\cos\alpha\, \cos\beta)$. In sum, for the elementary force we have

$$dF = q_e q_0 \frac{r_1}{r_2} \frac{d\alpha\, d\beta}{\sin\alpha} = \frac{q_e q_0}{\cos\alpha}\, d\alpha\, d\beta,$$

and for its vertical component

$$dF_{\perp} = dF \cos\beta \cos\alpha = q_{\mathrm{e}} q_0 \cos\beta \, d\alpha \, d\beta \, .$$

Integrating the expression for $F_{\perp}$ by β from $\beta = 0$ up to $\beta = \pi/2$, we obtain the force of attraction of the electron by the elementary "belt", located in the quadrant $r_1 > 0$, $r_3 > 0$:

$$dF_{\alpha}^{+} = q_{\mathrm{e}} q_0 \, d\alpha \, .$$

Summing dF_{α}^{+} over α from $\alpha = 0$ up to $\alpha = \pi/2$, i.e. over all belts of the quadrant $r_1 > 0$, $r_3 > 0$, we determine the attraction force induced by charges located in this quadrant:

$$dF^{+} = \frac{\pi}{2} q_{\mathrm{e}} q_0 \, .$$

Taking into account the action of all four quadrants of the surface of the lower plate and conducting similar considerations for the upper plate, we shall obtain the resulting attraction (repulsion) force of the electron to all charges of the capacitor

$$F = 4\pi q_{\mathrm{e}} q_0 \, . \tag{2}$$

The force F is directed along the axis r_2 to the lower plate (components of F on axes r_1, r_3, are obviously equal to zero by virtue of symmetry; to be convinced of this, it is enough to consider the action of a charge of the surface element located in quadrant $r_1 < 0$, $r_3 < 0$ and symmetrical to the surface element ds).

Since the force F does not depend on r_2, and on horizontal axes the particle is moving with a constant velocity v, we come to the situation of the previous section; applying Newton's second law, it is easy to derive formulae similar to (1), describing the motion of the electron via a parabolic trajectory and enabling one to calculate all its parameters. However as distinct from the case with the submarine, the direct application of the fundamental law of Coulomb for constructing a model of a moving electron, appears impossible. Based on the fundamental law, one has first to describe the elementary act of interaction of charges, and then, combining all those actions, to find out the resulting force.

The analogous situation and the sequence of operations are rather typical in the construction of models, so far as many fundamental laws install mutual relations just between elementary parts of the initial object. This is certainly valid not only for electrical forces, but, for example, for gravity.

3. Oscillations of the rings of Saturn. We shall construct a model of motion of a point mass M_0 in a field of gravity created by a matter ring of radius R_0 and linear density ρ_0. The ring is considered to be indefinitely

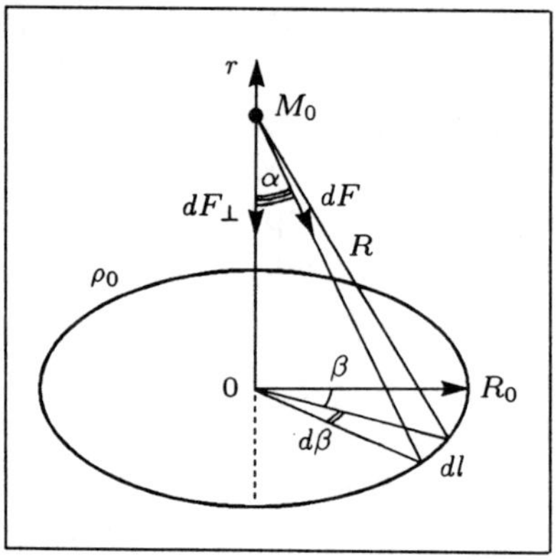

Fig.10.

thin, the motion is along the axes of the ring (Fig. 10). This scheme can be considered as an idealization of the process of the oscillations of the rings of Saturn. Nevertheless, despite essential simplification, the direct use of the law of world gravitation

$$F = \gamma \frac{m_0 m_1}{r^2},$$

where F is the force of attraction of two bodies of masses m_0 and m_1, and r is the distance between them, γ is the the constant of gravitation, cannot give a final model of the motion of the rings of Saturn, as the masses m_0, m_1 should be point-like.

Therefore, we first calculate the attraction force between a point mass M_0 and a mass dm, contained in the small element of the ring dl, which can be considered as a point

$$dF = \gamma \frac{M_0\, dm}{R^2}.$$

Here R, r are, correspondingly, the distance from a mass M_0 up to the ring and up to the center of the ring. Obviously, at $0 \le \alpha \le \pi/2$ (for $\pi/2 \le \alpha \le 2\pi$ the calculations are similar)

$$\frac{R_0}{R} = \sin\alpha = \frac{R_0}{\sqrt{r^2 + R_0^2}}, \qquad \frac{r}{R} = -\cos\alpha = \frac{r}{\sqrt{r^2 + R_0^2}}.$$

So far as $dm = \rho_0\, dl = \rho_0\, R_0\, d\beta = -\rho_0 r\, \tan\alpha\, d\beta$, then

$$dF = -\gamma \frac{M_0 \rho_0}{R^2} r\, \tan\alpha\, d\beta = -\gamma \frac{M_0 \rho_0}{r} \sin\alpha\, \cos\alpha\, d\beta.$$

Let us estimate the projection of the force dF on the axis r (just this projection determines the sought motion):

$$dF_\perp = dF \cos\alpha = -\gamma \frac{M_0 \rho_0}{r} \sin\alpha\, \cos^2\alpha\, d\beta.$$

Summing the forces of gravity induced by all elements of the ring, i.e. by taking the integral of $dF_\perp$ on β from $\beta = 0$ up to $\beta = 2\pi$, we shall estimate the resulting force:

$$F = -2\pi\gamma \frac{M_0\rho_0}{r} \sin\alpha \cos^2\alpha = -\gamma M_0 M_1 \frac{r}{(r^2 + R_0^2)^{3/2}}, \qquad (3)$$

where $M_1 = 2\pi R_0\rho_0$ is the total mass of the ring. As in the previous problem, the horizontal projection of the resulting force is equal to zero because of the symmetry of the ring with respect to the mass M_0.

The gravitational force (3) essentially differs from the expression given by the law for point masses, coinciding with it only at $r \gg R_0$, when the ring can be identified with a point mass due to the distance between the gravitating bodies which in great as compared with the size of the ring. If $r \ll R_0$, then

$$F = -\gamma \frac{M_0 M_1}{R_0^3} r,$$

and the force of attraction, contrary to the case of point masses, decreases by distance between the objects (one more limiting case is considered in exercise 3).

Applying Newton's second law to the mass M_0, we shall obtain the equation of its motion along the axis r:

$$\frac{d^2r}{dt^2} = -\gamma M_1 \frac{r}{(r^2 + R_0^2)^{3/2}},$$

which, as distinct from subsections 1 and 2, is essentially nonlinear, and becomes linear only at $r \ll R_0$:

$$\frac{d^2r}{dt^2} = -\gamma \frac{M_1}{R_0^3} r. \qquad (4)$$

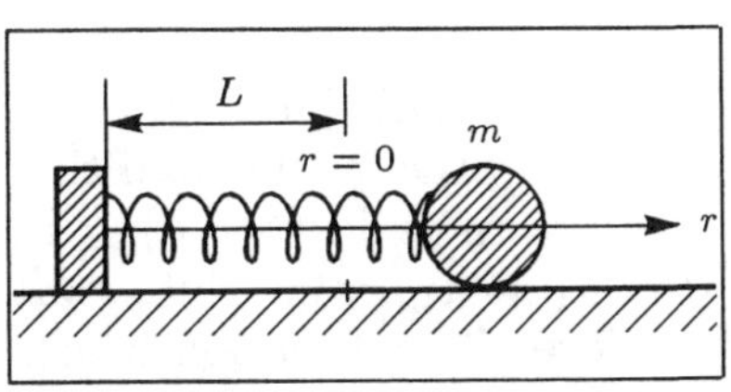

Fig.11.

4. Motion of a ball attached to a spring. In deriving models of subsections 1–3 the fundamental laws had the principal role, defining the origin and the magnitude of forces acting on the object, while Newton's second law was an auxiliary and was applied at the last stage of constructing the model. Certainly, such division is entirely conditional. Concerning the problems of dynamics, one can use another scheme, namely, first to connect the projection of acceleration of the body with the projections of acting forces (using Newton's law), and then, proceeding from certain considerations to estimate those forces as functions of coordinates, getting a closed model. We will demonstrate this approach to an example of a model of the motion of a ball connected to a spring, with a rigidly fixed edge (Fig. 11).

Let r be the coordinate of the ball along the axis of the spring lying on a horizontal plane, and the direction of the motion of the ball coincides with its axis. Then by the second law of dynamics

$$F = ma = m\frac{d^2r}{dt^2} ,$$

where m is the mass of the ball, a is its acceleration. We consider the plane ideally smooth (so that motion is without friction), and will also discount air resistance, and take into account that the weight of the ball is balanced by the reaction of the plane. The only force acting on the ball in the direction of the axis r, is obviously is the elasticity force of the spring. We will determine it, using Hook's law, stating that for the expansion (compression) of a spring it is necessary to have a force

$$F = -kr ,$$

where the coefficient $k > 0$ characterizes the elastic properties of the spring, and r is the magnitude of expansion or compression with respect to an unloaded position $r = 0$. The equation of motion of the ball has the form (equation of an elementary oscillator)

$$m\frac{d^2r}{dt^2} = -kr , \qquad t > 0 . \tag{5}$$

It describes the simple harmonic oscillations and has a general solution

$$r = A \sin\omega t + B \cos\omega t , \tag{6}$$

where $\omega = \sqrt{k/m}$ is the eigen-frequency of oscillations of the system "spring–ball". The values of A and B are easily determined from the initial condition of the object, i.e. through the magnitudes $r(t = 0) = r_0$ and $v(t = 0) = v_0$ ($v(t)$ is the velocity of the ball), and $r(t) \equiv 0$ at $r_0 = v_0 = 0$. Note that the equation (4) in fact coincides with (5), therefore the subsection 3 also concerns oscillations, but with respect to the system "Saturn–ring".

The approaches used to construct the models in the present section, should certainly not contradict other fundamental laws of nature. The appropriate check for consistency (if it is possible) is rather useful for the reliability of models. Let us clarify this, using the energy conservation law to derive of equation (5) rather than Newton's law. In so far as the attaching point of the spring is motionless, the wall does not perform any work on the system "spring–ball" (and the contrary), and its total mechanical energy E remains constant. Let us calculate it. The kinetic energy is determined by the motion of the ball (spring is considered weightless):

$$E_k = \frac{mv^2}{2} = \frac{m(dr/dt)^2}{2}.$$

The potential energy of the system "is contained" in the spring, it is easy to estimate it by defining the work necessary for the expansion (compression) of the spring on magnitude r:

$$E_n = -\int_0^r F\,dr' = \int_0^r kr'\,dr' = k\frac{r^2}{2}.$$

For the quantity $E = E_k + E_n$ (integral of energy) which is constant in time, we obtain

$$E = \frac{m(dr/dt)^2}{2} + \frac{kr^2}{2}.$$

In so far as $dE/dt \equiv 0$, differentiating the integral of energy by t, we arrive at the expression

$$m\frac{dr}{dt}\frac{d^2r}{dt^2} + k\frac{dr}{dt}r = \frac{dr}{dt}\left(m\frac{d^2r}{dt^2} + kr\right) = 0,$$

i.e. equation (5), thus checking the validity of its derivation. A similar procedure is easy to perform for the examples in subsection 1–3.

5. Conclusion.

1. Even in the elementary situations for constructing a model the use of not one, but several fundamental laws can be required.

2. The direct formal application of fundamental laws to an object considered as a whole, is not always possible (subsections 2, 3). In these cases one has to add the elementary acts of interaction between its parts, taking into consideration the properties of the object (for example, its geometry).

3. The same models can describe objects of completely different natures, described by different fundamental laws, and, on the contrary, a given law can correspond to different models (for example, to linear and nonlinear ones, see subsection 3).

4. It is necessary to use all means to check the validity of construction of the model (see the limiting cases in subsections 2, 3, the other fundamental laws in subsection 4, etc).

EXERCISES

1. In the problem of floating a submarine, the resistance of water is taken into account. Adopting the force of resistance $F_1 = -k_0 u$, where $k_0 > 0$ is the coefficient depending on the properties of water and the form of the submarine, u is the vertical velocity of the boat, estimate the maximum depth H, when the force F_1 can be neglected at any moment of time $t \leq t_k$ (the condition $F_1 \ll F - P$) should be fulfilled).

2. Repeating the considerations of subsection 2, obtain the force of attraction of an electron to the plate of a capacitor of dimensions R_1, R_3. Be convinced, that at $R_1 \to \infty$, $R_3 \to \infty$ the obtained expression turns to the formula (2).

3. In the problem in subsection 3 introduce the thickness of the ring d, and obtain the force F and show that the obtained expression at $d \to 0$ coincides with the formula (3).

4. Let the distance between the neutral position of the spring $r = 0$ and the wall to which it is connected equal L (see Fig. 11). Obtain, using the formula (6), the condition on the magnitudes r_0, v_0, where the ball cannot hit the wall (otherwise the model (5) is incorrect, in so far as at the impact with the wall the ball undergoes the action of the force which is not taken into account in equation (5)).

3 Variational Principles and Mathematical Models

Let us give a simplified formulation of Hamilton's variational principle for a mechanical system. Based on it we shall derive the equations of motion of a ball with a spring and of a pendulum in the gravity field. We will compare the results of deriving the models from the fundamental laws and from the variational principle.

1. The general scheme of the Hamiltonian principle. Consider a mechanical system, without giving its formal and rigorous definition, taking into account that the interaction between all its elements is determined by the laws of mechanics (one of the simplest examples is the "ball–spring" system considered in section 2.4). Introduce the concept *of a generalized coordinate* $Q(t)$, completely defining the position of the mechanical system in space. The quantity $Q(t)$ can coincide with the Cartesian coordinate (for example, the coordinate r of the system "ball–spring"), the radius-vector, the angular coordinate, the set of coordinates of mass points, etc. It is natural to label the quantity dQ/dt as *generalized velocity* of the mechanical system in the moment of time t. The set of magnitudes $Q(t)$ and dQ/dt determines the state of a mechanical system at all moments in time.

To describe the mechanical system the *Lagrange function* is introduced; its derivation is a separate problem, considered in more detail in chapter III. In the simplest cases the Lagrange function has a clear content and is denoted as

$$L(Q,\, dQ/dt) = E_k - E_p\,, \tag{1}$$

where E_k, E_p are the kinetic and potential energies of the system respectively. For the purposes of the present section there is no need to give the general definition of quantities E_k, E_p, in so far as they are calculated in obvious manner in the considered examples.

Let us introduce the quantity $S[Q]$, named *an action*:

$$S[Q] = \int_{t_1}^{t_2} L\left(Q,\, \frac{dQ}{dt}\right) dt\,. \tag{2}$$

The integral (2), is obviously a functional from the generalized coordinates $Q(t)$, i.e. it puts a correspondence between the function $Q(t)$ defined within the interval $[t_1,\, t_2]$ and some number S (action).

The Hamilton principle for the mechanical system states: if the system is moving in accordance with the laws of mechanics, then $Q(t)$ is a stationary function for $S[Q]$, or

$$\frac{d}{d\varepsilon}\, S[Q + \varepsilon\, \varphi]_{\varepsilon=0} = 0\,. \tag{3}$$

The function $\varphi(t)$ in the *principle of least action* (3) is a test function turning to zero at the moments t_1, t_2 and satisfying the condition that $Q(t) + \varepsilon\varphi(t)$ is the possible coordinate of the given system (in the rest $\varphi(t)$ is arbitrary).

The content of principle (3) is that from all a priori possible trajectories (motions) of the system between the moments t_1 and t_2, the one with the minimum of the functional of action is chosen (hence the name of the principle). The function $\varepsilon\varphi(t)$ is called *variation* of the quantity $Q(t)$.

Thus, the scheme of applying the Hamiltonian principle (3) for the constructing models of mechanical systems is as follows: the generalized coordinates $Q(t)$ and generalized velocities dQ/dt of the system are defined; the Lagrange function $L(Q,\, dQ/dt)$ and the functional of action $S[Q]$ are constructed; the minimization of the latter over variations $\varepsilon\varphi(t)$ of the coordinates $Q(t)$ gives the sought model.

2. The third way of deriving the model of the system "ball–spring". We use the Hamiltonian principle to construct the model of the motion of the ball connected with a spring (section 2.4). As generalized coordinates of the system it is natural to chose the usual Eulerian coordinate of the ball $r(t)$. Then the generalized velocity $dr/dt = v(t)$ is the usual velocity of the ball. The Lagrangian function (1), $L = E_k - E_p$, is rewritten using values of kinetic and potential energies of the system (already found in section 2.4):

$$L = \frac{m(dr/dt)^2}{2} - k\frac{r^2}{2}.$$

For the action we obtain

$$S[r] = \int_{t_1}^{t_2} L\left(r, \frac{dr}{dt}\right) dt = \int_{t_1}^{t_2} \left[\frac{m}{2}\left(\frac{dr}{dt}\right)^2 - \frac{k}{2}r^2\right] dt.$$

Now, in view of the correspondence with the scheme seen in subsection 1, we calculate the action over the variations $\varepsilon\varphi(t)$ of the coordinates $r(t)$:

$$S[r+\varepsilon\varphi] = \int_{t_1}^{t_2} \left[\frac{m}{2}\left(\frac{d(r+\varepsilon\varphi)}{dt}\right)^2 - \frac{k}{2}(r+\varepsilon\varphi)^2\right] dt.$$

The last formula has to be differentiated by ε (taking into account that the functions r, φ, dr/dt, $d\varphi/dt$ do not depend on ε):

$$\frac{d}{d\varepsilon}S[r+\varepsilon\varphi] = \frac{d}{d\varepsilon}\frac{1}{2}\int_{t_1}^{t_2} \left[m\left\{\left(\frac{dr}{dt}\right)^2 + 2\varepsilon\frac{dr}{dt}\frac{d\varphi}{dt} + \varepsilon^2\left(\frac{d\varphi}{dt}\right)^2\right\} - \right.$$

$$\left. -k\{r^2 + 2\varepsilon\, r\varphi + \varepsilon^2\varphi^2\}\right] dt =$$

$$= \int_{t_1}^{t_2} \left[m\left\{\frac{dr}{dt}\frac{d\varphi}{dt} + \varepsilon\left(\frac{d\varphi}{dt}\right)^2\right\} - k\{r\varphi + \varepsilon\varphi^2\}\right] dt,$$

and inserting $\varepsilon = 0$:

$$\frac{d}{d\varepsilon}S[r+\varepsilon\varphi]\,|_{\varepsilon=0} = \int_{t_1}^{t_2} \left[m\frac{dr}{dt}\frac{d\varphi}{dt} - k\,r\,\varphi\right] dt = 0.$$

The right hand side of this expression (equal to zero in accordance with the Hamiltonian principle – see (3)) after, integrating of its first member by parts and in view of $\varphi = 0$ in the moments t_1, t_2, will take the form

$$\frac{d}{d\varepsilon}S[r+\varepsilon\varphi]\bigg|_{\varepsilon=0} = -\int_{t_1}^{t_2} \varphi\left[m\frac{d^2r}{dt^2} + kr\right] dt = 0.$$

So far as the test function $\varphi(t)$ in the formulation of the principle of least action is arbitrary, then the part of the expression under the integral in square brackets should be equal to zero in all instances $t_1 < t < t_2$:

$$m \frac{d^2 r}{dt^2} = -kr ,$$

i.e. the motion of the system should be described the equation (5), derived in section 2 from Newton's law (first way) and the law of conservation of energy (second way). All three approaches appear to be equivalent (since there is a deep relation between them studied in more detail in Chapter III).

3. Oscillations of a pendulum in a gravity field. Consider a slightly more complicated example of applying the Hamiltonian principle, along with the detailed consideration of the initial stage of constructing a model – the description of a mechanical system.

Let a pendulum be hung on a fixed (motionless) hinge. The pendulum is a mass m attached on the edge of a rod of length l (Fig. 12). The hinge is considered to be ideally smooth, in the sense that there are no losses by friction. The motionless hinge means that no energy is passed from it to the system "rod–weight", i.e. no work is performed by the rod. The hinge is considered weightless and absolutely rigid, i.e. its kinetic and potential energies are equal to zero, and the weight cannot move along it. The weight is small in comparison with the length of the rod (material point), the acceleration of gravity g is constant, the air resistance is neglected, the oscillations occur in the fixed vertical plane (obviously, the vector of the initial velocity of the weight lies in this plane).

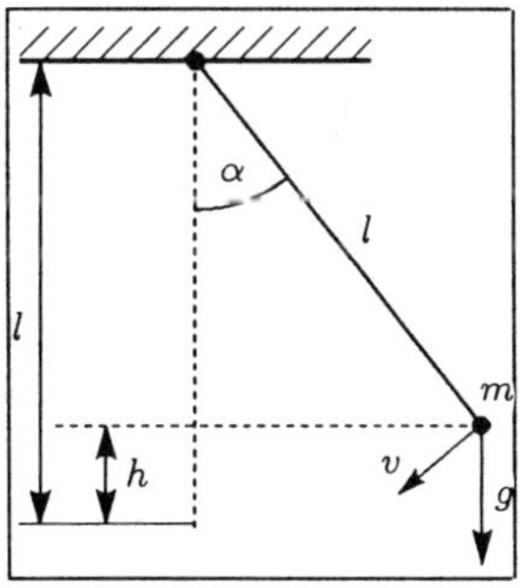

Fig.12.

It is clear that after these simplifying assumptions, the position of a pendulum is determined only by one of the generalized coordinates; we select the angle $\alpha(t)$ of the deviation of the rod from the vertical. A generalized velocity in this case is the angular velocity $d\alpha/dt$.

The kinetic energy of the system is given by the formula

$$E_k = \frac{1}{2} m v^2 = \frac{1}{2} m \left(l \frac{d\alpha}{dt} \right)^2 = \frac{1}{2} m l^2 \left(\frac{d\alpha}{dt} \right)^2 ,$$

and the potential energy is given by the expression

$$E_p = mgh = -mg(l \cos\alpha - 1) ,$$

where h is the deviation of the pendulum from the lowest position by the vertical. In further calculations we omit the quantity mgl in E_p, in so far as the potential energy is determined to be within a constant.

Now it is not difficult to estimate the Lagrange function (1) and the action (2):

$$L \left(\alpha, \frac{d\alpha}{dt} \right) = ml \left[\frac{1}{2} l \left(\frac{d\alpha}{dt} \right)^2 + g \cos\alpha \right] ,$$

$$S[\alpha] = ml \int_{t_1}^{t_2} \left[\frac{1}{2} l \left(\frac{d\alpha}{dt} \right)^2 + g \cos\alpha \right] dt .$$

Deriving the action over variations $\alpha + \varepsilon\varphi(t)$:

$$S[\alpha + \varepsilon\varphi] = ml \int_{t_1}^{t_2} \left[\frac{1}{2} l \left(\frac{d\alpha}{dt} + \varepsilon \frac{d\varphi}{dt} \right)^2 + g \cos(\alpha + \varepsilon\varphi) \right] dt =$$

$$= ml \int_{t_1}^{t_2} \left[\frac{1}{2} \left\{ \left(\frac{d\alpha}{dt} \right)^2 + 2\varepsilon \frac{d\alpha}{dt} \frac{d\varphi}{dt} + \varepsilon^2 \left(\frac{d\varphi}{dt} \right)^2 \right\} + g \cos(\alpha + \varepsilon\,\varphi) \right] dt ,$$

and differentiating it by ε and assuming $\varepsilon = 0$, we obtain

$$\frac{d}{d\varepsilon} S[\alpha + \varepsilon\varphi] \bigg|_{\varepsilon=0} = ml \int_{t_1}^{t_2} \left[l \frac{d\alpha}{dt} \frac{d\varphi}{dt} - \varphi\, g \sin\alpha \right] dt = 0 .$$

As in subsection 1, the first term of expression in the brackets is integrated by parts, taking into account, that $\varphi(t) = 0$ in moments t_1, t_2. Then we come to the following equation:

$$ml \int_{t_1}^{t_2} \varphi \left[l \frac{d^2\alpha}{dt^2} + g \sin\alpha \right] dt = 0 ,$$

which by virtue of an arbitrary $\varphi(t)$ can be satisfied only if for all $t_1 < t < t_2$ true

$$\frac{d^2\alpha}{dt^2} = -\frac{q}{l}\sin\alpha\,. \tag{4}$$

Note that the equation of oscillations of the pendulum (4) is nonlinear as distinct from equation (5) in subsection 2. This circumstance is connected with the more complicated geometry of the system "rod–weight", namely: the acceleration of the weight is not proportional to the coordinate, as in Hook's law, but is a complex function of the deviation from the position of equilibrium (angle α). If these deviations are small, then $\sin\alpha \approx \alpha$, and the model of small oscillations is linear:

$$\frac{d^2\alpha}{dt^2} = -\frac{g}{l}\alpha\,.$$

They are described by a formula similar (6) from subsection 2, where $\omega = \sqrt{g/l}$ is the eigen-frequency of small oscillations, and the magnitudes A, B are determined through $\alpha(t=0)$, $\frac{d\alpha}{dt}\,(t=0)$.

4. Conclusion. The examples of this use of the Hamiltonian principle for constructing models of mechanical systems enable one to draw up a rather precise program of actions, in a general form described in subsection 1. The universality of the the successive procedures not depending on the details of the concrete system, certainly, are the attractive features of variational principles. In the simple cases above mentioned models can easily be constructed in other ways. However for many other, more complicated objects, the variational principles appear to be actually the only method of constructing models. So, for example, the mechanical parts of the majority of robotic devices consist of a great number of various elements mutually connected in various ways. Their mathematical models include a large number of equations, obtained with the help of the variational principles. This approach is also successfully applied to other kinds of systems (physical, chemical, biological), for which the appropriate general statements about a character of their evolution (behavior) are formulated.

The circumstance that the Hamiltonian principle and other approaches give identical models is natural, in so far as they describe the same initial objects. Certainly, such a coincidence is guaranteed only when initial assumptions about the object are the same. If its idealization (as one of the initial phases of constructing a model) will be carried uniformly, the different modes of constructing models should give identical results. For example, in the system "ball–spring", let an additional constant external force acts on the ball F_0. Then from Newton's second law it is easy to derive the equation of motion of the ball

$$m\,\frac{d^2r}{dt^2} = -kr + F_0$$

(compare with (5) of section 2). Applying the Hamiltonian principle to such a system, it is necessary to take into account the presence of this force. Obviously, the definitions of generalized coordinates, velocity and kinetic energy E_k remain the same. At the same time, the expression for the potential energy is modified essentially (compare with subsection 2) – on the quantity equal to work performed by this force:

$$E_p = k\,\frac{r^2}{2} + \int\limits_0^r F_1\,dr = k\,\frac{r^2}{2} + F_0 r\,.$$

Using steps similar to those in subsection 2 involving the quantities L and Q, it is easy to be convinced that the Hamiltonian principle gives the above-derived equation with an external force F_0.

EXERCISES

1. Check the validity of the construction of the model (4), deriving it using Newton's second law.

2. Using the results of section 2.2 and the Hamiltonian principle, construct a model of the oscillations of a pendulum in an electrical field of a charged horizontal plane, above which the pendulum is hung. The charge of the weight is q, the surface charge density is $-q_0$ (the gravity is neglected). Why is the model obtained analogous to (4), despite the varied nature of the acting forces?

4 Example of the Hierarchy of Models

We are constructing a hierarchical chain of models based on the principle "from bottom to top" to describe the motion of a ball connected to a spring. We will now introduce step by step new complicating factors and will give their mathematical descriptions.

1. Various modes of action of the given external force. Let a known external force $F(r, t)$ acts on a ball, depending on time and the position of the ball. It can be generated by gravitational field (see exercise 1), or have electrical or magnetic origin, etc. From Newton's second law we readily obtain that in comparison with the basic model of oscillations

$$m\,\frac{d^2r}{dt^2} = -kr \tag{1}$$

in the right hand side of equation (1) there is an additional term

$$m\,\frac{d^2r}{dt^2} = -kr + F(r, t)\,. \tag{2}$$

The simplest version of equation (2) corresponds to a case of a constant force $F(r,\, t) = F_0$. Substituting $\bar{r} = r - F_0/k$, we obtain for $\bar{r}$

$$\frac{d^2\bar{r}}{dt^2} = -kr\,,$$

i.e. the constant force does not lead to any modifications in the process of oscillations with the exception that the coordinate of the neutral point, where the force acting on the ball is equal to zero, is shifted on the magnitude F_0/k.

Much more complex motion can be originated by a time-dependent force $F(t)$. Consider, for example, a periodic external force $F(t) = F_0 \sin\omega_1 t$:

$$m\,\frac{d^2r}{dt^2} = -kr + F(t) = -kr + F_0 \sin\omega_1 t\,. \tag{3}$$

The solution of the linear equation (3) is obtained as the sum of a general solution of the homogeneous equation (formula (6) in section 2) and of a partial solution of the inhomogeneous equation (3), which we search in the form

$$r_1(t) = C \sin\omega_1 t\,. \tag{4}$$

By substitution of this expression into (3), we obtain

$$C = \frac{F_0}{k - m\omega_1^2} = \frac{F_0}{m(\omega^2 - \omega_1^2)}\,,$$

where $\omega = \sqrt{k/m}$ is the frequency of oscillations of the spring in the absence of external forces, or the eigen-frequency of the system. To sum up, for the general solution of (3) we have

$$r(t) = A \sin\omega t + B \cos\omega t + \frac{F_0}{m(\omega^2 - \omega_1^2)} \sin\omega_1 t\,.$$

Thus, the external force $F(t)$ leads not only to the emergence in the system of additional oscillations with frequency ω_1, but also to a resonance – the unlimited growth of oscillation amplitude at $\omega_1 \to \omega$.

2. Motion of an attaching point, the spring on a rotating axis. The resonance in a system can be determined also by the action of forces of inertial origin. Let the attaching point of a spring move by a given law $r_0(t) = f(t)$. Then in coordinates connected to this point, apart from the tension of a spring, a force $ma(t)$ is acting on the ball, where $a(t)$ is the acceleration stipulated by the motion of the coordinates, $a(t) = d^2f/dt^2$. In this coordinates the motion of the ball is described by the equation

$$m\,\frac{d^2r}{dt^2} = -kr + F(t)\,,$$

where $F(t) = -ma(t) = -m\,d^2f/dt^2$ is a given function of time. As in the previous case, in the presence of the appropriate periodic motion of the attaching point a resonance is appearing in the system.

In more complicated geometry the inertial forces of the system can depend not only on time t, but also on the coordinate r. If the spring is mounted over an axis moving with an angular velocity $\omega(t)$, the centrifugal force of inertia is $F = mv^2(t)/R = m\omega^2(t)R$, where $v(t) = \omega(t)\,R$, $R = R{+}r$, R_0 is the length of the spring in unloaded state, r is the shift of the ball from the neutral position, $r > -R_0$. The equation of motion of the ball has the form

$$m\,\frac{d^2r}{dt^2} = -kr + F(r,\,t)\,, \tag{5}$$

where $F(r,\,t) = m\omega(t)\,(R_0 + r)$, or

$$m\,\frac{d^2r}{dt^2} = -(k - m\omega^2(t))\,r + m\omega^2(t)\,R_0\,,$$

and, obviously, at $r \ll R_0$ the linear equation (5) (its general solution is not represented here by virtue of its complicated form), turns to the equation (3) with $F(t) = -m\omega^2(t)\,R_0$.

However, a resonance in this case is impossible, in so far as the external force is always aimed in one direction and cannot shake the system.

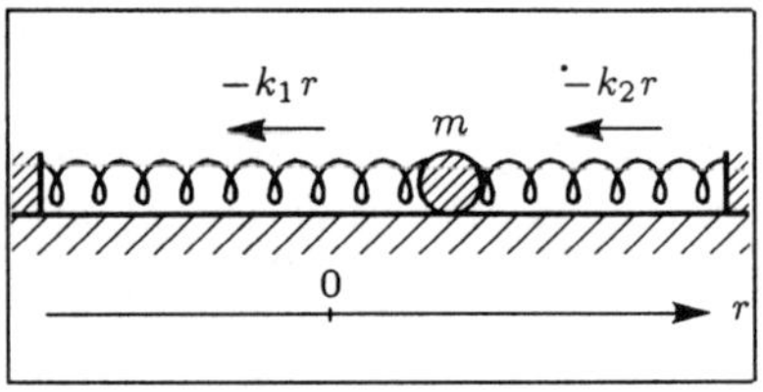

Fig.13.

Note that more complicated geometry (compared with the initial one), by no means implies more complicated behavior of the object. Consider, for example, a ball attached to two springs with rigidity k_1 and k_2 (Fig. 13). The beginning of coordinates we shall locate at a point where the forces acting on the ball by both springs, counterbalance each other (then, some condition on the parameters of the system should be fulfilled, so that the ball could not touch one of the attaching points – see exercise 2). By the Hook's law at deviation r the left spring is acting on the ball with the force $-k_1r$, and the right spring is acting with the force $-k_2r$ (both forces are acting in the same direction, in so far as at the expansion of the first spring the second one is contracting). To sum up, we come to the same equation, as in the case of one spring,

$$m\,\frac{d^2r}{dt^2} = -k_1r - k_2r = -kr\,,$$

but with a greater rigidity $k = k_1 + k_2$, combined from the rigidity of both springs.

3. Accounting for the forces of friction. In the considered system the forces of friction can occur for at least two reasons. The first is the non-ideality of the surfaces of the ball and the plane. In this case the force of friction is equal $F = k_1 P$, where k_1 is the coefficient of friction, $P = mg$ is the weight of the ball. It is always directed against motion of the ball, and its sign is opposite to the sign of the velocity of the ball $v = dr/dt$, $F = -k_1\, mg\ \mathrm{sign}(dr/dt)$. The motion of the ball is governed by the equation

$$m\,\frac{d^2r}{dt^2} = -kr - k_1\, mg\ \mathrm{sign}\,\frac{dr}{dt}\,, \tag{6}$$

which seems similar to equation (2) with constant force $F(r,\,t) = F_0$. However, because of the variation of the sign of the force it is not reduced to the standard equation of oscillations. This circumstance indicates that the equations (1) and (6) describe essentially different processes. Particularly, the amplitude of the ball oscillations in the latter case decreases with time. It is easy to show this, rewriting (6) in the form

$$m\,\frac{dv}{dt} + kr = -k_1\, mg\ \mathrm{sign}\,v\,,$$

multiplying both parts of this expression by $v/2$:

$$m\,\frac{v}{2}\,\frac{dv}{dt} + kr\,\frac{v}{r} = -k_1\, mg\ \mathrm{sign}\,v\,\frac{v}{2}\,,$$

and in view of $v = dr/dt$, obtaining

$$\frac{m}{2}\,\frac{dv^2}{dt} + \frac{k}{2}\,\frac{dr^2}{dt} = -\frac{1}{2}\,k_1\, mg\ \mathrm{sign}\,v\cdot v\,.$$

The latter equation is equivalent to the equation

$$\frac{d}{dt}\left(\frac{mv^2}{2} + k\,\frac{r^2}{2}\right) = -\frac{1}{2}\,k_1\, mg\ \mathrm{sign}\,v\cdot v\,. \tag{7}$$

Taking into consideration that the left hand side of (7) under the operation of a derivative includes the sum of kinetic and potential energies $E(t) = E_r(t) + E_p(t)$, while the right hand side of (7) at $v \neq 0$ is negative, we have

$$\frac{dE(t)}{dt} < 0\,, \qquad v \neq 0 \qquad \left(\frac{dE(t)}{dt} = 0\,, \qquad v = 0\right),$$

i.e. the total energy $E(t)$ is decreasing with time. In so far as at the moments the maximum amplitude $|r_m(t)|$ is reached the ball's velocity (and the kinetic energy E_k) is equal to zero, then at this moments $E_p = kr_m^2(t)/2 = E(t)$, and due to the decrease of $E(t)$ the amplitude $|r_m(t)|$ is also decreasing a function of time.

Consider in more detail the consequences of the action of the force of friction of another origin, originating due to the resistance of the medium in which the ball is moving (air, water etc.). In this case the force of friction is not constant, but essentially depends on the velocity of motion. This dependence is described by the well-known Stokes' formula

$$F = -\mu v = -\mu \frac{dr}{dt},$$

where the coefficient $\mu > 0$ is determined by the size of the ball, the density of the medium, its viscosity and so on. The equation of motion in a viscous medium has the form

$$m \frac{d^2 r}{dt^2} = -kr + F(v) = -kr - \mu \frac{dr}{dt}. \tag{8}$$

Let us obtain the general solution of the linear equation (8), first getting rid of the term with the first derivative. The substitution in (8) of $r(t) = \bar{r}(t)\, e^{\alpha t}$ gives the equation for a new function $\bar{r}(t)$

$$m \left(e^{\alpha t} \frac{d^2 \bar{r}}{dt^2} + \alpha e^{\alpha t} \frac{d\bar{r}}{dt} + \alpha e^{\alpha t} \frac{d\bar{r}}{dt} + \alpha^2 e^{\alpha t} \right) =$$

$$= -k\bar{r}\, e^{\alpha t} - \mu e^{\alpha t} \frac{d\bar{r}}{dt} - \mu\, \alpha\, e^{\alpha t}\, \bar{r}.$$

Reducing the multiplier $e^{\alpha t}$ and denoting $\alpha = -\mu/(2m)$, we arrive at the equation

$$m \frac{d^2 \bar{r}}{dt^2} = -\left(k - \frac{\mu^2}{4m}\right) \bar{r} = -k_1 \bar{r}. \tag{9}$$

As distinct from equation (1), the first multiplier on the right hand side of (9) can change its sign depending on the values of parameters k, μ, m; in view of the relation $r(t) = e^{\alpha t}$, this leads to an essentially different behavior as compared with the standard case.

At low viscosity, i.e. when $k - \mu^2/(4m) = k_1 > 0$ the solution $\bar{r}(t)$ is given by the formula (6) of section 2, and for $r(t)$ we have

$$r = \bar{r}\, e^{\alpha t} = e^{-t\mu/(2m)} \left(A \sin \omega t + B \cos \omega t\right),$$

where $\omega = (k_1/m)^{1/2}$, and the constants A, B are obtained via r_0, v_0. Damping oscillations of frequency ω (see also exercise 3) occur in the system.

If $k_1 = 0$, then the magnitude $d\bar{r}/dt$ is constant, or, alternatively, $\bar{r}(t) = ct + c_1$. For $r(t)$ in view of initial conditions, we obtain

$$r(t) = e^{-t\mu/(2m)}(ct + c_1) = e^{-t\mu/(2m)}\left[\left(v_0 + \frac{\mu r_0}{2m}\right) t + r_0\right].$$

In this case oscillations are absent due to the overwhelming action of the forces of viscous friction. The system can pass only once the point $r = 0$, when the necessary and sufficient conditions $v_0 < -\mu r_0/(2m)$, $r_0 > 0$ or $v_0 > -\mu r_0/(2m)$, $r_0 < 0$ are fulfilled, i.e. the initial velocity of the ball should be rather high and directed to $r = 0$. Thus, the velocity of the ball $v(t) = dr/dt$ can obviously change its sign only once.

Finally, at high viscosity the action of a friction force is so significant, that for any r_0, v_0 the ball "sticks" in the medium, never passing the point $r = 0$, only approaching it at $t \to \infty$. Indeed, at $k_1 < 0$ the solution of equation (9) (see also exercise 4) is of a constant sign (the opposite assumption immediately leads to a contradiction with the equation), therefore, $r(t)$ does not change its sign either. The behavior of the function $\bar{r}(t)$ at $t \to \infty$ can be understood from the properties of the first integral of equation (9)

$$m\left(\frac{d\bar{r}}{dt}\right)^2 = -k_1\bar{r}^2 + \text{const},$$

which can be obtained easily by multiplying both sides of (9) on $d\bar{r}/dt$ and integrating once by t. The assumption that $\bar{r}(t) \to \infty$ or $\bar{r}(t) \to C_1 \neq 0$ at $t \to \infty$ contradicts the latter equality. Only one possibility remains, $\bar{r}(t) \to 0$, $t \to \infty$, and, thus, $r(t) \to 0$, $t \to \infty$.

Thus, the motion of a system in a viscous medium is distinguished by essential diversity as compared with the ideal situation, and in all cases occurs with damping.

4. Two types of nonlinear models of the system "ball–spring". Strictly speaking Stokes' formula is valid only for steady motions, when the action of constant external force is balanced by the force of viscous friction in such a way that the body is moving with a constant velocity. While it is possible to conceive of situations in which the resistance of the viscous medium at low velocities is smaller, and at higher velocities is greater than that given by the Stokes formula; for example, $F(v) = -\mu v\,|\,v\,|^{\alpha}$, where $\mu > 0$, $\alpha > -1$. Then, the sought quantity $r(t)$ is determined from the equation

$$m\frac{d^2r}{dt^2} = -kr + F(v) = -kr - \mu v\,|\,v\,|^{\alpha}. \qquad (10)$$

The equation (10), as distinct from all the models considered above, is nonlinear, and its solution, generally speaking, cannot be rewritten (though it is also possible to study in detail the system in a nonlinear case, in particular, to establish the damping character of the motion as in the case of equation (6)). Therefore we shall confine ourselves by the approximate analysis of the behavior of the system to two limiting positions – in the neighborhood of points $v = 0$ and $r = 0$. Obviously, both positions cannot be achieved simultaneously, otherwise it should mean that the system is at rest.

If $v(t_0) = 0$ (here the moment t_0 is one of the moments at which maximum amplitude r_0 is reached), the second term in the right hand side of equation (10) can be neglected as compared with the first one, and it takes the form

$$m\,\frac{d^2r}{dt^2} = -kr_0\,.$$

In so far as the small neighborhood Δt of moment t_0 is considered, one can also neglect the deviation Δr as compared with r_0. Taking into account that $v(t_0) = 0$, we obtain

$$\Delta r = r - r_0 = -\frac{1}{2}\,\frac{k}{m}\,r_0\,(t - t_0)^2\,,$$

i.e. the ball is moving with a constant (in the first approximation) acceleration, since only the force of tension of the spring is acting constantly in the vicinity of the point $r = r_0$, and the force of friction is zero.

At $r(t_0) = 0$ (t_0 is one of the moments of passage by the system of the beginning of coordinates, if, certainly, the point $r = 0$ is reached at least once) the first term in the right hand side is small compared with the second one, and

$$m\,\frac{d^2r}{dt^2} = -\mu v_0\,|\,v_0\,|^\alpha\,.$$

Here also we neglect the deviation Δv from the value $v_0 = v(t_0)$ in view of its small size. So far as $r(t_0) = 0$, from the last equation we have

$$\Delta r = r = \frac{-\mu v_0\,|\,v_0|^\alpha}{2m}\,(t - t_0)^2 + v_0\,(t - t_0)\,.$$

This means that in this position the system undergoes a constant (in the first approximation) acceleration, determined only by the force of friction, as the tension of the spring is equal to zero. This conclusion is quite obvious and is valid for all positions of the system, though the acceleration of the ball at $v \neq 0$, $r \neq 0$ is determined already by the joint action of both forces. The only exception is the point where $k_0 = -\mu v_0\,|\,v_0\,|^\alpha$, when the right hand side of the equation (10) turns to zero and the first term in the acceleration in the moment $t = t_0$ equals zero. Expanding $r(t)$ via Taylor series in the neighborhood of the point $t = t_0$

$$r(t) = r(t_0) + \frac{dr}{dt}(t_0)(t - t_0) +$$
$$+\frac{1}{2}\frac{d^2r}{dt^2}(t_0)(t - t_0)^2 + \frac{1}{6}\frac{d^3r}{dt^3}(t_0)(t - t_0)^3 + \dots,$$

where the points denote terms of higher order of smallness, and taking into consideration, that $\frac{d^2r}{dt^2}(t - t_0) = 0$, we obtain

$$\Delta r = r(t) - r_0 = \frac{dr}{dt}(t_0)(t - t_0) + \frac{1}{6}\frac{d^3r}{dt^3}(t_0)(t - t_0) + \dots,$$

i.e. in the principal term the acceleration of the system in the neighborhood of the point $t = t_0$ is not constant, and is a linear function of time (see also exercise 5).

One more type of nonlinearity can be determined by the variation of the mechanical properties of a spring. Hook's law is valid, generally speaking, only for small deviations of the spring from the unloaded neutral position. For noticeable deformations the spring, depending on its matter and the magnitudes of deformation can behave as "soft", and then the tension will be less than that given by Hook's law (in the case of "rigid" springs – the opposite is true). The rigidity of the spring in such situations becomes a function of a coordinate, i.e. $k = k(r)$, and the equation of motion becomes

$$m\frac{d^2r}{dt^2} = -k(r)\,r, \tag{11}$$

where obviously $k(r) > 0$. For example, if $k(r) = k_0/(1 + |r|))$, then the spring is soft. Equation (11) is as nonlinear as equation (10), though at least two differences between (10) and (11) are obvious. As distinct from (10), it is possible to write down the implicit solution of (11) with repeated use quadratics. Besides, rewriting (11) in the form

$$m\frac{dr}{dt}\frac{d^2r}{dt^2} = -\frac{dr}{dt}k(r)\,r = -\frac{dr}{dt}\frac{d}{dr}\int_0^r k(r')\,r'\,dr' = -\frac{d}{dt}\left(\int_0^r k(r')\,r'\,dr'\right),$$

and taking into account that the left hand side of this expression is $mv\,dv/dt = \frac{1}{2}m\,dv^2/dt$, and integrating it by t, we obtain

$$E_k + E_p = \frac{1}{2}m\left(\frac{dr}{dt}\right)^2 + \int_0^r k(r')\,r'\,dr' = \text{const} > 0. \tag{12}$$

This denotes the *conservatism* of the motion described by the model (11), or constancy of the total energy of the system. The existence of the first integral (12) allows us to reveal something in the common with the case of a linear system – the oscillatory character of the motion. Indeed, from (12) the boundedness of functions $v(t) = dr/dt$ and $r(t)$ follows at any $t > 0$. The solution has no limit at $t \to \infty$, and in so far as at $v(t \to \infty) \to v_\infty \neq$ it would contradict the boundedness of function $r(t)$ at $t \to \infty$. For $v(t \to \infty) \to v_\infty = 0$ it is impossible that $r(t \to \infty) \to r_\infty \neq 0$ because from (11) the absence of limitation of $v(t)$ would follow at $t \to \infty$ (the case $v_\infty = r_\infty = 0$ contradicts (12)). Thus, the ball oscillates. It passes the point $r = 0$ an infinite number of times (otherwise $r(t)$ would be sign-definite, along with the acceleration d^2r/dt^2 (see (11)) and $v \to \infty$, $t \to \infty$).

5. Conclusion. The constructions considered in this section demonstrate the hierarchical chain of models of the system "ball-spring", obtained one from another via successive refusals from the assumptions idealizing the investigated object. In some cases the complications do not introduce anything new into the behavior of the system (constant external force, ball on two springs), in others – its properties change essentially. The path "from the simple to the complex" enables one to study more and more realistic models and to compare their properties.

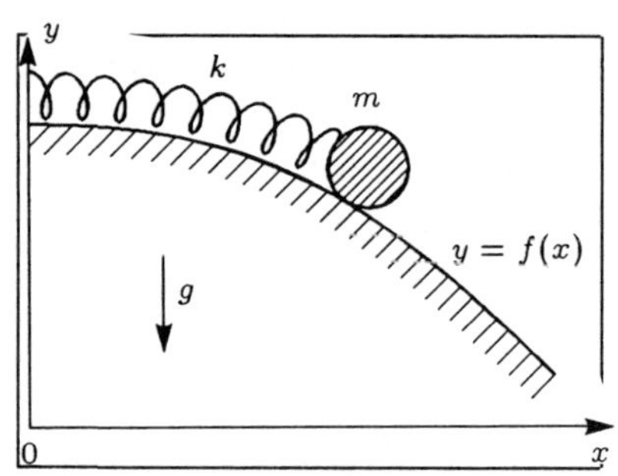

Fig.14.

There is another way to construct and study models – "from the general to the specific". From the results of the present section it is obvious: the sufficiently general equation of motion of the system "ball-spring" is given as follows

$$m\frac{d^2r}{dt^2} = -k(r,\,t)r + F\left(r,\,t,\,\frac{dr}{dt}\right), \qquad k > 0\,,$$

where k and F can be various functions of the arguments. Based on this general model, it is possible, using appropriate concrete assumptions, to obtain and study more simple models. For example, the dependence of k on r and t is given via the equation following equation (5) and equation (11). The dependence of F on r, t implies the presence of an external force or force of inertia (equations (2), (3) and (5)), while dependence on dr/dt implies the resistance of a medium (equations (6), (8) and (10)). The given approach is also widely applied, because it enables one to promptly establish the general features of the object, defining them concretely in more particular situations.

EXERCISES

1. Derive the equation of motion of a ball on a spring, moving on an ideal surface with non-uniform declination under the action of the force of tension of the spring and gravity. The equation of the surface is: $y = f(x)$, $y' < 0$ (Fig. 14).

2. Select the values of k_1, k_2, m, r_0, v_0 in the system in Fig. 13 to exclude the ball touching the attaching points.

3. Using the technique used at the analysis of the equation (6), show that in the case of equation (8) the motion occurs with damping.

4. Write down the solution of equation (9) at $k_1 < 0$ through hyperbolic functions, and be convinced that the solution of equation (8) tends to zero at $t \to \infty$.

5. Using expansion via Taylor series of the function $r(t)$ in the neighborhood of a point $t = t_0$, where the right hand side of equation (10) is equal to zero, obtain the quantity $d^3r/dt^3\,(t_0)$.

5 The Universality of Mathematical Models

Consider the processes of oscillations for different bodies. We will show that despite different essence of bodies, the same mathematical models correspond to them.

1. Fluid in a U-shaped flask. The fluid fills part of a U-shaped flask; it represents a bent pipe of a radius r_0 (Fig. 15). The mass of the fluid is M_0, its density is ρ_0. The walls of the flask are ideally smooth, the surface tension is neglected, atmospheric pressure P_0 and acceleration of gravity g are constant.

At the equilibrium the fluid, is obviously motionless, its height at either side of the U-bend is identical. If it is removed from the equilibrium, the motion will start with a character we will below establish with the help of energy conservation law, as long as our assumption that no energy loss exists in the system is correct.

We will calculate the potential energy of a system through work, which is necessary to shift it from the equilibrium state (where $h_1 = h_2$) to a position represented in Fig. 15. It is

$$E_p = -\int_{\bar{h}}^{h_2} P\,dh_2 = -\int_{\bar{h}}^{h_2} \rho_0\, s_0\,(h_1 - h)\,g\,dh\,, \qquad \bar{h} = \frac{h_1 + h_2}{2}\,, \qquad s_0 = \pi\, r_0^2\,,$$

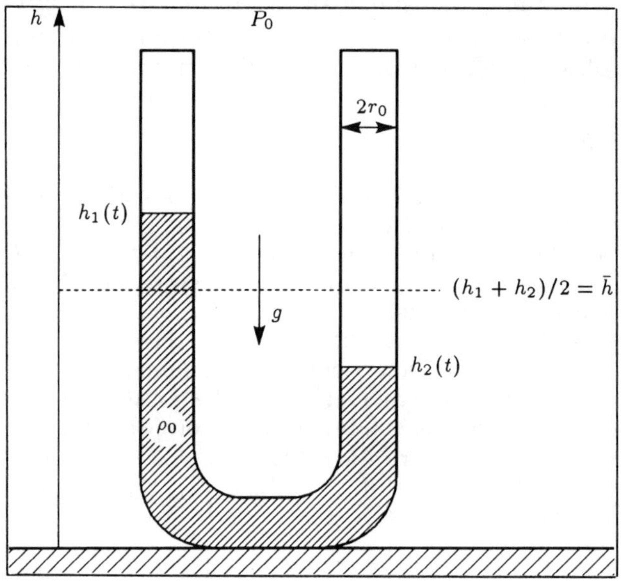

Fig.15.

where P is the weight of the part of the fluid to the left side of the bend; its level exceeds the magnitude h_2. The work of the forces of atmospheric pressure is equal to zero, in so far as for different bends the corresponding displacements have different directions.

The unknown quantities $h_1(t)$ and $h_2(t)$ are connected by an obvious relation $h_1(t) + h_2(t) = \text{const} > 0$, expressing a constance of total length of the pillar of the fluid with constant cross-section. Substituting the last equality into the expression for E_p, we obtain after integration

$$E_p = -\rho_0 \, s_0 \, g \left[-h_2^2(t) + Ch_2(t) + C_1\right].$$

At the estimation of kinetic energy we shall take into account the constance of the cross-section of the tube and the incompressibility of the fluid. It means that the pillar of fluid is moving as a whole, and its velocity $v(t)$ is identical in all cross-sections. Adopting for $v(t)$ the quantity $dh_2(t)/dt$, then

$$E_k = \frac{1}{2} M_0 \left(\frac{dh_2}{dt}\right)^2,$$

and from the energy conservation law it follows

$$E(t) = E_k(t) + E_p(t) = \frac{M_0}{2} \left(\frac{dh_2}{dt}\right)^2 - \rho_0 \, s_0 \, g \left(-h_2^2 + Ch_2 + C_1\right).$$

In so far as $dE/dt = 0$, differentiating this expression, we obtain

$$M_0 \frac{d^2 h_2}{dt^2} \rho_0 \, s_0 \, g \, (-2h_2 + C) \, ,$$

which in view of the same relation for the $h_1(t)$ gives the equation

$$M_0 \frac{d^2 h}{dt^2} = -\rho_0 \, s_0 \, g \, h = -\pi \, \rho_0 \, r_0^2 \, g \, h \, ,$$

where $h = (h_2 - h_1)/2$ is the deviation of the level of the fluid from the equilibrium position. It completely coincides within equation (1) section 4 for the system "ball-spring" (in this case the pillar of fluid is analogous to a ball, while the gravity takes the role of the spring).

The successive refusal from the idealization of the object gives more models (as in subsection 4). For example, accounting for the force of surface tension $\sigma \, 2\pi \, r_0$ (σ_0 is the coefficient of the surface tension) always directed against the motion of the fluid, leads to the equation of the type (7) (section 4) for the quantity h (see also exercise 1).

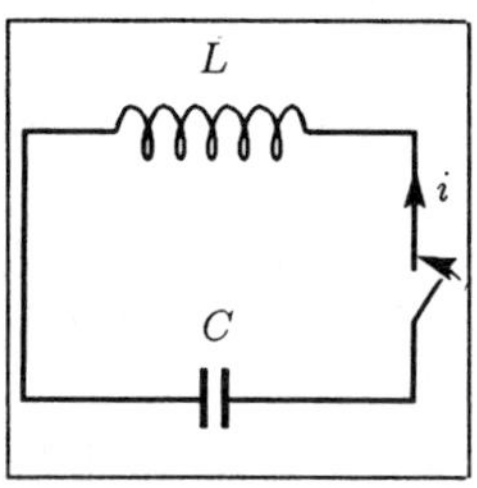

Fig.16.

2. An oscillatory electrical circuit. This device represents a capacitor, connected with wires to an inductive coil. At the moment $t = 0$ the circuit is closed and the charge of the plates of the capacitor begins to propagate over the circuit (Fig. 16).

The resistance of wires is considered equal to zero, the capacity of the capacitor is C, induction of the coil is L. For changing in time quantity $q(t)$, where $q(t)$ is the charge on the plates of the capacitor, is necessary to obtain the appropriate equation. Obviously, the current $i(t)$ and the voltage $v(t)$ are also functions of time.

By the physical content of the quantity C at any moment of time we have the equality $v(t) = q(t) \, C$ (the capacity is equal to the magnitude of the charge on the plate of the capacitor required for the increase of a potential difference by unity).

In so far as the electrical resistance in the circuit is absent, no loss of voltage in the wires exists, and the difference of potentials $v(t)$ of the capacitor, is immediately passed to the coil. At variable current in a coil an electromotive force of self-induction appears, equal to $\varepsilon = -L\,di/dt$. The Ohm law for the circuit in the absence of a resistance is as follows:

$$v(t) = -\varepsilon(t)\,,$$

or

$$q(t)\,C = -\varepsilon(t) = L\,di/dt\,.$$

So far as by definition $i = -dq/dt$ (at a change of the charge on the capacitor the current appears in a circuit), from the last relation we obtain

$$L\,\frac{d^2q}{dt^2} = -Cq\,,$$

describing the oscillations of $q(t)$ (and consequently of $i(t)$, $v(t)$) in the simplest electrical circuit identical to (1) of section 4. In the system "capacity–induction" the oscillations occur in the same manner and in the system "ball–spring" (and analogously the models become more complicated when additional processes are taken into account – see exercise 2).

3. Small oscillations at the interaction of two biological populations. Let two biological populations of numbers $N(t)$ and $M(t)$ coexist in the same territory, with the first ones being vegetarians, and the second ones being fed by representatives of the first population.

The rate of variation of $N(t)$ is determined by the first term on the right hand side of formula (10), section 1, describing the growth due to birth (effect of saturation is not taken into account; compare with (12), section 1) and from the rate of decrease due to the presence of the second population:

$$\frac{dN}{dt} = (\alpha_1 - \beta_1 M)\,N\,, \tag{1}$$

where $\alpha_1 > 0$, $\beta_1 > 0$, and the term $\beta_1\,M\,N$ describes the enforced decrease (the natural mortality of population is neglected).

The greater the number of the first population, the faster the second population expands, while at its absence decreases with a rate proportional to the number $M(t)$ (thus its birth rate is not taken into account, as well as the effect of saturation):

$$\frac{dM}{dt} = (-\alpha_2 + \beta_2 N)\,M\,, \tag{2}$$

where $\alpha_2 > 0$, $\beta_2 > 0$.

Obviously, the system is in an equilibrium at $M_0 = \alpha_1/\beta_1$ and $N_0 = \alpha_2/\beta_2$, when $dN/dt = dM/dt = 0$. Consider small deviations of the system from the equilibrium values, i.e. represent the solution as $N = N_0 + n$, $M =$

$M_0 + m$, $n \ll N_0$, $m \ll M_0$. Substituting N and M into the equations (1), (2), we obtain (neglecting the terms of higher order of smallness)

$$\frac{dn}{dt} = -\beta_1 N_0 m\,, \tag{3}$$

$$\frac{dm}{dt} = -\beta_2 M_0 n\,. \tag{4}$$

Differentiating (3) by t and substituting into the obtained equation the function dm/dt defined from (4), we shall come to the equation

$$\frac{d^2n}{dt^2} = -\alpha_1\alpha_2 n\,,$$

analogous to equation (1), section 4. Therefore, small oscillations of number with frequency $\omega = \sqrt{\alpha_1, \alpha_2}$ occur in the system, depending only on coefficients of birth and death α_1 and α_2.

Note that the value of $m(t)$ satisfies a similar equation, so that if the deviation $n(t)$ is zero in an initial moment $t = 0$, then $m(t = 0)$ has maximum amplitude, and vice versa (see the solution of the equation of oscillations (6), section 2). This situation, when the numbers $n(t)$ and $m(t)$ are in antiphase, is reproduced at all moments $t_i = iT/4$, $i = 1, 2 \ldots$, (T is the period of oscillations) and reflects the delay in the number of one population responding to a variation in the number of another (see also exercise 3).

4. Elementary model of variation of salary and employment. The trade market, where the employers and employees are interacting, is characterized by the salary $p(t)$ and occupation number $N(t)$. Let an equilibrium exist, i.e. a situation, when $N_0 > 0$ persons agree to work for salary $p_0 > 0$. If for any reasons this equilibrium is violated (for example, if some of the workers retire or the employers have financial difficulties), the functions $p(t)$ and $N(t)$ deviate from values p_0, N_0.

Consider that the employers change the salary proportionally to the deviation of number of employees from their equilibrium value. Then

$$\frac{dp}{dt} = -\alpha_1 (N - N_0)\,, \qquad \alpha_1 > 0\,.$$

Assume that the number of the workers also increases or decreases proportionally to the increase or decrease of the salary with respect the value p_0, i.e.

$$\frac{dN}{dt} = \alpha_2 (p - p_0)\,, \qquad \alpha_2 > 0\,.$$

Differentiating the first equation by t and excluding N from it with the help of the second equation, we come to a standard model of oscillation

$$\frac{d^2(p-p_0)}{dt^2} = \alpha_1\,\alpha_2\,(p-p_0)$$

of the salary relative the equilibrium (analogously for $N(t)$). From the first integral of this equation

$$\alpha_1\,(N-N_0)^2 + \alpha_2\,(p-p_0)^2 = \text{const} > 0$$

it is clear that in some moments $t = t_i$, $i = 1, 2, \ldots$, when $p = p_0$ (i.e. the salary becomes equal to the equilibrium value), we have $N > n_0$, i.e. the number of employees is more than the equilibrium value, while at $N = N_0$ we obtain $p > p_0$, i.e. the salary exceeds the equilibrium value. At these moments the salary fund, being equal to pN, exceeds the equilibrium value p_0N_0 (or is less than it), if the conditions $p > p_0$ or $N > N_0$ are fulfilled while approaching the moment t_i(and vice versa). Though on average over a period of oscillations the value of pN equals p_0N_0 (exercise 4).

5. Conclusion. The models constructed in the present section, in one case are based on exactly known laws (subsections 1 and 2), in others are based on the observable facts and on analogies (subsection 3), or on reasonable assumptions about the character of the object (subsection 4). Though the essence of the considered phenomena, and the approaches to deriving the adequate models are totally different, the constructed models appeared to be identical to each other. This reveals the crucial property of mathematical models – their *universality*, which is widely used at the study of systems of very different natures.

EXERCISES

1. In the problem about a U-shaped flask, let the left side have a variable cross-section, i.e. $r = r_0(h)$. Applying Newton's second law and assuming the absence of a horizontal velocity component of the fluid, show that for the quantity h an equation of type (11), section 4, is obtained.

2. Introducing a resistance R in the LC-circuit and using the Ohm law, prove that the model of oscillations in LCR-circuit is similar to the equation (8), section 4.

3. Reduce the nonlinear system (1) and (2) to a second order equation and show that it (as well as its linear analog (3) and (4)) has the first integral.

4. Using the formula (6) of section 2 for the general solution of the equation of oscillations, show that average value of the salary fund pN (subsection 4) during the period of oscillations is equal to its equilibrium value.

6 Several Models of Elementary Nonlinear Objects

Let us discuss the origin of nonlinearity and consider several consequences of it in the behavior of investigated objects. We will also illustrate the inevitability of application of numerical methods for their analysis.

1. On the origin of nonlinearity. As was mentioned in section 1.5, linear models obey the principle of superimposition. In this case, obtaining partial solutions and summarizing them, it is generally possible to construct a general solution (typical examples – formula (6), section 2 and the formula for the general solution of equation (3), section 4 in models of oscillations).

For nonlinear models the principle of superimposition is inapplicable, and the general solution can be found only in rare cases. Individual partial solutions of nonlinear equations may not reflect the character of the behavior of objects in a more general situation.

Nonlinearity can have many reasons. The fundamental laws of nature – the law of gravitation and the law of Coulomb – are nonlinear by origin (square-law relation of the force of interaction between masses or charges), and consequently the models based on them, generally speaking, are also nonlinear. Also contributing to the nonlinearity of models are the more complicated geometry of the phenomenon (see exercise 1, section 4 and exercise 1, section 5), various external actions (see equation (10) section 4) and, certainly, the change of the character of interaction in the object while changing its state (effect of saturation in models of populations, varying rigidity of the spring).

In essence only nonlinear models correspond to real phenomena, and linear models are valid only for descriptions of minor changes of parameters characterizing the object.

2. Three regimes in a nonlinear model of population. As distinct from the Malthus model (10), section 1, and model (12), section 1, the coefficient of birth we shall consider dependent on the population $N(t)$, i.e. $\alpha = \alpha(N)$. The coefficient of mortality β also depends on N. The equation of population dynamics

$$\frac{dN}{dt} = [\alpha(N) - \beta(N)]\, N \tag{1}$$

is nonlinear due to the variation of interaction characteristics within the population during the change of its state.

Assume for the sake of definiteness $\beta(N) = \beta_0 = \text{const}$, $\alpha(N) = \alpha_0(N)$, i.e. the birth rate is proportional to the population number (for example, because the population is interested in its own growth). Then the equation

(1) will be transformed to

$$\frac{dN}{dt} = \alpha_0 N^2 - \beta_0 N \tag{2}$$

with quadratic nonlinearity (peculiar also for some chemical reactions). Consider the behavior of function $N(t)$ at various initial populations $N(0) = N_0$ (Fig. 17).

a) At $N_0 < N_{cr} = \beta_0/\alpha_0$ the population monotonously decreases in time, tending to zero at $t \to \infty$. The solution is given by a formula similar to the solution of the equation (12), section 1, where t is replaced by $-t$ (inverse logistic curve; compare with Fig. 7, section 1).

b) At critical value $N_0 = N_{cr}$ the number of population does not depend on time.

c) At $N_0 > N_{cr}$ the nature of the solution drastically changes as compared with cases a) and b): the number increases so quickly that it tends to infinity at finite time $t = t_f$. The value of t_f is smaller, as N_0 increases (see exercise 1).

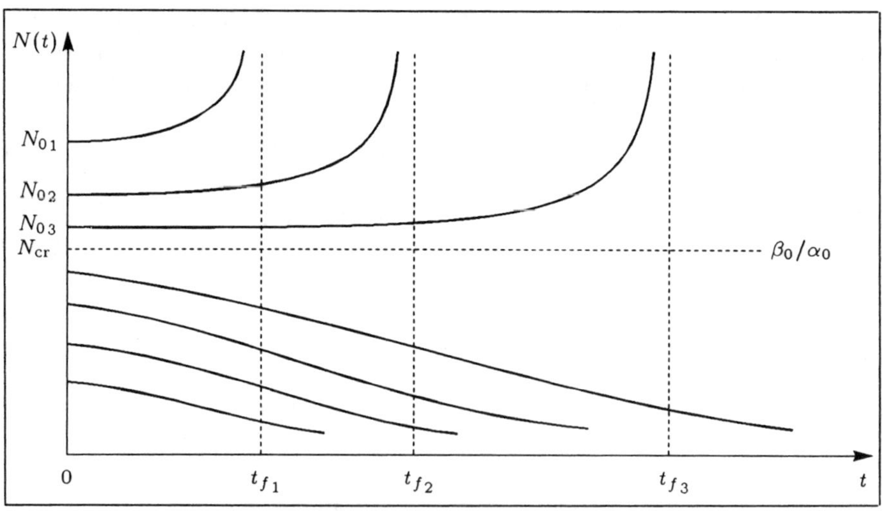

Fig.17.

The nonlinearity of equation (2) leads to a large variety of effects contained even in the elementary model: three possible regimes of variation of the population number in time; the instability of regime b) when at small deviations to the areas a) or c); the solution deviates from the line $N_{cr} = \beta_0/\alpha_0$; strong sensitivity of function $N(t)$ relative the initial conditions N_0; finally, the catastrophic growth of the number of population in finite time scale at $N_0 > N_{\text{cr}}$.

Note that the last property is not a particular result, but occurs for any models of the form

$$\frac{dN}{dt} = F(N), \qquad t > 0 \quad N(0) > 0, \quad F(N) > 0,$$

if at a large value of N the function $F(N)$ grows faster than the first order of N, or more precisely, if for $F(N)$ the following criterion is fulfilled

$$\int_{N(0)}^{\infty} \frac{dN}{F(N)} < \infty,$$

obtained by the direct integration of the equation.

3. Influence of strong nonlinearity on the process of oscillations. The equation of oscillations

$$m\,\frac{d^2r}{dt^2} = -k(r)\,r\,, \tag{3}$$

where the function $k(r) > 0$ describes the rigidity of a spring, is one of the few nonlinear equations for which one can write down the general solution. Introducing the velocity $v = dr/dt$, we rewrite (3) in the form

$$m\,\frac{dv}{dt} = -k(r)\,r\,, \qquad \frac{dr}{dt} = v\,;$$

dividing the first of these equations by the second one, we shall obtain the nonlinear equation of the first order

$$m\,\frac{dv}{dt} = -\,\frac{k(r)\,r}{v}\,. \tag{4}$$

Separating the variables in (4)

$$mv\,dv = -k(r)\,r\,dr\,,$$

and integrating the latter equation twice, we obtain

$$v^2 = \left(\frac{dr}{dt}\right)^2 = -2\int_0^r k'(r')\,r'\,dr' + C\,, \qquad k'(r) = \frac{k(r)}{m}\,,$$

$$\frac{dr}{dt} = \pm\sqrt{C - 2\int_0^r k'(r')\,r'\,dr'}\,,$$

$$t = \pm\int_0^r d\bar{r}\left(\sqrt{C - 2\int_0^{\bar{r}} k'(r')\,r',\,dr'}\right)^{-1} + C_1\,, \tag{5}$$

where in the implicitly written general solution (5) one can determine the constants C, C_1, knowing the initial conditions.

Note that this procedure is inapplicable in relation to the equations of type (10), section 4, (variables are not separated), and therefore it is impossible to find their general solution in this way.

In the linear case ($k(r) = k_0$) the integrated curve of the equation (4) represents concentric circles centered on the beginning of coordinates; the radius is determined by the initial energy of the system and "the motion" along circles describes a periodic process of oscillations (Fig. 18).

Consider now a strongly nonlinear system, when the spring behaves as "supersoft", for example, $k(r) = 1/(r^2 + \alpha)$, $\alpha > 0$. In a limiting case $\alpha = 0$ the equation (4) takes the form

$$m\,\frac{dv}{dr} = -\frac{1}{vr}\,,$$

and the solution drastically differs from solution (4) (see Fig. 19), since the energy is not conserved and, moreover, infinitely increases at $r \to \pm\infty$. When nonlinearity is weakened the process of oscillations takes the usual character (exercise 2).

4. On numerical methods. The examples considered here convincingly indicate the necessity of applying numerical methods to the modeling of nonlinear objects, because of clearly insufficient purely theoretical approaches and complicated and diverse behavior of the quantities describing those objects. However, this conclusion is fair also for linear models containing a large number of unknown parameters, independent variables and possessing complicated spatial structure. To construct appropriate numerical models, the methods and approaches developed at the creation of initial models are widely used, and there are specific problems requiring a deep investigation.

Let us explain this last statement by a simple example. For the equation (10), section 1

$$\frac{dN}{dt} = (\alpha - \beta)\,N = \gamma N\,, \qquad t > 0, \quad N(0) = N_0\,,$$

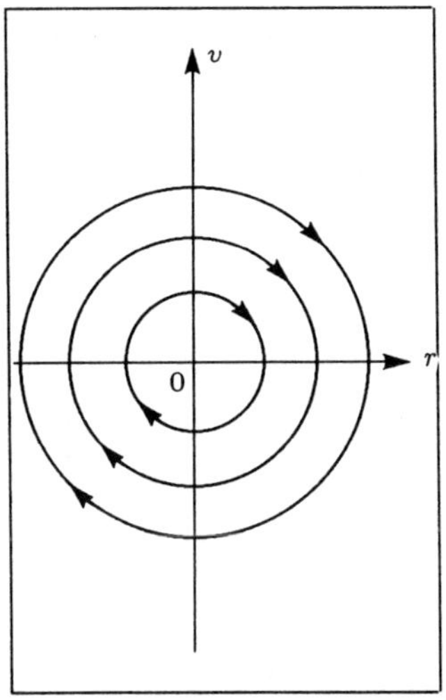

Fig.18.

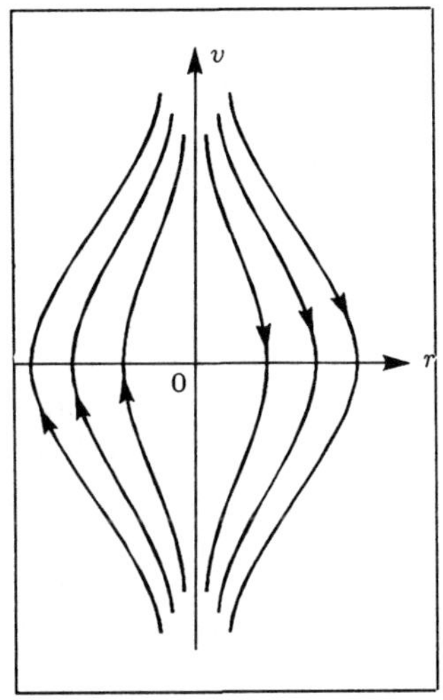

Fig.19.

where $(\alpha - \beta) > 0$, it is quite logical to suggest the following numerical procedure (by dividing the axis t into equal intervals $\tau = t_{i+1}$, $i = 0, 1, 2, \ldots$,; $t_0 = 0$ and by replacing the derivative with a finite difference):

$$\frac{N_{i+1} - N_i}{\tau} = \gamma N_i \qquad i = 0, 1, \ldots ; \quad N(t_0) = N_0 . \tag{6}$$

From (6) we obtain

$$N_{i+1} = (\tau\gamma + 1) N_i ,$$

this gives the solution

$$N_1 = (1 + \tau\gamma) N_0 , \qquad N_2 = (1 + \tau\gamma)^2 N_0 ,$$

$$N_i = (1 + \tau\gamma)^i N_0 = (1 + \tau\gamma)^{t/\tau} N_0 ,$$

i.e. at $t \to \infty$ the solution (6) may differ from the one sought as much as is possible. Therefore, to reach the required accuracy it is necessary to properly select the interval τ depending on the scale of integration T (exercise 3).

EXERCISES

1. Using the transformation applied at the analysis of the equation (8), section 4, obtain the solution of equation (2) at $N_0 > N_{\rm cr}$ and estimate t_f through N_0, α_0, β_0.

2. Find the restriction on the growth of function $k(r) \to \infty$, $r \to 0$ in equation (3), when the system "ball-spring" would become conservative, i.e. the total energy would conserve.

3. Using the representation of number e as a corresponding limit, show that for fixed values of γ, N_0, T the solution of equation (6) tends at $\tau \to 0$ to the solution of the initial problem.

In conclusion, based on the material of this chapter, we have selected the topics, which are crucial for the development and application of the methodology of mathematical modeling. They include: the problems of idealizing the initial object and formulating appropriate assumptions; applying both strict procedures (fundamental laws, variational principles) and the method of analogies and other approaches to the construction of mathematical models (including hardly formalizing ones); methods of qualitative research of nonlinear models; and constructing effective computing algorithms realizing the model by computers. These problems, along with the description of some important applications, form the basic content of subsequent chapters.

Bibliography for Chapter 1: [7, 16, 25, 40, 47, 57, 60, 66, 73, 76–79, 81, 83, 84]. i

Chapter II

DERIVATION OF MODELS FROM THE FUNDAMENTAL LAWS OF NATURE

1 Conservation of the Mass of Substance

Based on the considerations of balance of the mass of a substance and some additional assumptions, we shall construct models of a flow of non-interacting particles and motion of underground waters in a porous medium. We will describe the properties of the models obtained and discuss their possible generalizations.

1. A flow of particles in a pipe. In a cylindrical pipe with a cross-section S (Fig. 20) certain particles (dust, electrons) are moving. The velocity of their motion $u(t) > 0$ along the axis x, generally speaking, varies in time. For example, the charged particles can be accelerated or decelerated under the influence of an electrical field. To construct an elementary model of the considered motion we shall use the following assumptions:

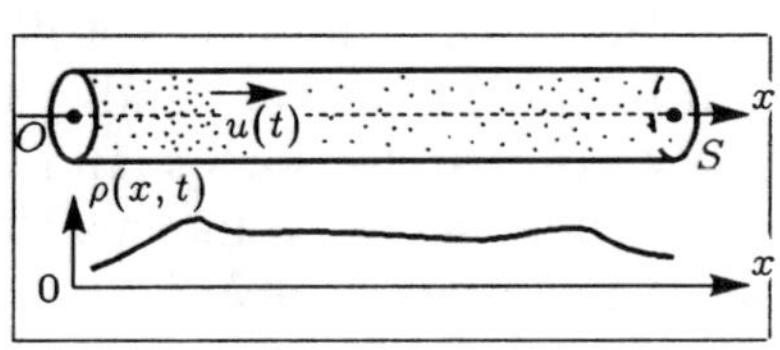

Fig.20.

a) the particles themselves do not interact (do not collide, attract each other and so on). For this to hold true the density of particles should clearly be rather small (then the charged particles not only fail to collide, but also they have no influence on each other due to the large distance between them);

b) the initial velocity of all particles in the same cross-section with a coordinate x is the same, and is directed along the axis x;

c) the initial density of particles depends only on the coordinate x;

d) the external forces acting on the particles are directed along the axis x.

Assumption (a) means that the velocity of the particles can vary only under the action of external forces; the assumptions (b–c) ensure the one-dimensional character of the transfer process, i.e. the dependence of the sought particle density only on the coordinate x and time $t \geq 0$.

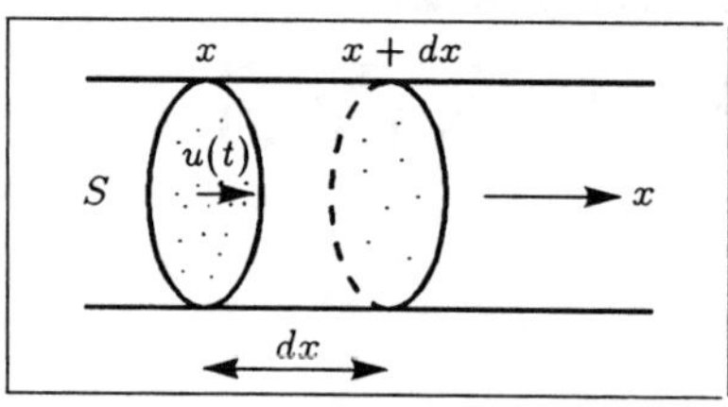

Fig.21.

Thus, given the initial density $\rho(x, t = 0) = \rho_0(x)$ one has to find the density of particles $\rho(x, t)$ at any moment of time for an arbitrary value of x (the velocity of motion $u(t)$ is given). Using the law of conservation of mass, by counting the balance of substance in a small element of the pipe from x up to $x + dx$ during dt (Fig. 21). From the left, the mass entering the elementary volume is

$$Su(t)\,dt\,\rho(x,\, t + \xi\, dt)\,, \qquad 0 \leq \xi \leq 1\,,$$

where $Su(t)\,dt$ is the volume of matter introduced during time interval dt. At the same time from the right cross-section the outflown mass equals

$$-Su(t)\,dt\,\rho(x + dx,\, t + \bar{\xi}\, dt)\,, \qquad 0 \leq \bar{\xi} \leq 1\,,$$

i.e. the resulting change in mass is

$$dm = Su(t)(\rho(x,\, t + \xi\, t) - \rho(x + dx,\, t + \bar{\xi}\, dt))\, dt\,.$$

By virtue of the small size of dt, the velocity $u(t)$ is considered as constant. Quantities $\rho(x,\, t + \xi\, dt)$ and $\rho(x + dx,\, t + \bar{\xi}\, dt)$ are the time averages of the density at cross-sections x and $x + dx$.

Another way of evaluating the balance in the fixed volume $S\,dx$ is obvious from the content of $\rho(x,\, t)$:

$$dm = S\,dx\,(\rho(x + \eta\, dx,\, t + dt) - \rho(x + \bar{\eta}\, dx,\, t)) \qquad 0 < \eta, \bar{\eta} < I\,,$$

where $\rho(x + \eta\, dx,\, t + dt)$ and $\rho(x + \bar{\eta}\, dx,\, t)$ are the spatial averages of the density at moments t and $t + dt$.

Equating both expressions obtained for dm and tending dx and dt to zero, we come to the equation for $\rho(x,\, t)$ implying the law of conservation of mass,

$$\frac{\partial p}{\partial t} + \frac{\partial p}{\partial x} u(t)\,, \qquad -\infty < x < \infty, \quad t > 0\,, \tag{1}$$

with initial conditions

$$\rho\,(x,\, 0) = \rho_0\,(x), \qquad -\infty < x < \infty\,. \tag{2}$$

The quantity ρu (*the matter flux* or *the mass flux*) is the amount of matter passing in a unit of time through a unit cross-section of the pipe. As can be seen from (1), the rate of density change of the matter in time in any cross-section is determined by the "rate" of the variation of the flux by the coordinate x. Similar properties are possessed by many models satisfying the conservation laws and describing absolutely different processes.

In the case of constant velocity $u(t) = u_0$, we come to a simplest linear equation in partial derivatives

$$\frac{\partial p}{\partial t} + u_0\,\frac{\partial p}{\partial x} = 0\,, \qquad -\infty < x < \infty, \quad t > 0\,. \tag{3}$$

It is not difficult to obtain its general solution by taking into consideration the fact that the equation (3) has characteristics – lines $x = u_0 t + C$, on which the sought function is constant in time, $\rho(x = u_o t + C,\, t) = \rho_c$, or equivalently

$$\rho(x,\, t) = \rho(x + u_0(t - t_0),\, t_0), \qquad t - t_0 \geq 0\,.$$

Selecting $t_0 = 0$, we obtain

$$\rho(x,\, t) = \rho(\xi) = \rho(x + u_0 t)\,. \tag{4}$$

The integral (4) is the general solution of equation (3). From the formula (4) and initial conditions (2) it is easy to find the sought function, so that it does not depend on separate variables x, t, but on their combination $\xi = x + u_0 t$ (*traveling wave*). The spatial profile of the density without distortions is transferred along the flow (Fig. 22) with constant velocity (the equation (3) is also known as *equation of transfer*). This basic property of the solution to equation (3) is slightly modified in the case where the velocity of particles depends on time (see exercise 1) – the density profile is transferred by different distances in equal time intervals. If, for some reason, the flow velocity depends on the density ($u = u(\rho)$), the equation (1)

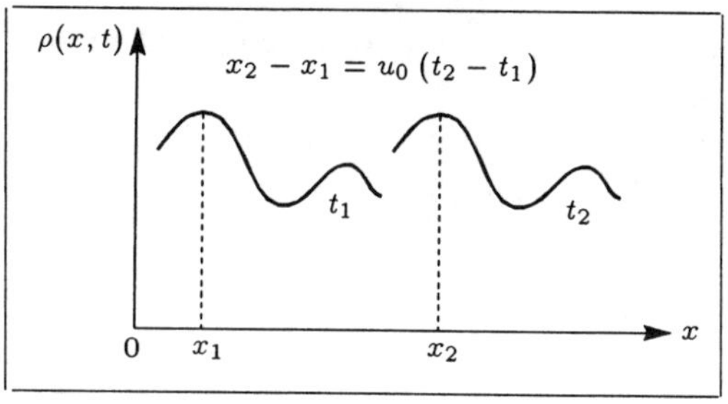

Fig.22.

becomes nonlinear, and the behavior of its solution can have a qualitatively different character (see subsection 7, section 4).

2. Basic assumptions on the gravitational nature of flows of underground waters. The porous medium represents a layer permeable by water (sand, clay), bounded below by an impermeable medium (granite), and from above by the surface of Earth (Fig. 23). If, due to intense action of artesian wells or heavy rains, the level of water somewhere varies, then gravity leads the water to fill the empty spaces.

To describe this process we have to make the following assumptions:

1) the water is considered to be an incompressible fluid with constant density ρ;

2) the thickness of the layer is much less than its width and length;

3) the underlying surface has no slits or gaps, so that the assigned function $H(x, y)$ is a sufficiently smooth function of the arguments;

4) the free surface of the water $h = h(x, y, t)$ varies smoothly by coordinates x, y;

5) the ground waters do not come to the surface at any point, so that on the free surface of the fluid the pressure is constant;

6) the soil is homogeneous, i.e. its physical and mechanical properties do not depend on arguments x, y, z.

The first assumption is quite natural, so far as in the considered processes it is not possible to reach pressures capable of noticeably changing the density of the water. The rest of the assumptions are of a simplifying nature. For example, the second assumption (the thin layer) means that flow is two-dimensional and all its characteristics do not depend on the coordinate z; the last two assumptions enable one to construct a model uniform in all points and so on. At the same time the assumptions 1–6 by no means distort the essence of the process, as they are fulfilled in plenty of realistic situations.

3. Balance of mass in the element of soil. Consider an elementary volume in the layer formed by the intersection of a vertical prism $ABCD$ of the underlying and free surfaces of the soil. In so far as the dimensions of the prism dx and dy are small, and the functions H and h are smooth (as-

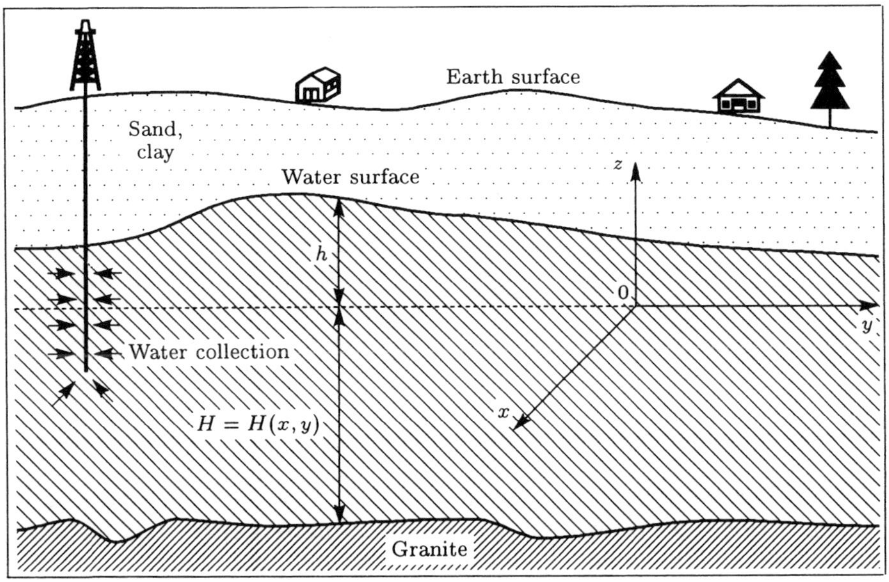

Fig.23.

sumptions 3 and 4), the resulting figure can be considered as parallelepiped. Let us introduce the unknown functions $v = v(x,\, y,\, t)$ and $u = u(x,\, y,\, t)$ – the components of velocity of the fluid along the axes x, y (Fig. 24).

We have to estimate the amount of fluid inflowing and outflowing from the parallelepiped within the time interval dt.

The mass of the water inflown through the facet DC is equal to the volume of the fluid multiplied by the density ρ:

$$\rho u(H + h)\, dy\, dt\,,$$

and through the facet AB the outflown mass is

$$\rho u(H + h)\, dy\, dt + \left\{ \frac{\partial}{\partial x}\, [\rho u(H + h)]\, dx \right\}\, dy\, dt\,.$$

In this expression as compared with the previous one, a term describing the increment of the function $\rho u(H + h)$ from the plane x to $x + dx$, is added. The quantity $\rho u(H + h)$ itself, as in subsection 1, has the content of the flux of mass (matter).

Thus, as the fluid moves along the axis x the mass accumulated within the element is

$$-\frac{\partial}{\partial x}[\rho u\, (H + h)]\, dx\, dy\, dt\,.$$

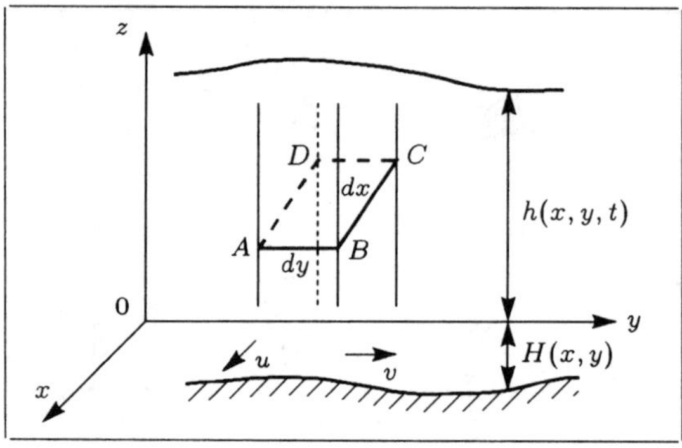

Fig.24.

Performing a similar procedure for the facets AD and BC, we obtain the variation of the mass of water due to its motion along the axis y

$$-\frac{\partial}{\partial y}[\rho v\,(H+h)]\,dx\,dy\,dt\,.$$

So far as along the axis z no water is inflown or outflown (below is the underlying boundary layer, while no matter flow exists through the free surface), the total change of the mass of water in the element of the soil is equal

$$-\left\{\frac{\partial}{\partial x}[\rho u\,(H+h)]+\frac{\partial}{\partial y}[\rho v\,(H+h)]\right\}\,dx\,dy\,dt\,. \tag{5}$$

The total amount of fluid in the parallelepiped is equal to its volume multiplied on the density ρ and the coefficient of porosity $m<1$ (since part of volume is occupied by soil):

$$m\rho\,(H+h)\,dx\,dy\,.$$

The variation of the mass of water within the element during dt, obviously, equals

$$\left\{\frac{\partial}{\partial t}\,[\rho m\,(H+h)]\,dx\,dy\right\}\,dt\,.$$

Taking into account the fact that $\partial H/\partial t\equiv 0$, $\partial\rho/\partial t\equiv 0$, from the latter expression we obtain

$$m\rho\,\frac{\partial h}{\partial t}\,dx\,dy\,dt \tag{6}$$

and equating (5) and (6), we come to the *equation of continuity*, expressing the law of conservation of mass in the considered process:

$$m\rho \frac{\partial h}{\partial t} = -\frac{\partial}{\partial x}\left[\rho u \left(H + h\right)\right] - \frac{\partial}{\partial y}\left[\rho v \left(H + h\right)\right]. \tag{7}$$

In equation (7) the rate of variation of the given quantity (in this case – the mass) in time is determined by a divergence of the flux of that quantity – a property peculiar to many models obtained from conservation laws (compare with equation (1)).

In view of $\partial\rho/\partial x \equiv 0$, $\partial\rho/\partial y \equiv 0$, equation (7) can be rewritten in more simple form:

$$m \frac{\partial h}{\partial t} = -\frac{\partial}{\partial x}\left[u \left(H + h\right)\right] - \frac{\partial}{\partial y}\left[v \left(H + h\right)\right]. \tag{8}$$

4. Closure of the law of conservation of mass. Equation (8) contains three unknown variables – h, u, v. Therefore, for the closure of a model it is necessary to use some additional considerations on the character of the process. They are given by the semi-empirical *Darsi law*

$$u = -\mu \frac{\partial p}{\partial x}, \qquad v = -\mu \frac{\partial p}{\partial y}. \tag{9}$$

where $p(x, y, z, t)$ is the pressure in the fluid, $\mu > 0$ is a coefficient factor determined by the properties of the soil. According to Darsi's law, the components of velocity of the flow of fluid are proportional to corresponding components of the pressure gradient. Note, that by its physical content the gradient of pressure is a force (referred to the unit volume). At the same time, in accord with Netwon's second law; the force acting on a body is proportional to the acceleration, instead of the velocity, as in Darsi law. However this contradiction is apparent, since the motion of fluid through the soil (filtration) implies the overcoming of the resistance of particles, as distinct of free flows (compare with the equation of fluid in section 4).

In the formulae (9) a new unknown quantity – pressure of fluid, is used. It is easy to find out its connection with quantities already introduced assuming the slow and almost horizontal character of motion. Then the dynamic component of the pressure can be to neglected and it can calculated via purely hydrostatic law as a pressure created by a column of a fluid:

$$p\left(x, y, z, t\right) = \rho\left(h\left(x, y, t\right)\right) - z) + \text{const},$$

where *const* is the pressure on the surface of fluid (for example, atmospheric), g is the acceleration of gravity.

Substituting the latter formula into (9), we obtain

$$u = -\mu\rho g \frac{\partial h}{\partial x}, \qquad v = -\mu\rho g \frac{\partial h}{\partial y}, \tag{10}$$

and, using (10) in the equation of continuity (8), we finally come to the equation of motion of underground waters

$$\frac{\partial h}{\partial t} = k \frac{\partial}{\partial x}\left[(H(x,\, y) + h)\frac{\partial h}{\partial x}\right] + k \frac{\partial}{\partial y}\left[(H(x,\, y) + h)\frac{\partial h}{\partial y}\right], \tag{11}$$

$$k = \frac{\mu\rho g}{m},$$

or to the *Bussinesque equation*, containing only one unknown function $h(x, y, t)$.

5. On some properties of the Bussinesque equation. The equation (11) is non-stationary (the sought function h depends on t), two-dimensional (h depends on x and y) and is of a parabolic type. It is inhomogeneous, as the function H depends on x, y, and is nonlinear, as its right hand side contains terms of the form $(hh_x)_x$ and $(hh_y)_y$. In comparison with the equation (1), the Bussinesque equation is mathematically much more complicated. In view of its nonlinearity the general solution cannot be found analytically, however it is relatively easy to obtain some reasonable partial solutions (see exercise 2), which also serve as tests of the development of numerical methods for equation (11).

To construct a complete model of motion of underground waters it is necessary to know input data for the equation (11): the form of a underlying surface $H(x, y)$, the coefficient k and boundary conditions assigning the function h in an initial moment of time and on the boundaries of the layer (and maybe in some chosen areas of the layer, for example, on an artesian well). A more detailed formulation of boundary conditions for the equations of a parabolic type is considered in section 2. Here we shall only mention that the simplest variant of formulating boundary conditions for the equation (11) is the formulating only of the initial condition of the function $h(x, y, t)$ at $t = 0$:

$$h(x,\, y,\, t = 0) = h_0(x,\, y), \qquad -\infty < x < \infty, \quad -\infty < y < \infty.$$

This formulation is known as *the Cauchy problem* for equation (11), solved evidently, also in area $-\infty < x < \infty$, $-\infty < y < \infty$. In the Cauchy problem, the function h for all $t > 0$ is obtained via the known distribution of the level of underground waters h_0.

The consideration of a layer of infinite dimensions is certainly is an idealization. However if the flow studied is in a small central area of a layer and during relatively short time interval, the influence of the boundaries of the layer can be neglected, and the solution of the Cauchy problem describes a quite realistic process.

Note that certain boundary conditions were actually already implicitly introduced into the model at the derivation of the Bussinesque equation. The assumption that the layer was inpermeable was used to derive the equation of balance, and without the assumption (5) on the "slit" between the surface of the soil and the surface of underground waters (i.e. when all the fluid is in a porous medium) it would be impossible to use the Darsi law for the considered area. Certainly, the fulfilling of these and other assumptions should be checked when studying the given object using the constructed model.

By introducing additional assumptions, the general model becomes simpler. Thus, if for any reason the solution does not depend on time t (stationary process), we come to an elliptical equation

$$\frac{\partial}{\partial x}\left[(H+h)\frac{\partial h}{\partial x}\right]+\frac{\partial}{\partial y}\left[(H+h)\frac{\partial h}{\partial y}\right]=0\,. \tag{12}$$

Its solution, obviously, does not require knowledge of the function h at the initial moment. In the simplest case (12) turns to the *Laplace equation* (see exercise 3). If the underlying surface is horizontal ($H(x,\, y) = H_0 = \text{const}$), the Bussinesque equation becomes homogeneous

$$\frac{\partial h}{\partial t}=k\,\frac{\partial}{\partial x}\left(h\,\frac{\partial h}{\partial x}\right)+k\,\frac{\partial}{\partial y}\left(h\,\frac{\partial h}{\partial y}\right)\,.$$

For additional assumptions on the one-dimensionality of the flow, when the sought solution depends only on one spatial variable, for example, on the coordinate x, we come to the equation

$$\frac{\partial h}{\partial t}=k\,\frac{\partial}{\partial x}\left(h\,\frac{\partial h}{\partial x}\right)\,, \tag{13}$$

also called *a one-dimensional equation of the type of nonlinear thermal conductivity* (see section 2) or one-dimensional equation of an isothermal filtration. For example, the flows in layers highly prolate in one direction are one-dimensional, so that one can neglect by the variation of quantities along the transversal section (if there is no flow through lateral areas). Finally, the most simple model of flow of underground waters is given by the *heat transfer equation* (or by *equation of diffusion of matter*)

$$\frac{dh}{dt}=kH_0\,\frac{d^2h}{dt^2}\,, \tag{14}$$

obtained at $h \ll H_0$, i.e. at small variations of the level of the fluid as compared with the thickness of a layer.

The last three equations concern a parabolic type, so that the equation (14) is linear and the methods to find its general solution are well known. Certainly, besides those mentioned, other simplifications of the initial model are possible as well, for example, the two-dimensional equation (13).

From the Bussinesque equation it is also relatively easy to obtain more complicated models, when some of the assumptions formulated in section 2 are not valid. In particular, in many cases the soil is inhomogeneous, $m = m(x, y)$, $\mu = \mu(x, y)$, and it is necessary to take into account the inflow of fluid into the layer due to rains. Then the generalization of the Bussinesque equation has the form

$$\frac{m(x, y)}{\rho g}\frac{\partial h}{\partial t} = \frac{\partial}{\partial x}\left[\mu(x, y)(H + h)\frac{\partial h}{\partial x}\right] + \\ + \frac{\partial}{\partial y}\left[\mu(x, y)(H + h)\frac{\partial h}{\partial y}\right] + q(x, y, t), \qquad (15)$$

where $q(x, y, t)$ characterizes the power of the rains at x, y in time t (see exercise 4).

So, the application of the fundamental law of conservation of mass has enabled us to derive various models of the considered processes. The difference between the models is determined by the type of the obtained equations (hyperbolic, parabolic, elliptic), by their spatial-temporal properties (stationary, non-stationary, one-dimensional, many-dimensional), the presence or absence of nonlinearities, as well as by the boundary conditions. Thus, depending on the concrete properties of the object and additional assumptions based on the same fundamental law, it is possible to derive completely different mathematical models. On the other hand, as we will see numerous times below, the same mathematical models can, thanks to their universality, describe objects of completely different natures.

EXERCISES

1. Find out the transformation of variables reducing equation (1) to equation (3), and show that the solution when $u = u(t)$ has the form (4), where $\xi = x + \int_0^t u(t)\,dt$.

2. The solution of equation (13) of the form $h(x, t) = u(t)\,\theta(x)$ (i.e. in separating variables) is called *regular regime of Bussinesque* in the case $u(t) \to 0$, $t \to \infty$. Show that $u(t)$ is a power function of time at large t.

3. Establish at what assumptions the equation (12) is reduced to the equation of Laplace.

4. Using the law of conservation of mass and the Darsi law, derive equation (15).

2 Conservation of Energy

We will apply the law of energy conservation together with some additional assumptions to construct models of heat propagation in a continuous medium. Also we will formulate the typical boundary problems for the equations of heat transfer and discuss some physical and mathematical properties of the models obtained.

1. Preliminary information on the processes of heat transfer. Thermal energy or heat is the energy of random motion of atoms or molecules of matter. The heat exchange between various parts of the object is called *heat transfer*, and materials possessing a well expressed property of heat transfer are said to be *thermal conductors.* They include the metals, where the thermal energy is transferred mainly by free electrons, some gases and so on. The processes of heat transfer are considered in conditions of so-called *local thermodynamic equilibrium* (LTE). The concept of LTE for gases is introduced at $\lambda \ll L$, i.e. when the length of the free path of particles of matter is much less than the characteristic sizes of the considered medium (*continuous medium*). LTE also implies, that the processes are studied at time scales larger than τ (the time scale between the collisions of particles) and on scales larger than λ. Then in volumes with dimensions exceeding λ (but much less than L), the equilibrium does exist in these cases one can introduce the concepts of mean density, velocity of thermal motion of particles, etc. These local magnitudes (different in various points of medium) at the formulated assumptions are obtained from the Maxwellian distribution of particles (see section 3 of chapter III). This includes the *temperature* T determining the mean kinetic energy of particles:

$$\frac{mv^2}{2} = \frac{3}{2}\,kT\,,$$

where m is the mass of a particle, v is the mean velocity of random motion, k is the Boltzmann constant (in case of so-called Boltzmann gas).

The energy connected with the random motion of particles (internal energy) is determined through the temperature with the help of the *specific heat* $c(\rho,\,T)$, namely

$$c(\rho,\,T) = \frac{\partial\varepsilon(\rho,\,T)}{\partial T}\,, \qquad c(\rho,\,T) > 0\,,$$

where $\rho = mn$ is the density of matter (n is the number of particles per unit volume), $\varepsilon(\rho,\,T)$ is the *internal energy of mass unit.* In other words, the thermal capacity is the energy which should be given to a unit mass in order to increase its temperature by one degree.

The most simple expression for thermal capacity is obtained in the case of an *ideal gas* (when the particles interact only via direct collisions like billiard balls without loss of total kinetic energy). If some volume of an ideal gas contains N particles, their total internal energy is

$$E = N\frac{mv^2}{2} = \frac{3}{2}NkT = \frac{3}{2}M\frac{k}{M}T\,,$$

where $M = Nm$ is the summarized mass of particles, and the specific internal energy or the energy per unit mass is given by the formula

$$\varepsilon = \frac{E}{M} = \frac{3}{2}\frac{k}{m}T\,,$$

i.e. the thermal capacity of ideal gas is $3k/(2m)$ and does not depend on ρ, T.

In the general case the relation between the internal energy and temperature is more complicated. For example, apart from the kinetic energy of moving particles, the internal energy contains a component connected with the potential energy of their interaction, depending on their mean distance r. In turn $r \approx (n)^{-1/3} = (\rho/m)^{-1/3}$, where n is the number of particles in a unit of volume, i.e. ε depends on the density ρ. Therefore, in the theory of heat transfer the quantities ε (or equivalently – c) are, generally speaking, functions of ρ and T. Their concrete form is determined by the properties of the considered medium.

2. Derivation of Fourier law from molecular-kinetic concepts. To derive a mathematical model of heat transfer, apart from of the concepts described in subsection 1, one has to introduce the important concept of heat flux. *The heat flux* (or flux of thermal energy) in the given point is the amount of heat transferred in a unit of time through a unit area at the given point of the matter (compare with the concept of a flux of mass in section 1). Obviously the heat flux is a vector (as in general it depends on the spatial orientation of the unit area).

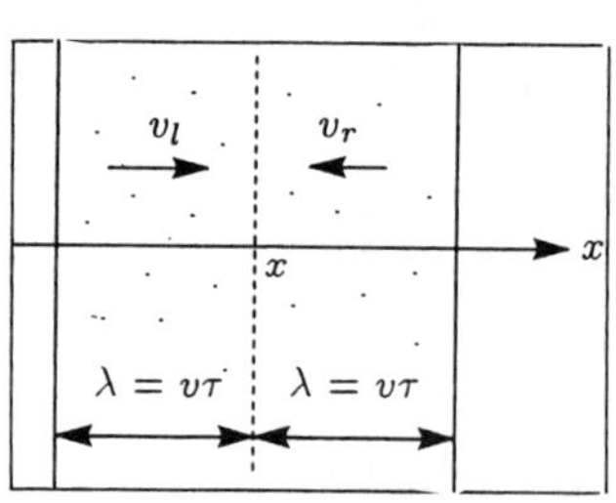

Fig.25.

Select a point with coordinates x, y, z within the medium and calculate the components of heat flux W on corresponding axes (the values of W_x, W_y, W_z). Let the unit area be located transversal to axis x (the dashed line in Fig. 25). The particles moving along the axis x intersect it from the right to the left and vice versa with equal probability. However if the temperature of particles (and hence their kinetic energy) is different on one of the sides, different energy should be transferred in a unit of time through that surface. The difference of these energies forms the heat flux along the axis x. On Fig. 25 select areas located at a distance $\lambda = v\tau$ to the right and to the left from the surface. Among the particles in the right hand side approximately $1/6$ are moving to the left, as all six directions (up – down, forward – back, to the right – to the left) are equally probable. During the time scale τ these particles will necessarily cross the surface and will transfer energy

$$\frac{1}{6} n\lambda \frac{mv_{\mathrm{r}}^2}{2},$$

where v_{r} is the velocity of particles in right hand side (the quantities n, λ are considered in the first approximation equal on both sides). Similarly, the particles from the left hand side transfer energy

$$\frac{1}{6} n\lambda \frac{mv_{\mathrm{l}}^2}{2},$$

where v_{l} is velocity of particles on the left side. The difference of these energies per time unit is

$$W_x = \frac{1}{6} nv \left(\frac{mv_{\mathrm{l}}^2}{2} - \frac{mv_{\mathrm{r}}^2}{2} \right) = \frac{mnv}{6} (\varepsilon_{\mathrm{l}} - \varepsilon_{\mathrm{r}}),$$

where ε_{l}, ε_{r} are the internal energy of matter in the left and right hand sides respectively, while v is the average between v_{l} and v_{r}. In the first approximation of quantities ε_{l}, ε_{r} one can express through ε (energy in the point x, i.e. on the surface) as follows

$$\varepsilon_{\mathrm{r}} = \varepsilon + \lambda \frac{\partial \varepsilon}{\partial x} = \varepsilon + \lambda c \frac{\partial T}{\partial x}$$

$$\varepsilon l = \varepsilon - \lambda \frac{\partial \varepsilon}{\partial x} = \varepsilon - \lambda c \frac{\partial T}{\partial x}$$

Substituting these formulae into the expression for W_x, we obtain

$$W_x = -\kappa \frac{\partial T}{\partial x}, \tag{1}$$

where $\kappa = \rho c \lambda v/3$. Analogously, we obtain for the components W_y, W_z

$$W_y = -\kappa \frac{\partial T}{\partial y}, \qquad W_z = -\kappa \frac{\partial T}{\partial z}. \tag{2}$$

The unification of (1) and (2) gives *the Fourier law*

$$W = -\kappa \operatorname{grad} T . \tag{3}$$

The quantity κ is called *thermal conductivity.*

Note that the thermal conductivity depends in general on the density and temperature of matter:

$$\kappa = \frac{\rho c \lambda v}{3} \geq 0 , \tag{4}$$

in so far as not only the thermal capacity c, but also the length of the free path λ can be functions of ρ, T. Thus, for example, in a gas under usual conditions the heat is transferred by molecules (molecular thermal conductivity). For λ in this case is fair $\lambda \sim 1/\rho$, and as $v \sim \sqrt{T}$, from (4) we have $\kappa_m \sim \sqrt{T}$ (thermal capacity is considered constant). In plasma (where the basic role in heat transfer is played by electrons) the free path of electrons depends on ρ, T, so that $\lambda \sim T^2\rho^{-1}$, and for κ_e is fair $\kappa_e \sim T^{5/2}$ (c is a constant).

Thus, the Fourier law states that the heat flux is proportional to the gradient of temperature. As the thermal energy is directly connected with the temperature, it is in some sense possible to consider that "flux" of temperature is proportional to a gradient of temperature itself. Absolutely the same property is peculiar to the process with close essence – to the diffusion of matter (Fick's law). A similar interpretation can be attributed also to the Darsi law (10) of section 1, though the motion of underground waters by its character differs drastically from heat diffusion (and the Darsi law has no such simple theoretical foundation, as Fourier's and Fick's laws).

3. The equation of heat balance. Apply the heat conservation law for mathematical description of the process of heat transfer. Consider that the internal energy of matter changes only due to the mechanism of thermal conductivity, i.e. other forms of energy are considered unimportant (for example, neglect the role of chemical reactions, the work of forces of pressure, contracting some volume of gas, and so on).

In a thermal conductive medium allocate an elementary cube with sides dx, dy, dz (Fig. 26), estimate the variation of the thermal energy contained in it during a small time interval dt. According to the assumptions made this change can be caused only by a difference in heat fluxes entering and outflowing through different facets of the cube. Thus, the fluxes along the axis x lead either to decrease or increase of the internal energy on the amount

$$[W_x(x, y, z, t) - W_x(x + dx, y, z, t)]\, dy\, dz\, dt ,$$

where $dy\, dz$ is the area of the facet perpendicular to axis x. In this formula it is considered that W_x as a function of time does not vary strongly during

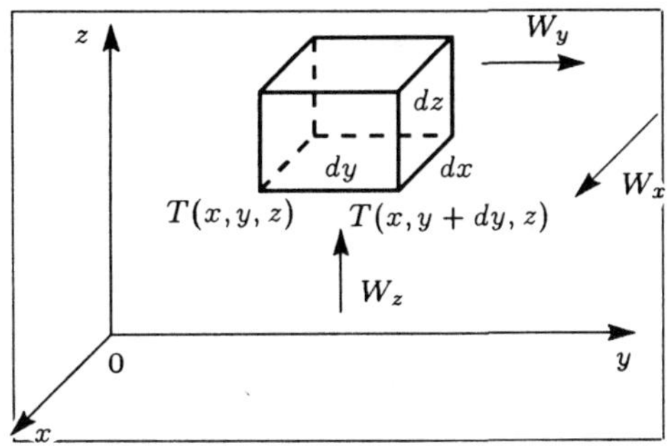

Fig.26.

dt, and it is possible to take its value at t. The variation of the internal energy along axes y, z is calculated in a similar way

$$[W_y(x,\, y,\, z,\, t) - W_y(x,\, y + dy,\, z,\, t)]\, dx\, dz\, dt\,,$$

$$[W_z(x,\, y,\, z,\, t) - W_z(x,\, y,\, z + dz,\, t)]\, dx\, dy\, dt\,,$$

The summarized variation of energy $\Delta E = E\,(t + dt) - E(t)$ is

$$\Delta E = -\mathrm{div}\; W\, dx\, dy\, dz\, dt\,.$$

On the other hand, the quantity ΔE can be expressed through the change in temperature of volume and through its thermal capacity via the formula

$$\Delta E = (T(t + dt) - T(t))\, c(\rho,\, T)\, \rho\, dx\, dy\, dz\,,$$

where in view of the small volume some average values of temperature and density are used.

Equating the two latter expressions and tending dt to zero, we obtain the *general equation describing the propagation of heat*

$$C\frac{\partial T}{\partial t} = \mathrm{div}\,(\kappa\, \mathrm{grad}\, T)\,, \tag{5}$$

and in explicit form

$$C\frac{\partial T}{\partial t} = \frac{\partial}{\partial x}\left(\kappa\frac{\partial T}{\partial x}\right) + \frac{\partial}{\partial y}\left(\kappa\frac{\partial T}{\partial y}\right) + \frac{\partial}{\partial z}\left(\kappa\frac{\partial T}{\partial z}\right), \tag{6}$$

where $C = \rho c$.

The equation (6) is a non-stationary, three-dimensional (function T depends on time t and three spatial variables x, y, z) equation of a parabolic type. It is inhomogeneous, in so far as the thermal capacity, thermal conductivity coefficient and density can generally be different in different points of the matter, and is nonlinear, in so far as the functions c and κ can depend on the temperature T (i.e. on the sought solution).

For additional assumptions on the character of the process of heat transfer the equation (6) can become simplified. Indeed, if the process is stationary, i.e. the temperature does not depend on time, (6) turns to the equation of an *elliptic type*

$$\frac{\partial}{\partial x}\left(\kappa\frac{\partial T}{\partial x}\right)+\frac{\partial}{\partial y}\left(\kappa\frac{\partial T}{\partial y}\right)+\frac{\partial}{\partial z}\left(\kappa\frac{\partial T}{\partial z}\right)=0\,, \tag{7}$$

and if the functions c, κ do not depend on temperature, (6) becomes a *linear parabolic equation*, which in the case of homogeneous medium (κ, c, ρ do not depend on x, y, z) has the form

$$\frac{\partial T}{\partial t}=k_0\,\Delta T\,, \tag{8}$$

where $k_0=\kappa/C$ is the *coefficient of thermal conductivity*. For equation (8) it is relatively easy to write down the general solution.

In a one-dimensional case (where temperature depends only on t and x) we obtain from (6)

$$C\frac{\partial T}{\partial t}=\frac{\partial}{\partial x}\left(\kappa\frac{\partial T}{\partial x}\right). \tag{9}$$

Equation (9) is reduced to an equation of the nonlinear thermal conductivity type

$$\frac{\partial u}{\partial t}=\frac{\partial}{\partial x}\left(k(u)\frac{\partial u}{\partial x}\right) \tag{10}$$

assuming that $\partial C/\partial x\equiv 0$ (compare with equation (13) from section 1). Finally, if $\kappa=\kappa_0$, $C=C_0$, where κ_0, C_0 are constants, from (10) we have the heat transfer equation – an elementary parabolic equation

$$\frac{\partial u}{\partial t}=k_0\,\frac{\partial^2 u}{\partial x^2}\,. \tag{11}$$

As in the case of Bussinesque equation from the basic equation (6) it is possible to obtain various generalizations corresponding to heat-transfer mechanisms which are more complicated than those considered above. So, for a anisotropic medium (when thermal conductivity is different in different directions) with energy release instead of (6) we have

$$C\frac{\partial T}{\partial t}=\frac{\partial}{\partial x}\left(\kappa_x\frac{\partial T}{\partial x}\right)+\frac{\partial}{\partial y}\left(\kappa_y\frac{\partial T}{\partial y}\right)+\frac{\partial}{\partial z}\left(\kappa_z\frac{\partial T}{\partial z}\right)+f(x,\,y,\,z,\,t,\,T)\,, \tag{12}$$

where κ_x, κ_y, κ_z are the coefficients in the Fourier law (3) along axes x, y, z, and the function f is the power of energy release. Anisotropic conditions, for example, in the case of an electronic thermal conductivity, can be caused by adequately strong magnetic field opposing the motion of heat carriers across the force lines of the field, and the energy release can be related to chemical reactions or the existence of an electric current.

All the equations in this subsection were derived with the help of the fundamental law of energy conservation and Fourier's law (compare with the derivation of the Bussinesque equation in section 1). Together with given functions c, κ, ρ and boundary conditions, they represent the closed mathematical models of the process of heat transfer.

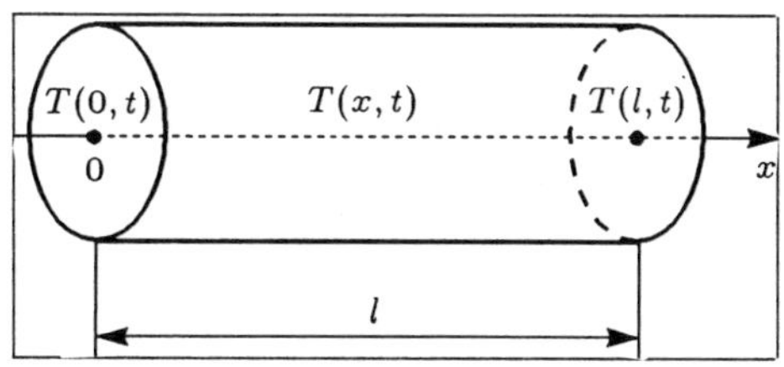

Fig.27.

4. The statement of typical boundary conditions for the equation of heat transfer. For simplicity we shall consider one-dimensional processes of thermal conductivity. They occur, for example, in a long and thin metal bar (Fig. 27), heated from one of its edges. Provided that the bar is isotropic, its initial temperature in arbitrary cross-section does not depend on y, z (the same property should be fulfilled also for the other edge), and the heat losses from the side surface can be neglected. Consider also that the thermal capacity of the bar is constant. Then the temperature depends only on x and t, and its distribution along the bar at various moments in time is described by the equation

$$\frac{\partial T}{\partial t} = \frac{\partial}{\partial x}\left(k(T)\,\frac{\partial T}{\partial x}\right), \tag{13}$$

valid at $0 < x < l$, $t > 0$. For the definition of the function $T(x, t)$), i.e. of the solution it is sufficient to know the initial temperature of the bar

$$T(x, t) = T_0(x)\,, \qquad 0 \le x \le l\,, \tag{14}$$

and the temperature on its edges at any moment of time:

$$T(0,t) = T_1(t)\,, \qquad T(l, t) = T_2(t)\,, \qquad T > 0\,. \tag{15}$$

The problem (13)–(15) is called *the first boundary problem* for the parabolic equation (13) *in interval* $x \in [0, l]$. Physically the condition (15) means that the edges of the bar are kept at a definite temperature with the help of some external heat sources, generally depending on time.

However, when on the edges of the bar instead of (15) heat fluxes are gicen as functions of time:

$$-k(T(0, t))\left.\frac{\partial T}{\partial x}\right|_{x=0} = W_1(t)\,, \quad k(T(l, t))\left.\frac{\partial T}{\partial x}\right|_{x=l} = W_2(t)\,, \quad t > 0\,, \tag{16}$$

then it is called *the second boundary problem on an interval* [0, l]. The given situation is realized, for example, when the edges of the bar are heated by laser beams of a known intensity.

A more complicated (nonlinear) version of conditions at the edges corresponds to a highly heated and hence radiating bar, not contacting with any other bodies. Then in a unit of time the bar will lose energy on its boundaries equal to $\sigma T^4(0, t)$ and $\sigma T^4(l, t)$, respectively, and instead of (15) or (16) the conditions will be

$$\sigma T^4(0, t) = k(T(0, t))\left.\frac{\partial T}{\partial x}\right|_{x=0}, \sigma T^4(l, t) = -k(T(l, t))\left.\frac{\partial T}{\partial x}\right|_{x=l}, \tag{17}$$
$$t > 0\,,$$

where $\sigma > 0$.

Other boundary conditions corresponding to other physical situations are possible as well. Certainly, various combinations of conditions (15)–(17) are admissible, for example, the temperature is given on the left edge, while the heat flux is given on the right one.

The diversity of boundary conditions for the equations of heat transfer is connected with various idealizations of the initial problem (13)–(15). In an analysis of the heat distribution near one of the edges of a long bar during a relatively short time scale one can neglect the influence of the other edge. Instead of (15) it is enough to set only one of the conditions (for definiteness on the left edge)

$$T(0, t) = T_1(t)\,, \qquad t > 0\,, \tag{18}$$

and to solve the equation at $x > 0$ ((13), (14), (18) – *the first boundary problem in semispace*).

The Cauchy problem discussed above in the example of the Bussinesque equation (section 1) is considered in the whole space $-\infty < x < \infty$. For the equation (13) only the initial distribution of temperature (14) is set. Such a statement is quite reasonable, when the processes in the central part of the bar are considered and the influence of both edges can be considered as insignificant.

For many-dimensional thermal conductivity equations the formulation of boundary conditions is not changed essentially as compared with the one-dimensional case: either the temperature, or the heat flux, or their more complicated combinations are given on the boundaries, and also (at $t = 0$) the initial distribution of temperature. Note that in the case of stationary equation (7) only the boundary conditions are set. The boundary conditions for the equation of motion of underground waters from section 1 are quite similar to described in these subsection (then the analogs of temperature and heat flux in the Bussinesque equation are the level of underground waters and the mass flow).

5. On the peculiarities of heat transfer models. The simplest of all the discussed problems of thermal conductivity is the problem of stationary process for the equation (11) within the interval $[0, l]$:

$$k_0 \frac{\partial^2 T}{\partial x^2} = 0\,, \qquad T(0) = T_1\,, \quad T(l) = T_2\,.$$

Its solution is a linear function of coordinate x:

$$T(x) = \frac{T_2 - T_1}{l}\, x + T_1\,, \qquad 0 < x < l\,. \tag{19}$$

The solution (19) has a quite obvious physical sense. Indeed, at the stationary process the heat fluxes entering and leaving any cross-section of the bar are equal (otherwise the temperature of the section would vary). Therefore the flux should be constant at any point x, which in accordance with the law of Fourier (3) at $\kappa = \kappa_0 =$ const is possible only at a linear "profile" of the temperatures.

At the same time the application of the Fourier's law leads to an effect having no physical sense, but being peculiar to equations of parabolic type. We explain this via a consideration of equation (11) to be solved in the whole space $-\infty < x < \infty$, the problem of so-called *instantaneous point source of heat.* It is required to find out the distribution of temperature at all $t > 0$, $-\infty < x < \infty$, caused by the release of a heat Q_0 at time $t = 0$ in a plane $x = 0$. The initial temperature is considered equal to zero: $T(x, 0) = T_0(x) \equiv 0$, $-\infty < x < \infty$. Such a formulation is an idealization of the actual process valid at fulfilling the corresponding conditions (for example, an intense transversal pulse of an electric current is moving over the center of a cold bar, acting for a very short time and affecting a small area of metal). The solution of this problem is given by the formula

$$T(x, t) = \frac{Q_0}{2\sqrt{\pi k_0 t}} \exp\left(-\frac{x^2}{4k_0 t}\right), \qquad t > 0 \quad C \equiv 1\,, \tag{20}$$

which can be checked by direct substitution into the equation (11). The symmetrical function (20) by virtue of a known equality

$$\int\limits_{-\infty}^{\infty} e^{-y^2}\, dy = \sqrt{\pi}$$

has the property

$$\int_{-\infty}^{\infty} T(x,\, t)\, dx = Q_0\,, \qquad t > 0\,,$$

so that the energy conservation law is fulfilled. At the same time in accordance with (20) the temperature in any point of space at any moment $t > 0$ is different from zero. Thus the model (11) and many other models of heat transfer describe processes with infinite velocity of propagation of perturbations (the temperature at $t = 0$ was equal to zero for all x).

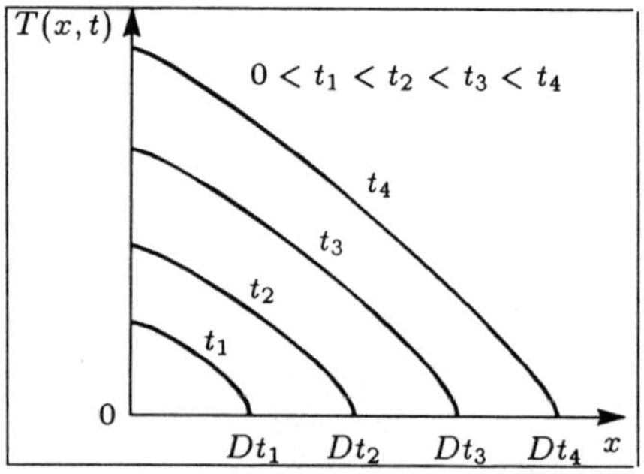

Fig.28.

This shortage is absent (but only under certain conditions) for nonlinear thermal conductivity type equations (10) (in particular, equation (13), section 1). For the model (10) with $k(T) = k_0 T^\sigma$, $\sigma > 0$ consider the process of heat propagation in a semispace $x > 0$ at given temperature on the boundary: $T(0,\, t) = T_1(t)$. The initial temperature of the medium is considered zero: $T(x,\, 0) = T_0(x) \equiv 0$. A partial solution to this problem, adequate for the boundary law

$$T_1(t) = \left(\frac{\sigma D^2}{k_0}\, t\right)^{1/\sigma}, \qquad t > 0\,,$$

has the form of a traveling wave (compare with the solution (4) section 1), propagating from the boundary downwards into the matter not with an infinite, but with a finite velocity $D > 0$ (Fig. 28):

$$T(x,t) = \begin{cases} \left(\frac{\sigma D}{k_0}\, t\right)^{1/\sigma} (Dt - x)^{1/\sigma}\,, & x \le Dt\,, \\ 0, & x > Dt\,, \end{cases} \qquad t > 0\,. \tag{21}$$

However, this property is realized only during the propagation of heat in a cold medium and is lost in the case of a non-zero initial temperature of matter (such problems are considered in more detail in Chapter V).

The shortage described generated by the inapplicability of Fourier's law (and the Darsi law in the case of Bussinesque equation) in the neighborhood of the front of propagation of thermal energy, do not prevent the broad application of parabolic equations (from (20) it is seen that the fraction of energy contained in matter at sufficiently high values of x is negligible in comparison with the total energy Q_0.). They are good examples of the universality of mathematical models, describing plenty of diverse processes of drastically different natures (see also section 1, Chapter IV).

EXERCISES

1. Find an integral replacement for the function T, when equation (9) will take the form of (10).

2. Using the same considerations as at the derivation of the law (3) and of equation (6), find equation (12).

3. Find out the solution (20) for the problem on the instantaneous point source of heat, representing it as $T(x, t) = f(t)\varphi(\xi)$, where $\xi = x/\sqrt{4k_0 t}$.

4. Construct the solution (21) for equation (10) with $k(T) = k_0 T^\sigma$, $\sigma > 0$ by presenting the temperature as $T(x, t) = Af(\xi)$, $A > 0$, $\xi = Dt - x$. Prove that the heat flux is zero at the wavefront $x = Dt$.

3 Conservation of the Number of Particles

Let us introduce some concepts of the theory of thermal radiation, transferred by means of light quanta in a medium. The conservation law of the number of quanta will be used to derive the kinetic equation, which the distribution function of photons obeys. We will discuss some properties of the constructed model of radiation heat exchange within matter.

1. Basic concepts of the theory of thermal radiation. In matter which has been heated to a sufficiently high temperature, an essential role is played by the processes of energy transfer by light quanta. Propagating within the medium, being reflected and absorbed by atoms and molecules of matter, as well as being emitted by them, the photons perform the radiative heat exchange between various parts of the medium. A heated fireplace heats the air in its vicinity using exactly this mechanism .

The radiation field can be considered as an electromagnetic radiation of frequency of oscillations ν and wavelength λ, connected through the speed of light ($\lambda = c/\nu$). While speaking about the radiation field as a system of a large number of particles – light quanta it is necessary to introduce the concept of energy of a quantum $h\nu$ (h is the Planck constant) moving with velocity c. As distinct from the temperature field characterized by coordinates x, y, z and time t, to describe radiation it is important to know also its frequency ν (which are generally different for different quanta) and the direction of motion of quanta in any point at any moment t.

To trace the trajectory of any of the huge number of photons is impossible. Therefore a statistical probability approach based on the concept of *the distribution function of particles* is used in the theory of radiation. This important concept is successfully used in the study of systems of large numbers of particles or other objects in various areas of research (see, for example, section 3, Chapter III).

The distribution function of photons $f = f(\nu, \vec{r}, \vec{\Omega}, t)$, depends on the frequency of quanta, the radius-vector $\vec{r}$ (i.e. on coordinates x, y, z), the direction of motion of particles $\vec{\Omega}$ and the time t. Its content is as follows. Consider in a moment t a volume element $d\vec{r}$ near a point $\vec{r}$ (Fig. 29). Then the quantity

$$f(\nu, \vec{r}, \vec{\Omega}, t)\, d\nu\, d\vec{r}\, d\vec{\Omega} \tag{1}$$

by definition is the number of quanta in the spectral interval $(\nu, \nu + d\nu)$ (i.e. their frequency lies between the values ν and $\nu + d\nu$), occupying the volume $d\vec{r}$ and having direction of motion between $\vec{\Omega}$ and $\vec{\Omega} + d\vec{\Omega}$ ($\vec{\Omega}$ – unit vector). The volume $d\vec{r}$ is supposed to be much greater than the wavelength λ, so that the wave effects are insignificant.

The distribution function (1) is one of the basic concepts of the theory of radiative heat exchange. It enables us to introduce and estimate all the other characteristics of this process. The quantity I_ν defined as

$$I_\nu(\vec{r}, \vec{\Omega}, t) = h\nu c f(\nu, \vec{r}, \vec{\Omega}, t)\,, \tag{2}$$

is called *the spectral intensity of radiation.* It represents the amount of radiated energy in a spectral interval from ν up to $\nu + d\nu$ transferred by photons per unit time through a unit area located at $\vec{r}$ and perpendicular to the directions of their motion (which lie within the angles from $\vec{\Omega}$ up to $\vec{\Omega} + d\vec{\Omega}$; Fig. 30).

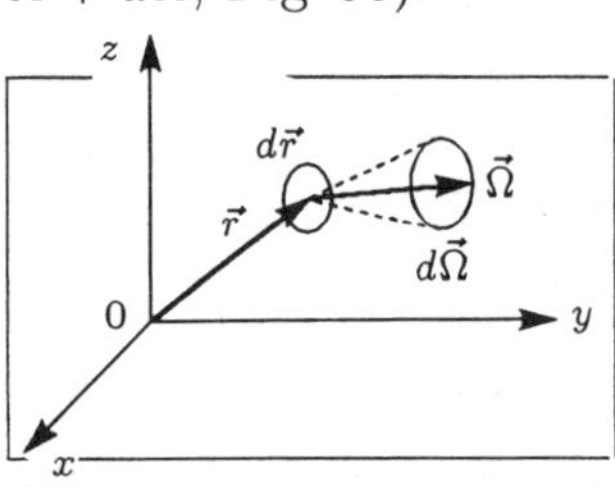

Fig.29.

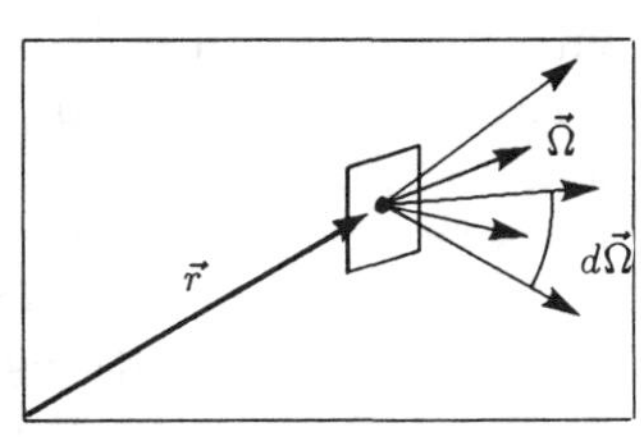

Fig.30.

Indeed, in so far as the energy of quantum is $h\nu$, and the total number of quanta with frequency from ν up to $\nu + d\nu$ and with motions directed from

$\vec{\Omega}$ up to $\vec{\Omega} + d\vec{\Omega}$ in unit volume equals $f\, d\nu\, d\vec{\Omega}$, then the energy transferred per second through the area element perpendicular to the motion in 1 cm^2, is equal to $h\nu c f\, d\nu\, d\vec{\Omega}$, in accordance with the definition (2).

Spectral density of radiation

$$U_\nu(\vec{r},\, t) = h\nu \int_{4\pi} f\, d\vec{\Omega} = \frac{1}{c} \int_{4\pi} I_\nu\, d\vec{\Omega} \tag{3}$$

represents the amount of energy of quanta contained in 1 cm^3 at a point $\vec{r}$ in a moment t in unit interval of frequencies and with frequency ν.

One more important descriptor is the *spectral flux of radiation* $\vec{S}_\nu$. The photons intersecting the unit area with direction of normal $\vec{n}$, transfer through it an energy (in 1 second in an interval ν and $\nu + d\nu$ equal to $h\nu c \int_{2\pi} \cos\theta\, \vec{\Omega}$ (Fig. 31)). The energy propagated through the area element from the right to the left is estimated in a similar way, but the integration is carried over the left hemisphere. Their difference just gives S_ν

$$S_\nu(\vec{r},\, t,\, \vec{n}) = h\nu c \int_{4\pi} f\, \cos\theta\, d\vec{\Omega}\,, \tag{4}$$

where θ is the angle between the direction of quanta and the normal. The quantity S_ν is the projection of vector $\vec{S}_\nu$ on the normal $\vec{n}$, while the vector itself is

$$\vec{S}_\nu = \int_{4\pi} I_\nu \vec{\Omega}\, d\vec{\Omega}\,. \tag{5}$$

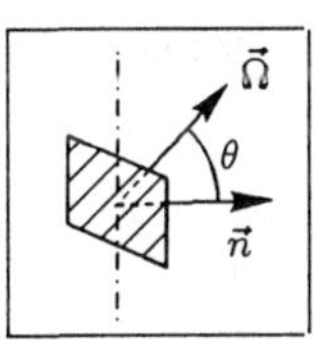

Fig.31.

Note that for isotropic radiation (not depending on direction $\vec{\Omega}$) the spectral density, as follows from (3), is

$$U_\nu = 4\pi\, h\nu\, f\,,$$

and from (5) it is seen that the flux $\vec{S}_\nu$ is equal to zero at any point in space.

The total intensity, density and radiation flux can be estimated using the spectral characteristics via integration over the whole spectrum of frequencies ν.

2. Equation of balance of the number of photons in a medium. We will derive an equation describing the transfer of radiation using the conservation law of the number of particles and following assumptions:

1) the process of propagating quanta is one-dimensional, i.e. $f = f(\nu, x, \vec{\Omega}, t)$;

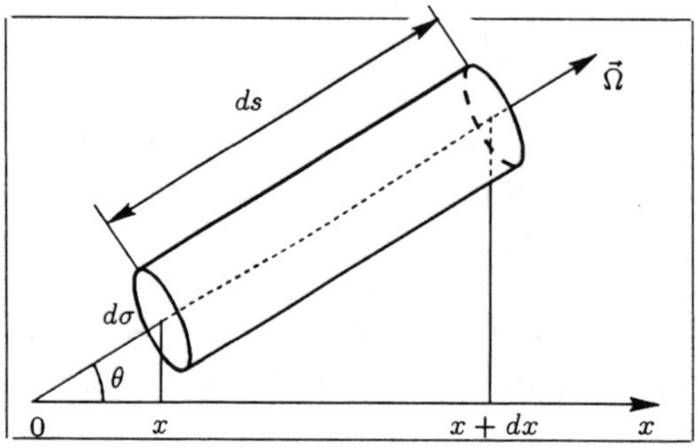

Fig.32.

2) the scattering of quanta on atoms or molecules (i.e. the change of their direction) can be neglected;

3) the character of absorption and emission of light by atoms and molecules is known;

4) the photons do not appear and disappear spontaneously.

Consider the balance of particles within an elementary cylinder with an axis directed $\vec{\Omega}$, length $ds = dx/\cos\theta$ and bases $d\sigma$ (Fig. 32), where θ is the angle between the axis x and vector $\vec{\Omega}$. We shall be interested in radiation of frequency ν in unit interval of frequencies propagated within unit solid angle in direction $\vec{\Omega}$.

In accordance with (1) (see also definition (2)), during dt a number of particles

$$cf(\nu, x, \vec{\Omega}, t)\, d\sigma\, dt\,,$$

enter the left base of the cylinder. During the same time scale, the following number of photons escape from it right base

$$(cf(\nu, x, \vec{\Omega}, t) + c\,df)\, d\sigma\, dt\,,$$

where df describes the increment of function f at passage from one base to other. So far as $f = f(\nu, \vec{\Omega}, x, t)$, this quantity can be represented as

$$df = \frac{\partial f}{\partial t}\, dt + \frac{\partial f}{\partial s}\, ds\,, \qquad ds = \frac{dz}{\cos\theta}\,,$$

where the first member described the increment in time during dt, and the second one described the increment of coordinate s.

Taking into account that the velocity of photons equals c and $dt = ds/c$, we obtain

$$df = \left(\frac{1}{c} \frac{\partial f}{\partial t} + \frac{\partial f}{\partial s} \right) ds .$$

So the number of photons in the cylinder during dt has varied on

$$- \left(\frac{\partial f}{\partial t} + c \frac{\partial f}{\partial s} \right) ds \, d\sigma \, dt = - \left(\frac{\partial f}{\partial t} + c \cos\theta \frac{\partial f}{\partial x} \right) ds \, d\sigma \, dt . \qquad (6)$$

Note that there is no scattering, and photons with direction $\vec{\Omega}$ do not escape through the side surface of the cylinder.

Thus, the variation of the number of quanta in the cylinder can be caused only by their absorption or emission by atoms and molecules of the matter of the cylinder. For the evaluation of its value the concept of *equilibrium radiation* is introduced, when the number of quanta absorbed by the matter is equal to the number of emitted particles (radiation and matter are in equilibrium) at any moment of time. The equilibrium distribution function f_{P} is (*Planck law*)

$$f_{\mathrm{P}} = \frac{2\nu^2}{c^3} \exp\left(1 - \frac{h\nu}{kT}\right) , \qquad (7)$$

where T is the temperature of matter (the medium is considered to be in local thermodynamic equilibrium, and in any point temperature, internal energy and other descriptors can be introduced).

In the absence of an equilibrium between radiation and matter the intensity of absorption (emission) of photons is proportional to the difference between f_{P} and f:

$$\kappa_\nu c(f - f_{\mathrm{P}}) ,$$

where $\kappa_\nu = \kappa'_\nu(1 - \exp(h\nu/kT))$, and κ_ν is the absorption coefficient determined by the state of the medium and its properties. The variation in the number of quanta in the volume of the cylinder during dt equals

$$\kappa_\nu \, (f - f_{\mathrm{P}}) \, d\nu \, ds \, dt . \qquad (8)$$

Equating (6) and (8), we obtain *the kinetic equation* for the distribution function, describing the transfer of radiation in a medium:

$$\frac{\partial f}{\partial t} + c \cos\theta \frac{\partial f}{\partial x} = \kappa_\nu \, (f_{\mathrm{P}} - f) , \qquad (9)$$

where f_{P} is given by formula (7). The equation (9), together with functions f_{P}, κ'_v and boundary conditions, represents the closed model of distribution of radiation energy for the above made assumptions.

3. Some properties of the equation of radiative transfer. The non-stationary one-dimensional inhomogeneous hyperbolic equation (9) obtained based on the conservation law of the number of particles can also be derived with the help of the energy conservation law. Indeed in the cylinder, the balance of particles of identical frequency ν and, therefore, of identical energy $h\nu$ was considered. Taking this into account, it is easy to rewrite (9) as a equation of relative spectral intensity $I_\nu = h\nu c f$:

$$\frac{1}{c}\frac{\partial I_\nu}{\partial t} + \cos\theta \frac{\partial I_\nu}{\partial x} = \kappa_\nu \left(I_{\nu \mathrm{p}} - I_\nu\right), \tag{10}$$

which is equivalent to (9), but has a more clear physical content.

Integrating (10) over a solid angle $\vec{\Omega}$ (i.e. in all directions of quanta) we obtain the equation connecting the density of radiation (3) and its flux (4):

$$\frac{\partial U_\nu}{\partial t} + \frac{\partial S}{\partial x} = c\kappa_\nu \left(U_{\nu \mathrm{p}} - U_\nu\right). \tag{11}$$

This equation can be treated as an equation of the continuity of radiation of a given frequency, reflecting the conservation law of radiation and being quite similar to equation (7), section 1, in the theory of the motion of underground waters and to equation (5), section 2, in the theory of thermal conductivity. This analogy is most obvious in three-dimensional cases, when equations (10) and (11) take the form

$$\frac{1}{c}\frac{\partial I_\nu}{\partial t} + \vec{\Omega} \nabla I_\nu = \kappa_\nu \left(I_{\nu \mathrm{p}} - I_\nu\right), \tag{12}$$

$$\frac{1}{c}\frac{\partial U_\nu}{\partial t} + \operatorname{div} \vec{S}_\nu = c\kappa_\nu \left(U_{\nu \mathrm{p}} - U_\nu\right), \tag{13}$$

Though equations (9)–(13) are linear, generally speaking, one cannot claim that the models of radiation exchange are simpler than the nonlinear models considered in sections 1 and 2. Indeed, solving (9)–(13), it is possible to obtain the spectral (i.e. for the given frequency ν) characteristics of the radiation propagated in a given direction $\vec{\Omega}$. For a complete picture it is necessary to find the required quantities for all values of ν, $\vec{\Omega}$ (or their certain integrals), which is a much more complicated task. Besides, in more complicated situations (during the scattering of photons, etc.) the models (9)–(13) can become considerably more complicated.

The most simple model of radiation transfer is obtained from (10), if one can neglect the absorption and emission of quanta and consider the case when all particles move in one direction. Then for any values of ν it is possible to adopt $\cos\theta = 1$ and to have the equation

$$\frac{1}{c}\frac{\partial I_\nu}{\partial t} + \frac{\partial I_\nu}{\partial x} = 0,$$

which is absolutely identical to equation (3), section 1, for the flux of non-interacting particles.

If the radiation intensity does not depend on time, (10) turns to an inhomogeneous linear differential equation

$$\cos\theta\,\frac{\partial I_\nu}{\partial x}+\kappa_\nu I_\nu=\kappa_\nu I_{\nu\mathrm{p}} \tag{14}$$

which has a general solution

$$I_\nu(x)=\int\limits_{x_0}^{x}\kappa_\nu I_{\nu\mathrm{p}}\,e^{-x(x')}\,dx'+I_{\nu_0}e^{-\kappa(x_0)}\,. \tag{15}$$

Here $\kappa(x')=\int_{x'}^{x}\kappa_\nu\,dx''$, $\kappa(x_0)=\int_{x_0}^{x}\kappa_\nu\,dx''$ (for simplicity in (15) we adopt $\cos\theta=1$) and I_{ν_0} is the constant of integration.

As we are not aiming to discuss in detail the physical content of solution (15), we mention that the first term is due to the radiation emitted in the medium in an interval x_0 and x (and weakened by absorption). The second term represents the radiation from any external sources entering the medium through its boundary x_0 (also weakened via absorption during propagation through the medium).

If $I_{\nu\mathrm{p}}$ and κ_ν are known functions of coordinate x (the temperature and density of matter have to be known along the trajectory of particles), the solution of equation (14) is reduced to a quadrature.

In the opposite case of a spatially homogeneous field of radiation from (10) we obtain

$$\frac{1}{c}\frac{\partial I_\nu}{\partial t}=\kappa_\nu\,(I_{\nu\mathrm{p}}-I_\nu)\,. \tag{16}$$

The process described by equation (16), corresponds to a situation, when in an infinite and initially cold medium (i.e. at $t=0$ no radiation exists) of constant density, quick heating of medium up to temperature T occurs, which later is kept constant in time. So far as no radiation losses occur from the boundaries, the spatial gradients of T are zero and κ_ν, $I_{\nu\mathrm{p}}$ do not depend on x, y, z. The radiation caused due to heating also has a zero gradient (i.e. $I_\nu=I_\nu(t)$) and exchanging energy with matter tends by time to its equilibrium value by exponential law.

EXERCISES

1. Check the validity of expression (5).
2. Repeating the considerations of subsection 2, derive the three-dimensional equation (9) or (10) in the case where the distribution function depends on x, y, z.
3. Using the definitions (3), (4) derive equation (12) from equation (13).

4. Obtain solution (15) of equation (14) and specify it in the case of constant κ_ν, $I_{\nu p}$.

5. Be convinced that the solution of equation of "saturation" (16) (compare with equation (12), section 1, Chapter I) has an exponential form, and obtain the index of the exponent.

4 Joint Application of Several Fundamental Laws

We will use laws of conservation of mass, momentum and energy to construct a mathematical model describing the motion of compressible gas. We will discuss the differences of the model obtained from models considered in sections 1–3, as well as some consequent properties for gas dynamical motions.

1. Preliminary concepts of gas dynamics. The noticeable variation in density of fluids and rigid bodies can be achieved only at huge pressure of tens and hundred thousands of atmospheres and higher. The gaseous media can be compressed much more easily: at a pressure variation of one atmosphere the density of a gas initially at atmospheric pressure decreases or increases by a magnitude comparable with its initial density.

In gas dynamics investigating the motion of compressible media under the action of any external forces or forces of pressure of the matter itself, the inequality $\lambda \ll L$, where λ is the length of free path, L is the characteristic scale of the considered motion is by assumption fulfilled (continuous medium). The LTE conditions (see subsection 1, section 2) are considered to be fulfilled as well. In LTE conditions the compressible medium can be considered as a configuration of a large number of liquid particles which are much bigger than λ, but much smaller than L. For each such particle connected with a small fixed mass of medium, its average descriptors – density ρ, pressure p, temperature T, internal energy ε, etc, as well as the velocity $\vec{v}$ of its macroscopic motion are introduced. All these quantities, in general, depend on three spatial variables x, y, z and time t.

Below we shall also assume the absence in the medium of the processes of heat transfer, viscous friction, sources and consumers of energy (for example, radiation), and external spatial forces and sources (consumers) of mass.

2. Equation of continuity for compressible gas. We use considerations similar to those used to derive continuity equation (7), section 1, (5), section 2, for the flow of underground waters and the process of heat transfer. Consider an elementary cube with sides dx, dy, dz in a spatial region filled by moving gas and estimate the mass balance during dt (Fig. 33). Here v_x, v_y, v_z are the velocity components by corresponding axes.

On axis x along the facet of coordinate x a mass of gas is flowing into

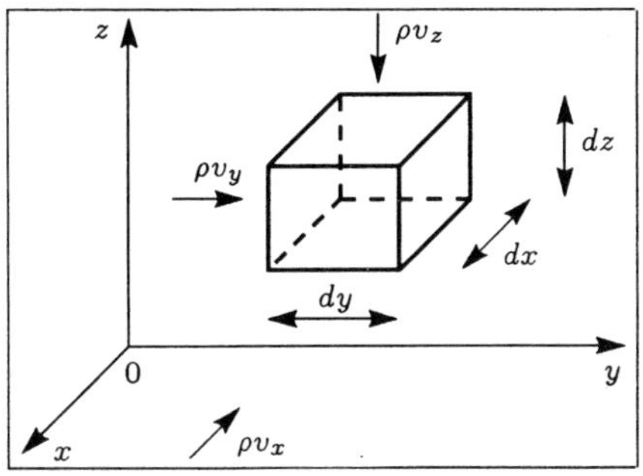

Fig.33.

the cube during dt equal to

$$\rho v_x \, dy \, dz \, dt \,,$$

where ρv_x is nothing other than the mass flow in the direction of axis x. During the same time a mass outflows from the facet with coordinate $x + dx$

$$[\rho v_x + d(\rho v_x)] \, dy \, dz \, dt \,,$$

where $d(\rho v_x)$ is the increment of mass flow due to transition from the coordinate x to $x + dx$. Summarizing the latter two expressions and taking into account that

$$d(\rho v_x) = \frac{\partial}{\partial x} (\rho v_x) \, dx \,,$$

we obtain the amount of variation of mass in the cube during dt due to gas motion along the x axis

$$dm_x = -\frac{\partial}{\partial x} (\rho v_x) dx \, dy \, dz \, dt \,. \tag{1}$$

In precisely the same way we derive the variation of mass due to motion along on axes y, z

$$\begin{aligned} dm_y &= -\frac{\partial}{\partial y} (\rho v_y) dx \, dy \, dz \, dt \,, \\ dm_z &= -\frac{\partial}{\partial z} (\rho v_z) dx \, dy \, dz \, dt \,. \end{aligned} \tag{2}$$

In the fixed volume of the cube, the variation of its mass is also expressed through the variation of its density in time

$$dm = \frac{\partial \rho}{\partial t} \, dt \, dx \, dy \, dz \,. \tag{3}$$

Summing dm_x, dm_y, dm_z and equating the result dm, we obtain from (1)–(3) the sought equation of continuity

$$\frac{\partial \rho}{\partial t} + \operatorname{div} \rho \vec{v} = 0\,, \tag{4}$$

expressing the law of conservation of mass of substance for compressible gas. In its form and content (the variation rate of a quantity is determined by divergence of its flux) it is quite similar to the equation of continuity (7), section 1 and equations (5), section 2 and (11), section 3.

However, the similarity to the flow of underground waters ends here. When a gas is moving freely its dynamics is determined only by forces of gas pressure, as distinct from the motion of a fluid undergoing the resistance of particles of soil.

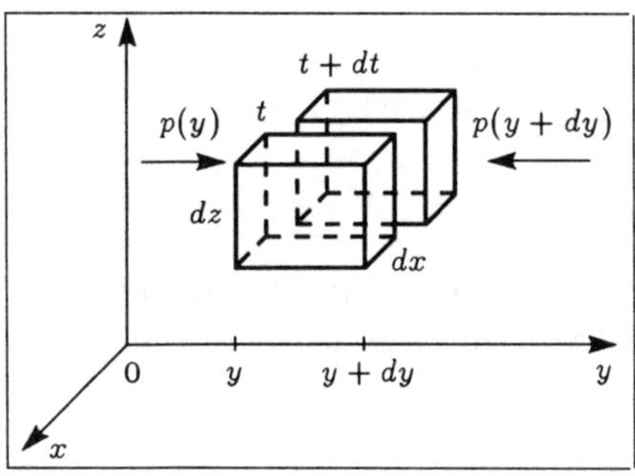

Fig.34.

3. Equations of gas motion. To derive these we use Newton's second law for an elementary liquid particle having in some moment t the form of a cube with edges dx, dy, dz (Fig. 34). The liquid particle is a volume moving in space and modifying its shape, and containing at different moments t the same atoms and molecules of gas. Thus its mass dm is constant. To simplify the derivation we consider that during the short time dt, the cube does not change its shape and is shifted in all directions at a distance which in much smaller than its dimensions.

First we define the force acting on the cube, for example, in the direction of axis y. Obviously, it is equal to the difference of pressure on the left and right edges multiplied on their square (no other forces are considered to be present)

$$F_y = [p(x,\, y,\, z,\, t) - p(x,\, y + dy,\, z,\, t)]\, dx\, dz\,.$$

The force F_y is equal to the acceleration of a liquid particle in direction y, multiplied by its mass $dm = \rho\, dx\, dy\, dz$

$$F_y = \frac{dv_y}{dt}\, \rho\, dx\, dy\, dz\,. \tag{5}$$

Substituting in the first expression for F_y the difference of pressures through the derivative of pressure on y and equating it to (5), we come to the equation describing the gas motion along the axis y

$$\rho \frac{dv_y}{dt} = -\frac{\partial p}{\partial y}. \tag{6}$$

In precisely this way obtain the equations of motion in directions x, z

$$\rho \frac{dv_x}{dt} = -\frac{\partial p}{\partial x}, \tag{7}$$

$$\rho \frac{dv_z}{dt} = -\frac{\partial p}{\partial z}, \tag{8}$$

having, as (6), an obvious physical content. In vectorial form equations (6)–(8) are as follows

$$\rho \frac{d\vec{v}}{dt} = -\mathrm{grad}\, p. \tag{9}$$

We have to explain that in (6)–(9) df/dt denotes the ordinary derivative (connected with fixed particles of gas) by time of some quantity describing the given constant mass of gas.

Opening df/dt through partial derivatives of x, y, z and t in correspondence with a rule $df/dt = \partial f/\partial t + (\vec{v}\,\mathrm{grad})\, f$, we come to *the Euler's equations of motion*

$$\frac{\partial \vec{v}}{\partial dt} + (\vec{v}\,\mathrm{grad})\, \vec{v} = -\frac{1}{\rho}\,\mathrm{grad}\, p. \tag{10}$$

In coordinate form, they are

$$\frac{\partial v_x}{\partial t} + v_x \frac{\partial v_x}{\partial x} + v_y \frac{\partial v_x}{\partial y} + v_z \frac{\partial v_x}{\partial z} = -\frac{1}{\rho}\frac{\partial p}{\partial x}, \tag{11}$$

$$\frac{\partial v_y}{\partial t} + v_x \frac{\partial v_y}{\partial x} + v_y \frac{\partial v_y}{\partial y} + v_z \frac{\partial v_y}{\partial z} = -\frac{1}{\rho}\frac{\partial p}{\partial y}, \tag{12}$$

$$\frac{\partial v_z}{\partial t} + v_x \frac{\partial v_z}{\partial x} + v_y \frac{\partial v_z}{\partial y} + v_z \frac{\partial v_z}{\partial z} = -\frac{1}{\rho}\frac{\partial p}{\partial z}. \tag{13}$$

As distinct from the flow of underground waters, the pressure gradients in equations of gas motion (6)–(13) define the components of acceleration of substance, instead of the components of its velocity (compare with the Darsi law (9), section 1).

The equations (4), (11)–(13) contain five unknown quantities – ρ, p, v_x, v_y, v_z. To make the set of equations complete it is most natural to use the energy conservation law.

4. The equation of energy. To derive this we use the same simplified scheme as in subsection 3: we shall consider the change of interior energy of a fixed mass of gas dm in a short time interval dt. In so far as the assumptions we have made no thermal conductivity, viscosity and sources (consumers) of energy exist in the substance, this variation is caused only by the work of forces of pressure on facets of cube at its compression or expansion.

The work of pressure connected with the motion of the facets of cube along the axis x, is obviously

$$dA_x = p(v_x(x) - v_x(x + dx))\, dt\, dy\, dz\,,$$

where the members in brackets, with cancelled terms of the second order, can be rewritten through the derivative $\partial v_x/\partial x$

$$dA_x = -p\,\frac{\partial v_x}{\partial x}\, dx\, dy\, dz\, dt\,.$$

Here p is the mean pressure in elementary volume. Similarly,

$$dA_y = -p\,\frac{\partial v_y}{\partial y}\, dx\, dy\, dz\, dt\,,$$

$$dA_z = -p\,\frac{\partial v_z}{\partial z}\, dx\, dy\, dz\, dt\,.$$

The total work accomplished upon the gas during dt, is

$$dA = dA_x + dA_y + dA_z = -p \operatorname{div} \vec{v}\, dx\, dy\, dz\, dt\,.$$

This is equal to the variation of internal energy of the volume, i.e.

$$dA = \rho d\varepsilon\, dx\, dy\, dz\,,$$

where ε is the specific internal energy. Equating both expressions for dA and tending dt to zero, we obtain

$$\rho\,\frac{d\varepsilon}{dt} + p \operatorname{div} \vec{v} = 0\,, \tag{14}$$

where $d\varepsilon/dt$ is the total derivative of internal energy by time.

Note that with the help of equations of continuity and motion, equation (14) is reduced, like (4), to *a divergent form*

$$\frac{\partial}{\partial t}\left(\rho\varepsilon + \frac{\rho v^2}{2}\right) = -\operatorname{div}\left[\rho\vec{v}\left(\varepsilon + \frac{v^2}{2}\right) + p\vec{v}\right]. \tag{15}$$

On the left hand side in (15) there is a derivative from total (internal and kinetic) energy of gas in the given spatial point.

In so far as the thermodynamic properties of substance are assumed to be known, ε is a known function of already defined quantities p and ρ, and the equation (14) or (15) give the missing link for the definition of unknown gas dynamics quantities.

5. The equations of gas dynamics in Lagrangian coordinates. In the models obtained the gas motion is characterized by the dependence of quantities on Cartesian coordinates x, y, z and time t. This manner of description (*Euler's approach*) treats the motion of a medium from the point of view of the motionless side observer and is convenient, for example, when studying gas flows in models of flight vehicles in aerodynamic tubes. In the *Lagrangian approach* the coordinate is linked not with a defined point in space, but with a definite fixed particle of substance – a liquid particle. The Lagrangian coordinates are convenient, for example, when analyzing certain internal processes within the particle, say, of chemical reactions with a rate determined not by spatial position of the particle, but by its temperature and density. The Lagrangian approach can also be useful for another reason: it was implicitly used in subsections 3 and 4 to simplify the derivation of equations of motion and energy (the liquid particle was considered as a cube).

Gas dynamical equations in Lagrangian coordinates are especially clear and simple in the one-dimensional case. Indeed, by taking a column of unit base in the direction of axis x, with moving left boundary and connected to fixed particles of substance, one can introduce the Lagrangian coordinates as follows

$$m(x) = \int_{x_0}^{x} \rho(x, t)\, dx\,, \qquad dm = \rho\, dx\,, \tag{16}$$

where x_0 is the coordinate of particles of the column on its left edge, $x(t)$ is the variable coordinate (as a particle with coordinate $x_0(t)$ it is possible to choose a particle either near the wall bounding the gas, or near a void, if one is available). The quantity $m(x)$ is the mass of column between x_0 and x.

The relation (16) is the most natural and simple way to introduce the Lagrangian variable (called in this case *a mass coordinate*). Then, all gas dynamic quantities are treated as depending not on x, t, but on m, t. We derive for them the equations of motion from the one-dimensional equations in Euler form

$$\begin{aligned} \frac{\partial \rho}{\partial t} + v\,\frac{\partial \rho}{\partial x} &= -\rho\,\frac{\partial v}{\partial x}\,, \\ \frac{\partial v}{\partial t} + v\,\frac{\partial v}{\partial x} &= -\frac{1}{\rho}\,\frac{\partial p}{\partial x}\,, \end{aligned} \tag{17}$$

$$\frac{\partial \varepsilon}{\partial t} + v\,\frac{\partial \varepsilon}{\partial x} = -\frac{p}{\rho}\,\frac{\partial v}{\partial x}\,,$$

The expressions on the left hand side of (17) are already defined substantive derivatives describing the variation in time of quantities concerning the fixed mass coordinate m. Then, using (16), we obtain from (17)

$$\frac{\partial}{\partial t}\,\frac{1}{\rho} = \frac{\partial v}{\partial m}\,, \tag{18}$$

$$\frac{\partial v}{\partial t} = -\frac{\partial p}{\partial m}\,, \tag{19}$$

$$\frac{\partial \varepsilon}{\partial t} = -p\,\frac{\partial v}{\partial m}\,. \tag{20}$$

Here $\partial/\partial t$ is the substantive derivative by time. The physical treatment of the equations of motion (19) and energy (20) is the same as in Euler coordinates (as distinct from the equation of continuity (18)). The latter represents an obvious property: the volume (and hence, the density) of fixed liquid particles varies in time due to the difference of velocities on its boundaries.

With the help of (19) the equation of energy can be rewritten in divergent form

$$\frac{\partial}{\partial t}\left(\varepsilon + \frac{v^2}{2}\right) = -\frac{\partial}{\partial m}\,(pv)\,, \tag{21}$$

where the time derivative of total energy of a particle is on the left hand side.

If the solution of Lagrangian equations is found, in particular, specific volume $V(m,\,t) = 1/\rho(m,\,t)$ is estimated, then the dependence of gas dynamical functions of Euler coordinates is given via a quadrature (see (16))

$$dx = V\,(m,\,t)\,dm\,, \qquad x\,(m,\,t) = \int\limits_0^m V(m,\,t)\,dm + x_0(t)\,.$$

Using the same considerations it is easy to derive Lagrangian equations in cases of cylindrical and spherical symmetries, when gas dynamical quantities depend only on one spatial coordinate r (r is the distance from the axis or from the center of symmetry) and from time t. They also have rather a simple and clear form, which is not the case for two-dimensional and three-dimensional problems.

Despite the different forms of representation the Euler and Lagrangian equations of gas dynamics clearly have similar properties, both being non-linear hyperbolic partial differential equations (non-stationary and usually many-dimensional). On the basis of these, more complicated models of motion of compressible media, including additional physical processes, are derived. Thus, for a gas possessing a thermal conductivity and performing transfer of energy, the equations of continuity (18) and motion (19) remain valid, while the energy equation has the form

$$\frac{\partial \varepsilon}{\partial t} = -p\,\frac{\partial v}{\partial m} - \frac{\partial W}{\partial m}\,, \tag{22}$$

where $W = -\kappa\rho\,\partial T/\partial m$ is the heat flux, $\kappa = \kappa(\rho,\, T)$ is the thermal conductivity. The internal energy of a liquid particle of such a gas varies not only due to the action of pressure forces, but also because of the presence of heat transfer. More simple models of gas dynamics than (18)–(20) are obtained at corresponding additional assumptions (see subsection 7).

6. Boundary conditions for the equations of gas dynamics. Their formulation is most obvious in the case of one-dimensional gas flow described by equations (18)–(20). Consider such motion in a pipe, in which the gas is limited from the right and the left by impenetrable rigid pistons (Fig. 35). For particles located near the left wall, we assign a coordinate $m = 0$; then, the coordinate of particles near the right wall is $m = M$, where M is the total mass of gas between pistons within the column of a unit cross-section. The internal particles have coordinates $0 < m < M$. The equations (18)–(20) are considered within the area $0 < m < M$ and at $t > 0$.

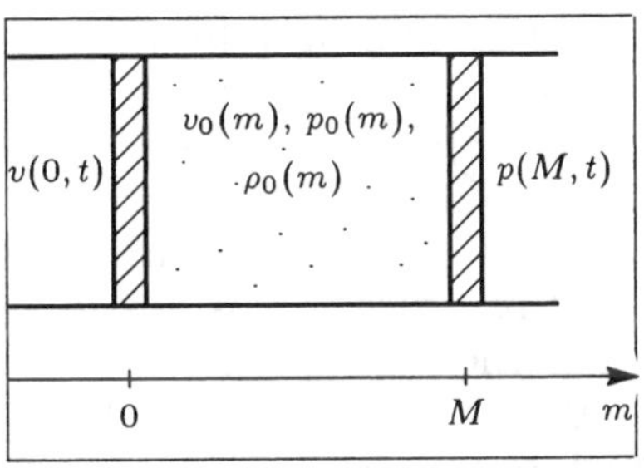

Fig.35.

To determine the motion at all $0 < m < M$ and $t > 0$ it is necessary to set

1) initial conditions, i.e. the condition of gas at moment $t = 0$,

$$v(m,\,0) = v_0(m)\,, \qquad p(m,\,0) = p_0(m)\,, \qquad \rho(m,\,0) = \rho_0(m)\,, \tag{23}$$

$$0 \le m \le M\,;$$

in (23) instead of p or ρ it is possible, using the equations of the state of the medium, to set the initial temperature $T(m,\,0) = T_0(m)$;

2) Boundary conditions, i.e. the dependence on time of gas dynamical quantities on the boundaries $m = 0$, $m = M$, for example, the law of pressure variation

$$p(0,\, t) = p_1(t)\,, \qquad p(M,\, t) = p_2(t)\,, \qquad t > 0\,, \tag{24}$$

or law of a change of the velocity of pistons (i.e. their trajectory in Eulerian coordinates, in so far as $\partial x/\partial t = v$)

$$v(0,\, t) = v_1(t)\,, \qquad v(M,\, t) = v_2(t)\,, \qquad t > 0\,. \tag{25}$$

The knowledge of boundary conditions (23) and (24) (or (25)) completely determines the unique solution of the considered *problem of a piston.*

As boundary conditions various combinations of (24) and (25) are possible as well, when the pressure $p_1(t)$ is fixed on the left boundary, while on the right boundary the velocity $v_2(t)$ is given (or vice versa, as in Fig. 35).

If one is interested in the flow only in a neighborhood of one of pistons, neglecting the influence of the second piston, it is enough to set only one of conditions (24) (or (25)), for example, at $m = 0$, and condition (23) at $m > 0$ (the gas fills the semi-space limited to the left by a piston).

Finally, the boundary conditions (24) or (25) are not set at all, when the influence of boundaries on the motion of gas in a central area can be neglected (considering the process during relatively short time scales). Then from the initial problem for equations (18)–(20) one comes to the Cauchy problem – on evolution in time of some initial distribution of gas dynamical quantities given in infinite volume. Then the initial data (23) are defined for all $-\infty < m < \infty$, and the equations are solved at $-\infty < m < \infty$, $t > 0$.

An important class of boundary conditions are the conditions on the boundary with a vacuum (void). Let a highly compressed gas at high pressure start to expand in a relatively rarefied medium at low pressure. Idealizing this process, it is possible to consider the pressure and density equal to zero in the space where the gas is expanding, i.e. to set a condition $p_1(t) = 0$, $t > 0$, or $p_2(t) = 0$), $t > 0$ (or $p_1(t) = p_2(t) = 0$, $t > 0$), depending on the concrete problem.

7. Some peculiarities of models of gas dynamics. To explain these peculiarities we will first simplify equations (18)–(20) using two circumstances. First, we assume the absence in the medium of energy change due to thermal conduction, viscosity, radiation, external sources and consumers of energy and so on. From thermodynamic point of view this means that the process is adiabatic and the entropy S of each fixed liquid particle is not changed in time. Then the equation of energy (20) can be rewritten in an equivalent form

$$\frac{\partial S}{\partial t} = 0\,. \tag{26}$$

It is easy to prove this, also formally applying the second law of thermodynamics

$$T\,dS = d\varepsilon + p\,dV \tag{27}$$

to a liquid particle.

The second circumstance is the special simplicity of expression of entropy through the pressure and density in the case of an ideal gas:

$$S = C\mathrm{v}\,\ln p\rho^{-\gamma} + S_0\,, \tag{28}$$

where $\gamma > 1$ is the adiabatic exponent equal to the ratio of specific heat at constant pressure (C_p) and constant volume (C_V), S_0 is an insignificant constant.

From (26) in view of (28), we have

$$\frac{\partial(p\rho^{-\gamma})}{\partial t} = 0\,,$$

which is equivalent to the expression

$$p\rho^{-\gamma} = \varphi(m)\,, \tag{29}$$

meaning the independence on time of entropy of gas particles. The function $\varphi(m)$ describes the distribution of entropy by mass of gas, being determined by given at the moment $t = 0$ functions $p_0(m)$, $\rho_0(m)$.

Using the integral (29) instead of the differential equation (20), we shall reduce (18) and (19) to a differential second order equation with respect to the density:

$$\frac{\partial^2\rho}{\partial t^2} = \frac{\partial}{\partial m}\left(\alpha_0\rho^{\gamma+1}\frac{\partial\rho}{\partial m}\right), \tag{30}$$

where $\alpha_0 = \gamma\varphi_0$, and constant φ_0 is entropy assumed to be independent also on the mass coordinate (the case of adiabatic flow).

The hyperbolicity of equation (30), and hence of equations of gas dynamics can easily be established without calculating its characteristics, but by deriving its linear analog. Then we consider small perturbations of gas dynamical quantities in the neighborhood of constant solution $\rho(m, t) \equiv \rho_0$. Representing the perturbed solution as $\rho(m, t) = \rho_0 + \bar{\rho}$ and assuming both the perturbations and their derivatives to be small, from (30) we obtain the equation for $\bar{\rho}$ (we omit the hyphen)

$$\frac{\partial^2\rho}{\partial t^2} = c_0^2\,\frac{\partial^2\rho}{\partial m^2}\,. \tag{31}$$

The linear equation (31) is absolutely similar to the equation of oscillations of a string from section 2, Chapter III, being, of a hyperbolic type. It describes the propagation of small (sound) perturbations in gas (*equation of acoustics*) with velocity of sound $c_0 = \sqrt{\gamma p_0/\rho_0}$ and, by virtue of linearity, its general solution can be easily found.

One more simplification of equations (18), (19), (29) is obtained on the assumption that the flow has a simple wave character, i.e. arbitrary gas dynamical quantities are functions of a certain chosen quantity, for example, of density. From (18), (19), (29) and in view of $v = v(\rho)$, we obtain

$$-\frac{1}{\rho^2}\frac{\partial \rho}{\partial t} = v_\rho \frac{\partial \rho}{\partial m}, \qquad v_\rho \frac{\partial \rho}{\partial t} = -\alpha_0 \rho^{\gamma-1}\frac{\partial \rho}{\partial m},$$

where v_ρ is the derivative of velocity by density. Excluding v_ρ from the latter equations, we come to the *Hopf equation*

$$\frac{\partial \rho}{\partial t} + \sqrt{\alpha_0}\,\rho^{\frac{\gamma+1}{2}}\frac{\partial \rho}{\partial m} = 0. \tag{32}$$

The equation (32) is of a first order, but it contains typical gas dynamical nonlinearity, and consequently is a good model for the study of nonlinear effects characteristic of flows of compressible gas. Known among them is the "gradient catastrophe" concerning the appearance of infinite gradients of gas dynamical quantities in compression waves, in spite of the fact that initially all functions are smooth.

We explain this concept by the following simple considerations. The Hopf equation can be rewritten in a characteristic form

$$\left(\frac{d\rho}{dt}\right)_s = 0. \tag{33}$$

Here script s means that the total derivative by time is taken along characteristics – the line in coordinates m, t, on which the value of solution (density) remains constant in all moments. Opening (33) we have

$$\frac{\partial \rho}{\partial t} + \frac{dm_s(t)}{dt}\frac{\partial \rho}{\partial m} = 0, \tag{34}$$

where $m_s(t)$ is the value of m for the given characteristics at different instants, and comparing (32) and (34), we obtain the equation of characteristics

$$m_s(t) = \sqrt{\alpha_0}\,\rho^{\frac{\gamma+1}{2}}\,t + m_s(0).$$

From this expression it is seen that the condition of high density is propagated over the mass of gas with a higher velocity, than the condition of lower density, and in certain moment of time "reach" the latter. An ambiguity appears in the solution, as its gradients increase infinitely to the point of merging. Schematically this process is represented in Fig. 36, where the evolution in time of the initial profile of density of a triangular form is shown: the top of the triangle in a certain time scale arrives in a point with the same coordinate m_k, as its front.

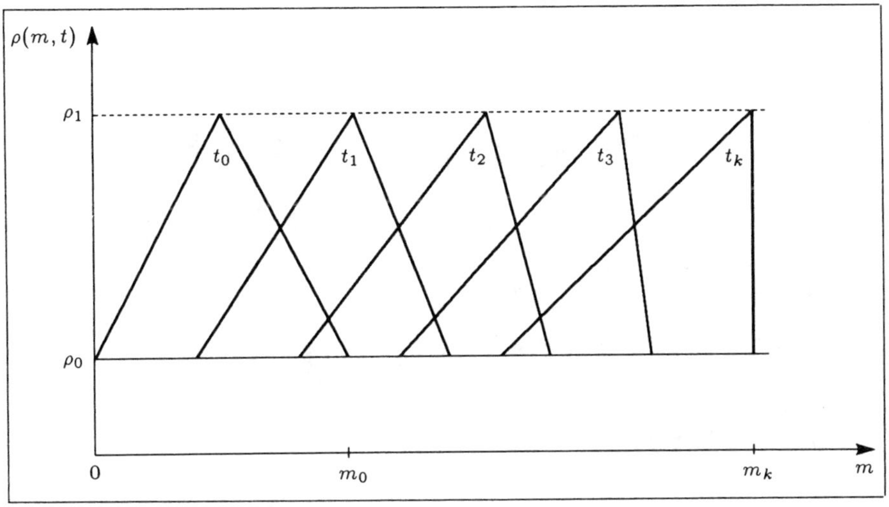

Fig.36.

"Gradient catastrophe" is a nonlinear effect (for more details see Chapter V). It does not arise in linear equations (31) and linear transfer equations (3), section 1, obtained from (32) at consideration of small perturbations in the vicinity of constant solution. The existence of this effect leads to the need to consider discontinuous solutions of equations of gas dynamics (note that at passage of the discontinuity through a liquid particle its entropy is changing). This represents an important difference between nonlinear hyperbolic equations and parabolic ones (models in subsections 1 and 2).

EXERCISES

1. Derive equation (21) from equation (20), using equation (19).

2. Using the energy conservation law with respect to the internal energy of a liquid particle, derive equation (22).

3. Prove the equivalence of equations (26) and (20), using (27) and (28).

4. Obtain equation (31) not from (30), but from (18), (19), (29), applying the same considerations.

5. Using the characteristic form of the Hopf equation estimate the moment t_k and point m_k of the "gradient catastrophe" for the situation in Fig. 36.

Bibliography for Chapter II: [30, 43, 44, 55, 69, 74].

Chapter III

MODELS DEDUCED FROM VARIATIONAL PRINCIPLES, HIERARCHIES OF MODELS

1 Equations of Motion, Variational Principles and Conservation Laws in Mechanics

We will now give elementary information from classical mechanics illustrating the connection between the equations of dynamics of Newton and Lagrange, variational Hamiltonian principle, conservation laws for mechanical systems and properties of space-time.

1. Equation of motion of a mechanical system in Newtonian form. The dynamics of a system consisting of N point masses influenced by some forces, are described by the equations

$$m_i \frac{d^2\vec{r}_i}{dt^2} = \vec{F}_i\,, \qquad i = 1, \ldots, N\,, \tag{1}$$

where m_i is the mass of material points, $t > 0$ is the time, $\vec{r}_i$ is its position vector, $\vec{F}_i = \vec{F}_i\,(t,\, \vec{r}_i,\, d\vec{r}/dt)$ is the resultant action of all forces. Through $\vec{r}$ we denote the set of coordinates of all points of the system. The quantities

$\vec{F}_i$ are considered known and can depend both on time and on spatial coordinates and velocities (and not only of chosen i-th point, but also of all considered points). The system (1) is the mathematical expression of Newton's second law applied to a totality of mass points. Written in coordinate form, it obviously consists of $3N$ second order equations with respect to $3N$ unknown coordinates $x_i(t)$, $y_i(t)$, $z_i(t)$, $i = 1, \ldots, N$. If the initial coordinates $x_i(0)$, $y_i(0)$, $z_i(0)$, $i = 1, \ldots, N$ of the points and their velocities dx_i/dt, dy_i/dt, dz_i/dt at a moment $t = 0$ are known, the system (1) enables to find the coordinates and velocities of all points at any $t > 0$, i.e. to solve the basic problem of classical mechanics.

The equations of dynamics in the form (1) are valid for an *inertial*, or *Galilean system of coordinates.* In this frame the free mass point, i.e. a point which is not experiencing any interaction, is moving uniformly and in a straight line. In other words, in this system Newton's first law is fulfilled. An example can be the coordinates rigidly connected to a train, moving with a constant velocity along a straight and horizontally smooth railway track. If a ball is moving on the smooth floor of a coach of such a train, it will perform a uniform and straight trajectory (this property is lost immediately if the train brakes or accelerates on turns and slopes when the system of coordinates becomes non inertial). Certainly, an inertial system of coordinates is an idealized concept justified, however, when considering many important mechanical phenomena. The obvious absence of an inertial system rigidly connected to the Earth (at least due to its rotation) does not prevent us from accurately describing several "terrestrial" motions, such as the motion of a train on its surface (for the flight of a ballistic missile a similar description can be not exact).

Any system of coordinates either at rest or moving in a straight line with constant velocity relative to some inertial system, is also an inertial one. The set of such systems "generated" by an initial system x, y, z, t is determined by the following transformation of coordinates and time

$$\begin{array}{l} x^* = x + a, \quad y^* = y + b, \quad z^* = z + c, \quad t^* = t \\ x^* = x, \quad y^* = y, \quad z^* = z, \quad t^* = t + a \\ x^* = x\cos\alpha + y\sin\alpha, \quad y^* = -x\sin\alpha + y\cos\alpha, \quad z^* = z, \quad t^* = t \\ x^* = x - v_x t, \quad y^* = y - v_y t, \quad z^* = z - v_z t, \quad t^* = t \end{array} \tag{2}$$

and also by any combinations of these transformations. Here a, b, c, v_x, v_y, v_z are arbitrary constants, as well as α the angle of rotation of a coordinate system relative to one of the axes of coordinates (in this case axis z).

The classical mechanics proceeds from the Galilean principle of relativity of the equality of all inertial systems (2), i.e. the laws of mechanics are identical in any of them. This principle reflects, in particular, the properties of homogeneity of space and time (first and second transformations (2)) and isotropy of space, i.e. the absence of chosen directions (third transformation).

The latter transformation (2) – the passage to coordinates moving with constant velocity relative the initial one – is called *a Galilean transformation.*

Note that if for any reasons it is necessary to consider motion in an essentially non inertial system of coordinates, Newton's second law in the form (1) is not fulfilled. Thus, a ball moving on a smooth horizontal surface of a rotating disk from its center to the edge, in a system rigidly connected to the disk, is moving not in a straight line, but via a curved trajectory, as if under the influence of an external force. For non inertial systems the equations (1) and other laws of mechanics become valid after relatively simple modification. The essence of this consists in adding "false" external *forces of inertia* to the acting, forces their magnitude is determined by the character of motion of a non inertial system relative the chosen inertial system. Usually the initial system of coordinates in mechanics is considered to be inertial, and its non-inertiality is specially mentioned.

The invariance of the laws of mechanics with respect to the transformations (2) can be expressed in various ways. If, for the equations corresponding to these laws, at passage to a new system

1) their structure does not vary,

2) the form of functions of coordinates, velocities and accelerations , appearing in the equations (these are the forces, the energy, the momentum and other mechanical quantities), does not vary then the equations are *invariant* with respect to given transformations. An example are the equations

$$m_1 \frac{d^2 r_1}{dt^2} = -k(l + r_1 - r_2)\,, \qquad m_2 \frac{d^2 r_2}{dt^2} = -k(r_2 - r_1 - l)\,, \tag{3}$$

invariant at any transformation (2) (see also exercise 1).

The equation describing the oscillations of the rings of Saturn, from subsection 3, section 2, Chapter 1

$$M_2 \frac{d^2 r}{dt^2} = \gamma M_1 M_2 \frac{r}{(r^2 + R_0^2)^{3/2}}\,,$$

is invariant at the shift of time $t^* = t+a$, but varies at the shift of coordinates $r^* = r + a$:

$$M_2 \frac{d^2 r^*}{dt^2} = \gamma M_1 M_2 \frac{r^* - a}{((r^* - a)^2 + R_0^2)^{3/2}}\,,$$

in the sense that the expression on the right hand side (i.e. the force), is rewritten differently as a function of a coordinate, than in the initial equation. The form of equations of mechanics violating the property 1), but fulfilling the property 2) at transformations (2), is called *covariant.* From the coordinate form of equations (1) their covariance with respect to transformation (2) follows immediately (however in the case of more complicated transformations of an initial system, for example, when passing to cylindrical or spherical coordinates, the covariance of equations of motion in Newton's form is lost).

2. Equations of motion in Lagrangian form. Lagrange equations are a convenient representation of Newton's second law, and are covariant relative to a much broader class of transformations of coordinates system, as compared with (2) (the covariance is understood in the sense, as with transformations of inertial systems). We explain how these equations are deduced using an example of a most simple mechanical system consisting of one point mass performing one-dimensional motion. In an initial inertial system of coordinates it is described by the equation

$$m\frac{d^2r}{dt^2} = F\left(r,\, \frac{dr}{dt},\, t\right). \tag{4}$$

Consider the transformation of coordinates $r(t) = r(q(t),\, t)$, where $q(t)$ is the new coordinate of point mass. For its velocity we obtain the following expression through functions $r(q)$ and $q(t)$:

$$v = \frac{dr}{dt} = \frac{\partial r}{\partial q}\dot{q} + \frac{\partial r}{\partial t}. \tag{5}$$

Here $\dot{q} = dq/dt$ is the "velocity" of a point in new coordinates (in general, it is no more an actual kinematic velocity, and only a characteristic of rate of change of coordinate q of the point in time). The "real" velocity v in the new system, as distinct from the old one, in accordance with (5) becomes a function of coordinate q and velocity $\dot{q}$, i.e. $v = v(q,\, \dot{q},\, t)$. We now calculate the acceleration of a point ω in new coordinates, first representing it in an equivalent form

$$\omega = \frac{d^2r}{dt^2} = \frac{dv}{dt} = \frac{1}{\partial t/\partial q}\frac{dv}{dt}\frac{\partial r}{\partial q} = \frac{1}{\partial r/\partial q}\left[\frac{d}{dt}\left(v\frac{\partial r}{\partial q}\right) - v\frac{d}{dt}\frac{\partial r}{\partial q}\right].$$

Differentiating (5) by $\dot{q}$, we obtain

$$\frac{\partial r}{\partial q} = \frac{\partial v}{\partial \dot{q}},$$

i.e. the expression for transformation of the first term in square brackets. For the second term, taking into account (5), we have a chain of equalities

$$\frac{d}{dt}\frac{\partial r(q(t),\, t)}{\partial q} = \frac{\partial^2 r}{\partial q^2}\dot{q} + \frac{\partial^2 r}{\partial q \partial t} = \frac{\partial}{\partial q}\left[\frac{\partial r}{\partial q}\dot{q} + \frac{\partial r}{\partial t}\right] = \frac{\partial v}{\partial q}.$$

Using the two last equalities, we come to a final expression

$$\omega(q,\, \dot{q},\, t) = \frac{1}{\partial r/\partial q}\left[\frac{d}{dt}\frac{\partial(v^2/2)}{\partial \dot{q}} - \frac{\partial(v^2/2)}{\partial q}\right].$$

Substituting it into (4), we obtain the equation

$$\frac{d}{dt}\frac{\partial(mv^2/2)}{\partial \dot{q}} - \frac{\partial(mv^2/2)}{\partial q} = F(q,\, \dot{q},\, t)\,\frac{\partial r}{\partial q}. \tag{6}$$

In (6) the quantity $F(q,\, \dot{q},\, t)$ on the right hand side of equation (4) is the force, but rewritten in new coordinates, its dependence on arguments $q,\, \dot{q},\, t$ in general is not the same as on $r,\, \dot{r},\, t$. Equation (6) is the simplest equation of motion in the Lagrangian form. On its left hand side under the operation of derivation is obviously kinetic energy of a point, and on the right is the acting force multiplied by $\partial r/\partial q$, describing the relation between old and new coordinates.

For three-dimensional motion of a point instead of coordinates $x,\, y,\, z$ it is necessary to introduce three new coordinates $q_1,\, q_2,\, q_3$. We repeat the considerations leading to conclusion (6), but instead of $v,\, \omega,\, F$ we shall take the variables $v_x,\, v_y,\, v_z;\, \omega_x,\, \omega_y,\, \omega_z;\, F_x,\, F_y,\, F_z$, i.e. the projection of velocity, acceleration and force, respectively. Consider them as functions of arguments $q_1,\, \dot{q}_1,\, t;\, q_2,\, \dot{q}_2,\, t;\, q_3,\, \dot{q}_3,\, t$ and substitute the expressions obtained into Newton's equations in coordinate form $m\,dv_x/dt = F_x$, $m\,dv_y/dt = F_y$, $m\,dv_z/dt = F_z$.

As a result we come to three equations which are similar to (6). For example, the first of them has a form

$$\frac{d}{dt}\frac{\partial T}{\partial \dot{q}_1} - \frac{\partial T}{\partial q_1} = F_x\frac{\partial x}{\partial q_1} + F_y\frac{\partial y}{\partial q_1} + F_z\frac{\partial z}{\partial q_1}, \tag{7}$$

where $T = m\,(v_x^2 + v_y^2 + v_z^2)/2$ is the kinetic energy of a point.

From these examples the procedure for the general case of motion of N points determined by three coordinates $x_i(t),\, y_i(t),\, z_i(t),\, i = 1,\, \ldots,\, N$ becomes clear (the calculations are quite similar to those used at derivation (6), (7), though they are too cumbersome to be represented here). The transition to new coordinates is set by arbitrary transformation

$$\begin{aligned} x_i &= x_i(q_1,\, \ldots,\, q_n;\, t), \\ y_i &= y_i(q_1,\, \ldots,\, q_n;\, t), \\ z_i &= z_i(q_1,\, \ldots,\, q_n;\, t), \end{aligned} \tag{8}$$

where $i = 1,\, \ldots,\, N$, $n = 3N$. In (8) q_j, $1 \le j \le 3N$ are so-called *generalized coordinates.* Their total number, as in an old system of coordinates, is naturally $3N$. At the transition to coordinates q_j each old coordinate can generally depend, on all the new ones (as, for example, in the case of transition from Cartesian coordinates x, y, z to spherical coordinates ρ, φ, ψ). Therefore on the right hand side of (8), dependence on all magnitudes q_j is implied.

After the consideration of "projections" (more precisely, of components) of all vectors $\vec{v}_i$, $\vec{\omega}_i$, $\vec{F}_i$, $i = 1, \ldots, N$, upon all new "coordinate axes" q_j, substitution of obtained expressions into the coordinate equations (1) and their summing by index i at fixed value of j, we come to general *Lagrange equations*

$$\frac{d}{dt}\frac{\partial T}{\partial \dot{q}_i} - \frac{\partial T}{\partial q_i} = \Phi_j \qquad j = 1, \ldots, n \tag{9}$$

In (9), as at $N = 1$, the quantity T is the kinetic energy of the system

$$T = \sum_{i=1}^{N} \frac{m_i v_i^2}{2} = \sum_{i=1}^{N} \frac{m_i \vec{v}_i \cdot \vec{v}_i}{2}, \qquad i = 1, \ldots, N,$$

written in coordinates q_j. The quantities

$$\Phi_j = \sum_{i=1}^{N} \left(F_{ix}\frac{\partial x_i}{\partial q_j} + F_{iy}\frac{\partial y_i}{\partial q_j} + F_{iz}\frac{\partial z_i}{\partial q_j} \right) = \sum_{i=1}^{N} \vec{F}_i(q, \dot{q}, t)\frac{\partial \vec{r}_i}{\partial q_j} \tag{10}$$

can be understood as expressed in coordinates q_j the "projections" of forces $\vec{F}_i$ on axes q_j – *generalized forces* (functions $|\partial \vec{r}_i / \partial q_j|$ are *Lame coefficients*; $\vec{A} \cdot \vec{B}$ denote the scalar product of vectors $\vec{A}$ and $\vec{B}$). In this treatment equations (9) represent Newton's second law (1) in "projections" onto axes q_j.

They become noticeably simpler in the case where all forces are potential, i.e. there is a function $\Pi(x_1, \ldots, x_N; y_1, \ldots, y_N, z_1, \ldots, z_N; t)$, such that

$$F_{ix} = -\frac{\partial \Pi}{\partial x_i}, \quad F_{iy} = -\frac{\partial \Pi}{\partial y_i}, \quad F_{iz} = -\frac{\partial \Pi}{\partial z_i}, \quad i = 1, \ldots, N.$$

Substituting these expressions into (10), we obtain

$$\Phi_j = -\sum_{i=1}^{N} \left(\frac{\partial \Pi}{\partial x_i}\frac{\partial x_i}{\partial q_j} + \frac{\partial \Pi}{\partial y_i}\frac{\partial y_i}{\partial q_j} + \frac{\partial \Pi}{\partial z_i}\frac{\partial z_i}{\partial q_j} \right).$$

Replacing the old arguments in the last equality in function Π with new ones, we come to the conclusion that its right hand side represents a partial derivative of some function $V(q,\, t)$, called *potential* over argument q_j, that is

$$\Phi_j = -\frac{\partial V}{\partial q_j}\,.$$

In other words, if the initial forces are potential, then the generalized forces are potential as well, and the equations (9) have the form

$$\frac{d}{dt}\frac{\partial L}{\partial \dot{q}_j} - \frac{\partial L}{\partial q_j} = 0\,, \qquad j = 1,\, \ldots,\, n\,, \tag{11}$$

$$L = T - V\,. \tag{12}$$

At derivation of (11), (12) is taken into account, that the function $V(q,\, t)$ does not depend on $\dot{q}$, and consequently $\partial L/\partial q_j = \partial T/\partial \dot{q}_j$. In Lagrange equations (11) the quantity L is the difference between kinetic and potential energies of a system expressed in new coordinates. It is known as *Lagrange function* or *Lagrangian* or *kinetic potential of a system.*

The equations (9),(11) obtained from Newton's equations (1), are covariant, as follows from their form with respect to the arbitrary transformation (8) while its functions have clear mechanical content. To write them down for a concrete system it is necessary to perform the following operations: to select independent coordinates $q_1,\, \ldots,\, q_n$; to find generalized forces (10) as functions of new coordinates (if forces are potential, then it is not necessary); to express in new coordinates the kinetic T and potential Π energies and to find the Lagrangian (12) (in case of potential forces); to substitute the obtained expressions into (9) or (11).

This standard procedure is called *Lagrangian formalism.* After its realization a system of $3N$ differential second order equations for coordinates $q_1,\, \ldots,\, q_n$ is obtained, which always solvable relative to its second derivatives

$$\ddot{q}_j = G_j(q,\, \dot{q},\, t)\,, \qquad j = 1,\, 2,\, \ldots,\, n\,. \tag{13}$$

At known initial values $q_j(0)$, $\dot{q}_j(0)$, $j = 1,\, \ldots,\, n$, the equations (13) enable one to describe the motion of a system at any moment $t > 0$, i.e. to solve the basic problem of mechanics.

The advantages of the Lagrangian approach are in its uniformity (covariance) in deriving a relatively simple mathematical description of the motion of very complicated mechanical devices (for example, robotic) in any coordinate system (including, as seen from (8) non inertial systems without introduction of additional forces of inertia). Note that the equations (1) also can be rewritten in covariant form relative to transformations (8). However they are not solvable with respect to higher derivatives and contain many more functions from new coordinates (as compared with equation (9)), without a simple mechanical content.

One more advantage of the described formalism is revealed when studying systems with *mechanical connections* (in all previous considerations only free systems have been taken into account for the sake of simplicity). The connections pose certain restrictions on the motion of points of the system, caused by the presence of material objects, not directly included in it. As examples we can use the connections caused by the presence of a rigid surface bounding the motion of the a ball, or a weightless rigid rod connecting two point masses, etc.

The connections are posed by a set K of expressions $f_k(\vec{r}_i, \dot{\vec{r}}, t) = 0$, $0 \leq k \leq K \leq 3N$, which the solution of Newton's equations (1) have to satisfy. The equations (1) themselves have to be modified, on their right hand sides one has to substitute the quantity $\vec{F}_i + \vec{R}_i$ instead of $\vec{F}_i$. A force $\vec{R}_i$ is the *reaction of connections*, estimated from certain considerations. Therefore, using Newton's approach, the description of unfree systems as compared with the free ones can become noticeably complicated.

Using the Lagrangian approach the reaction of connections are automatically taken into account for a broad class of functions $f_k(\vec{r}_i, \dot{\vec{r}}, t)$, therefore the generalized forces (10) do not contain the quantities $\vec{R}_i$, (the procedure of derivation of Lagrangian equations is absolutely similar to those of the free systems). Moreover, the number of Lagrangian equations is equal to the so-called degrees of freedom $l = 3N - K$ and can be significantly less than for a system without connections. The number of new independent coordinates q_j, naturally also equals $3N - K$.

Thus, Lagrangian equations are deduced from Newton's equations, and vice versa (exercise 2). Thus, the Lagrangian approach, as well as Newton's approach, may serve as a basis of mechanics.

3. Variational Hamiltonian principle. As the basis of mechanics one can consider not only the differential equations (1), (9) or (11) connecting the mechanical parameters in the given instant t, but also some general properties describing the motion of a mechanical system in the whole, within an arbitrary time interval from t_0 up to t_1. We shall prove it analyzing the quantity

$$Q = \int_{t_0}^{t_1} L(q(t), \dot{q}(t), t)\, dt\,, \tag{14}$$

called a *Hamiltonian action in interval* $[t_0, t_1]$. Obviously, (14) is a functional depending on the motion of the system within time interval $t_0 \ll t \ll t_1$.

In $(n+1)$-dimensional space q, t we select two points $M_0(q(t_0), t_0)$ and $M_1(q(t_1), t_1)$ by fixing thus, the moments t_0, t_1 and the position of the system at these moments ("velocities" $\dot{q}$ in moments t_0, t_1 are not fixed). The system can reach from a point M_0 to a point M_1 while moving in space q, t, generally speaking, via arbitrary kinematically possible trajectories ("paths"), i.e. via paths allowed by existing connections (see Fig. 37 for the case of space q_1, q_2, t). Let there exist a so-called *straight path* (solid line). On this path the functions $q_j(t)$, $j = 1, \ldots, n$ at any moment of time satisfy Lagrangian equations (11). The rest paths are called *indirect* (dashed lines).

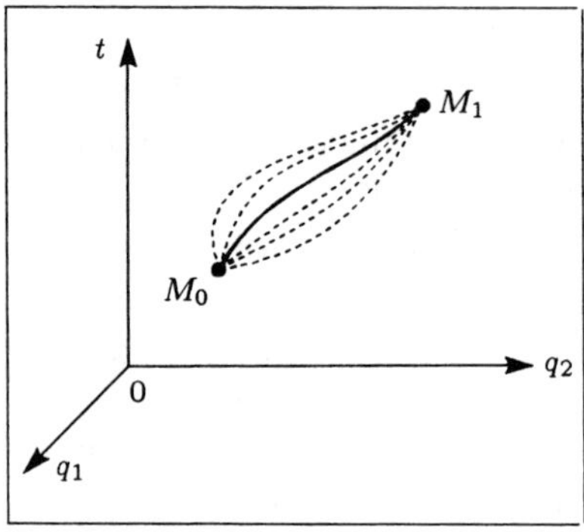

Fig.37.

The Hamiltonian principle is formulated as follows: the action Q has an extreme value on the straight path as compared with the indirect paths.

We characterize all possible paths by a one-parameter family of functions

$$q_j = q_j(t, \alpha), \qquad t_0 \le t \le t_1, \qquad |\alpha| \le \beta \le \infty, \qquad j = 1, \ldots, n,$$

where the value $\alpha = 0$ corresponds to the straight path, and values $\alpha \neq 0$ to indirect paths. Then the action (14), is obviously a function of parameter α:

$$Q(\alpha) = \int_{t_0}^{t} L[q_j(t, \alpha), \dot{q}_j(t, \alpha), t]\, dt.$$

The variation of Q at variation of the parameter α is

$$\delta Q = \frac{\partial Q}{\partial \alpha}\, d\alpha = \int_{t_0}^{t_1} \delta L\, dt = \int_{t_0}^{t_1} \sum_{j=1}^{n} \left(\frac{\partial L}{\partial q_j}\, \delta q_j + \frac{\partial L}{\partial \dot{q}_j}\, \delta \dot{q}_j \right) dt, \tag{15}$$

i.e. is equal to the sum of increments caused by the variation of coordinates $\delta q_j(t, \alpha)$ and velocities $\delta \dot{q}_j(t, \alpha)$.

Integrating by parts the second term on the right hand side of (15), we obtain

$$\delta Q = \sum_{j=1}^{n} \frac{\partial L}{\partial \dot{q}_j}\, \delta(q_j)\Big|_{t_0}^{t_1} - \int_{t_0}^{t_1} \sum_{j=1}^{n} \left(\frac{d}{dt} \frac{\partial L}{\partial \dot{q}_j} - \frac{\partial L}{\partial q_j} \right) \delta q_j\, dt,$$

or, using the commutativity of operations of variation on α and differentiation by t:

$$\delta(\dot{q}_j) = \delta\left[\frac{dq_j(t,\alpha)}{dt}\right] = \frac{\partial}{\partial\alpha}\frac{d}{dt}\,q_j(t,\,\alpha)\,\delta\alpha = \frac{d}{dt}\left[\frac{\partial}{\partial\alpha}\,q_j(t,\,\alpha)\,\delta\alpha\right] = \frac{d}{dt}\,\delta q_j\,,$$

we come to equality

$$\delta Q = \sum_{j=1}^{n}\frac{\partial L}{\partial\dot{q}_j}\bigg|_{t_0}^{t_1} - \int_{t_0}^{t_1}\sum_{j=1}^{n}\left(\frac{d}{dt}\frac{\partial L}{\partial\dot{q}_j} - \frac{\partial L}{\partial q_j}\right)\delta q_j\,dt\,. \tag{16}$$

In accordance with the definition the variations $\delta q_j(\alpha)$ are equal to zero at moments t_0, t_1, i.e. the first term on the right hand side of (16) is equal to zero. For the straight path by definition the Lagrangian equations (11) are valid, therefore the second member is also equal to zero.

Thus, on a straight path $\delta Q = 0$. This is the mathematical representation of the Hamiltonian principle. It is still called the *principle of least action*, in so far as the action along the straight path, as it is possible to show at some additional assumptions, has a least value as compared with the indirect paths. For more general mechanical systems (not governed by equations (11)) the analogous statement is called the *Hamilton-Ostrogradsky principle.*

Recall that in terms of variational calculus equations (11) are called *Euler's differential equations* for the variational problem

$$\delta Q = \delta\int_{t_0}^{t_1} L(q,\,\dot{q},\,t)\,dt = 0\,.$$

The straight path is called *extremal*, and the corresponding number $Q(\alpha = 0)$ – a *stationary value of functional* Q.

It is easy to prove the validity of the statement, inverse to the Hamiltonian principle: if for some path the condition $\delta Q = 0$ is fulfilled, this path is a straight one (exercise 3). Therefore all three approaches considered above are equivalent and can be the basis of a mathematical description of mechanical systems.

4. Conservation laws and space-time properties. The analysis of integral mechanical characteristics is not reduced to a formulation of the Hamiltonian principle and its various generalizations. It enables one to formulate other fundamental properties of mechanical systems. We will demonstrate some of those properties in the case of motions in potential fields, when both external and internal forces acting on points of a system are potential ones. Then it is required to calculate a variation of action Q on a set which

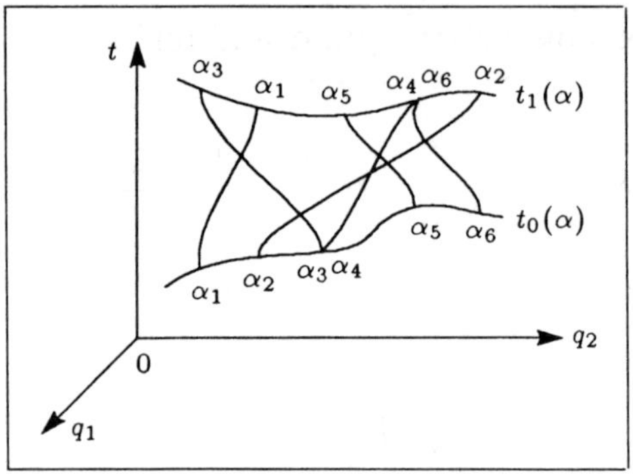

Fig.38.

is more general than the set of straight and indirect paths from subsection 3, among curves in $(n+1)$-dimensional $q_1, \ldots, q_n, t$ space.

Consider this one parametric family

$$q_1 = q_1(t, \alpha), \qquad \ldots, \qquad q_n = q_n(t, \alpha)\,, \tag{17}$$

where α is the parameter uniquely defining the curve, in particular its "lower" and "upper" edges in space q, t (Fig. 38). No restrictions are posed for the bundle (17), therefore curves can "start" and "terminate" in different instants $t_0(\alpha)$ and $t_1(\alpha)$, they can intersect, the different curves can coincide in initial and final moments, etc. The variation of action

$$\delta Q = \delta \int\limits_{t_0(\alpha)}^{t_1(\alpha)} L(q, \dot{q}, t)\, dt$$

on bundle (17) is obtained by the help of considerations similar to those used in the derivation of (16), however corresponding calculations are much more complicated, in so far as the initial and final positions of the family of curves in space q, t depend on the parameter α. Therefore, we represent only the final result: the general formula for δQ in the case of one parameter family (17) has the form

$$\delta Q = \left(\sum_{j=1}^{n} p_j\, \delta q_j - H\, bt \right) \Bigg|_{t_0(\alpha)}^{t_1(\alpha)} - \int\limits_{t_0(\alpha)}^{t_1(\alpha)} \sum_{j=1}^{n} \left(\frac{d}{dt} \frac{\partial L}{\partial \dot{q}_j} - \frac{\partial L}{\partial q_j} \right) \delta q_j\, dt\,. \tag{18}$$

The content of functions p_j and H will be explained below.

Obviously, if the beginnings and the ends of all curves of a bundle coincide, then $\delta q_j = \delta t = 0$, and from (18) the equation (16) or the Hamiltonian principle follows.

The functions p_j and H are called, respectively, *generalized momentums* and the *Hamiltonian of a system*. The quantities p_j are partial derivatives of the Lagrangian by velocities $\dot{q}_j$

$$p_j = \frac{\partial L}{\partial \dot{q}_j} = p_j(q,\, \dot{q},\, t)\,, \qquad j = 1,\, \dots,\, n\,, \tag{19}$$

and in the simplest case of motion of a point in Cartesian coordinates in a fixed potential field (the function Π from subsection 2 does not depend explicitly on time) coincide with projections of momentum on axes x, y, z. By virtue of general properties of a Lagrangian (not considered here), the relations (19) can be solved with respect to the generalized velocities

$$\dot{q}_j = h_j(q,\, p,\, t)\,,$$

i.e. there is a one-to-one transition from Lagrangian variables q, $\dot{q}$, t to so-called *Hamiltonian variables* q, p, t. The function H is defined by the equality

$$H(q,\, p,\, t) = \sum_{j=1}^{n} p_j\, \dot{q}_j - L\,, \tag{20}$$

where the Lagrangian L and the generalized velocities are written in variables q, p, t. The Hamiltonian plays particularly important role in the study of potential motions, since from Lagrangian equations (11) the equations of motion in Hamiltonian form (canonical equations) are obtained

$$\dot{q}_j = \frac{\partial H}{\partial p_j}\,, \qquad \dot{p}_j = \frac{\partial H}{\partial q_j}\,, \qquad j = 1,\, \dots,\, n\,, \tag{21}$$

and vice versa (exercise 4). The function H has a clear mechanical content: if the transformations of an initial system (8) are stationary (do not depend explicitly on t), at any moment the Hamiltonian is numerically equal to the total energy of the system, that is

$$H = T + \Pi = E\,.$$

After these explanations we will establish connection of conservation laws in mechanics with properties of space and time. Consider one parametric transformation of a coordinates system q, t

$$q_j^* = \varphi_j(q,\, t,\, \alpha), \qquad j = 1,\, \dots,\, n\,, \qquad t^* = \psi(q,\, t,\, \alpha) \tag{22}$$

such that it is an identity at $\alpha = 0$ and its inverse transformation does exist.

Let the Lagrangian of a given mechanical system moving in a potential field be invariant relative transformations (22). This implies that the new

Lagrangian $L^*(q^*, \dot{q}^*, t^*)$ does not depend on α and as a function q^*, $\dot{q}^*$, t^* has the same form as the initial Lagrangian in variables q, $\dot{q}$, t. Then there exists a function of p, q, t

$$\phi(q, p, t) = \sum_{j=1}^{n} p_j \left(\frac{\partial \varphi_j}{\partial \alpha}\right)_{\alpha=0} - H \left(\frac{\partial \psi_j}{\partial \alpha}\right)_{\alpha=0}, \tag{23}$$

not changing its value by time on straight paths (first integral of motion). In (23) p_j, H are the generalized momentums (19) and the Hamiltonian (20) respectively, written in variables p, q, t.

The scheme of the formulated statement (Nether's theorem) is as follows. In space q, t a curve $q(t)$ is chosen, on which $\delta Q = 0$, i.e. corresponding to a part of some straight path in an interval $[t_0, t_1]$. In accordance with (22) this curve "generates" in space q^*, t^* a family of curves $q^*(t^*, \alpha)$. For them, by virtue of invariance of the Lagrangian, the variation of an action is equal to zero at all values of α or in view of (18), applied for the space q^*, t^*,

$$\begin{aligned}\delta Q = &\left(\sum_{j=1}^{n} p_j^*\, \delta q_j^* - H^*\, \delta t^*\right)\Bigg|_{t_0^*(\alpha)}^{t_1^*(\alpha)} - \\ &- \int_{t_0^*(\alpha)}^{t_1^*(\alpha)} \left(\sum_{j=1}^{n} \frac{d}{dt^*}\frac{\partial L^*}{\partial \dot{q}^*} - \frac{\partial L^*}{\partial q_j^*}\right) \delta q_j^*\, dt = 0\,.\end{aligned} \tag{24}$$

If in (24) we put $\alpha = 0$, i.e. consider an identical transformation, the integrand expression will turn to zero, in so far as in coordinates q, t the curve $q(t)$ is a straight path, where the Lagrangian equations (11) for $L = L^*$ are satisfied. Thus, the following condition has to be fulfilled

$$\left[\left(\sum_{j=1}^{n} (p_j^*\, \delta q_j^* - H^*\, \delta t^*)\right)\Bigg|_{t_0^*(\alpha)}^{t_1^*(\alpha)}\right]_{\alpha=0} = 0\,.$$

From here in view of properties of transformation (22), it is easy to calculate δq_j^* and δt^* and, tending α to zero we deduce the formula (23). In so far as the straight path and the points t_0, t_1 have been selected arbitrarily, the statement of the theorem is valid for any straight path (for all real motions) of the system.

From Noether's theorem the conservation law of total mechanical energy follows for systems, the Lagrangian (as well as the Hamiltonian) of which does not depend explicitly on time; these systems are called *conservative systems*. Really, taking the transformation (22) as the shift in time

$$q_j^* = q_j\,, \qquad j = 1, \dots, n\,, \qquad t^* = t + \alpha\,,$$

it is easy to prove the invariance of Lagrangian and to (23) we obtain

$$-\phi = H = T + \Pi = \text{const}\,.$$

The conservation law of momentum for closed (i.e. not experiencing actions of external forces) system (exercise 5) and series of other laws of mechanics, are formulated in the same way. When considering motions not on one straight path, but on some selected sets of straight paths, the more general first integrals of mechanical systems – integrated invariants are obtained. Some of them can be considered along with the Hamiltonian principle to be at the basis of mechanics.

The deep connection of equations of motion, conservation laws, variational principles and properties of symmetry enables one to use various approaches to construct mathematical models of mechanical systems. Note that the invariant properties of objects are effectively applied not only to the construction, but also to the analysis (see, for example, section V) of the models of various phenomena.

EXERCISES

1. Show that equations (3) describe the motion of balls of masses m_1, m_2, connected with a weightless spring of rigidity k (l is the length of a unloaded spring, $r_1(t) \leq r_2(t)$). Check an invariance of equations at the passage from one inertial system to another.

2. Using in (8) the identical transformation of an initial system of coordinates, be convinced that from equations (9) the coordinate form of equations (1) is obtained.

3. Using the equality (16), check that for a path along which $\delta Q = 0$, the Lagrangian equations (11) are satisfied.

4. Establish the equivalence of the Hamiltonian equations (21) and Lagrangian ones (11) in an example of the motion of a single point mass in a stationary potential field, considering it in Cartesian coordinates x, y, z.

5. In the absence of external forces, the potential energy of a system and hence its Lagrangian function do not vary at the shift of beginning of coordinates. Taking in (22) the transformations of Cartesian coordinates $x_i^* = x_i + \alpha$, $y^* = y_i$, $z^* = z_i$, $(i = 1, \ldots, N)$, $t^* = t$, prove that (23) has a form $\phi = \sum_{i=1}^{N} m_i \dot{x}_i = \text{const}$, i.e. for such a system the momentum conservation law in projection on axis x is valid.

2 Models of Some Mechanical Systems

The Lagrangian equations and the Hamiltonian principle can be used to describe various types of motion of a pendulum and small oscillations of a string, and also oscillations of an electric current in a circuit, for which we used electromechanical analogy. We will discuss some properties of the investigated processes.

1. Pendulum on the free suspension. A system consists of two point masses, m_1 and m_2, connected with a weightless rigid rod of length l (Fig. 39). The motion occurs in a gravity field and is considered to be in a plane, i.e. is considered in the coordinates x; y, t. The location of point of a mass m_1 (suspension) is not fixed, and can move along the axis x (compare with subsection 3, section 3, Chapter 1).

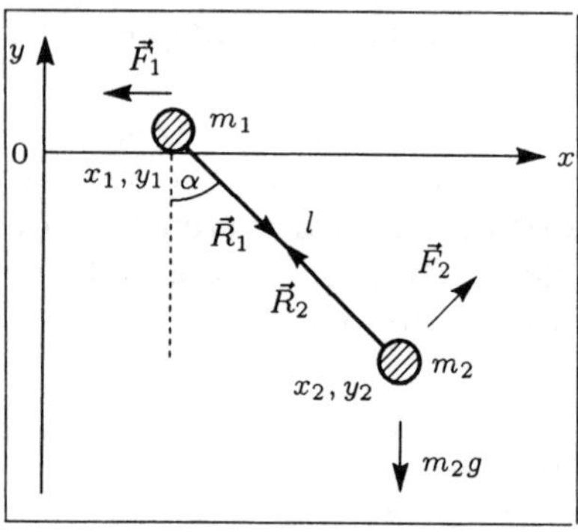

Fig.39.

To describe the plane motion of two points in an initial system of coordinates it is usually necessary to find out from equations (1), section 1, four functions of time, $x_1(t)$, $y_1(t)$, $x_2(t)$, $y_2(t)$, i.e. the Cartesian coordinates of the first and second points. However the investigated system is not free, in so far as it contains two mechanical connections (subsection 2, section 1). One of them is described by the equation $y_1 \equiv 0$ (suspension cannot move vertically), while the second is described by equation $(x_1 - x_2)^2 + y_2^2 = l^2$ (distance between points at any t is equal to the length of the rod). Therefore at transition to Lagrange equations it is enough to select (in accordance with the number of degrees of freedom) only two new independent coordinates. We choose the generalized coordinates as $q_1(t) = x_1(t)$ and $q_2(t) = \alpha(t)$, where α is the angle between the vertical and the axis of rod. This choice corresponds to transformation (8), section 1, having a form

$$x_1 = q_1, \qquad x_2 = q_1 + l \sin q_2, \qquad y_2 = -l \cos q_2 .$$

First, we express the kinetic energy of system $T = T_1 + T_2$ in coordinates q_1, q_2. For the suspension we have

$$T_1 = \frac{m_1 v_1^2}{2} = \frac{m_1 v_{1x}^2}{2} = \frac{m_1 \dot{x}_1^2}{2} .$$

For the pendulum we obtain

$$T_2 = \frac{m_2 v_2^2}{2} = \frac{m_2}{2} \left(v_{2x}^2 + v_{2y}^2\right)$$

With the help of equalities $v_{2x} = \dot{x}_1 + l\,\dot{\alpha}\,\cos\alpha$, $v_{2y} = l\,\dot{\alpha}\,\sin\alpha$, the first of which takes into account the composite motion of a mass m_2 along an axis x as a sum of motions together with the suspension and relative it, we rewrite T_2 as function of x_1 and α

$$T_2 = \frac{m_2\dot{x}_1^2}{2} + \frac{m_2}{2}\left(2l\dot{\alpha}\dot{x}_1\,\cos\alpha + l^2\dot{\alpha}^2\right).$$

Consider now the forces acting on the points m_1 and m_2. Gravitational force and the vertical projection R_{1y} of the reaction of a rod $\vec{R}_1$ acted on the suspension, are balanced by the force of the base, and hence, the vertical balancing force is equal to zero. The force $\Im_{1x}$ obviously represents the horizontal projection of reaction of a rod (connection) R_{1x}. In the Lagrangian approach it is not necessary to take into account the force of reaction of a rod on the motion both of the suspension and the pendulum, i.e. the forces $\vec{R}_1$ and $\vec{R}_2$ (a concrete example is given in subsection 2). Therefore from all acting forces it is enough to take into consideration only gravity $\vec{\Im}_2$ acting on the pendulum. For its projection we have

$$\Im_{2x} = 0\,, \qquad \Im_{2y} = -m_2 g = -m_2\,\frac{\partial\Pi}{\partial y_2}\,,$$

where $\Pi(y_2) = m_2 g y_2$ is the potential energy of the pendulum. In coordinates q_1, q_2 $\Pi(y_2)$ is expressed by the formula

$$V(q_2) = -m_2 l g\,\cos\alpha\,.$$

In so far as the considered motion is potential, it is necessary to use the Lagrangian equations (11), section 1, where $j = 1, 2$, $i = 1, 2$, and

$$L = T - V = T_1 + T_2 - V\,,$$

or in explicit form

$$L = \frac{m_1 + m_2}{2}\,\dot{x}_1^2 + \frac{m_2 l}{2}\left(l\,\dot{\alpha}^2 + 2\dot{x}_1\dot{\alpha}\,\cos\alpha\right) + m_2 l g\,\cos\alpha\,. \tag{1}$$

Differentiating (1) by q_1, $\dot{q}_1$, q_2, $\dot{q}_2$ (recall, that $q_1 = x_1$, $q_2 = \alpha$) we obtain

$$\frac{\partial L}{\partial q_1} = 0, \qquad \frac{\partial L}{\partial \dot{q}_1} = (m_1 + m_2)\dot{x}_1 + m_2 l\dot{\alpha}\,\cos\alpha\,,$$

$$\frac{\partial L}{\partial q_2} = -m_2 l\,\sin\alpha(\dot{x}_1\dot{\alpha} + g)\,, \qquad \frac{\partial L}{\partial \dot{q}_2} = m_2 l\left(l\dot{\alpha} + \dot{x}_1\,\cos\alpha\right).$$

Substituting the obtained expressions into Lagrangian equations and differentiating them by t, we come to two equations with respect to x_1, α

$$(m_1 + m_2)\ddot{x}_1 + m_2 l \cos\alpha \cdot \ddot{\alpha} = m_2 l \sin\alpha \cdot \dot{\alpha}^2 ,$$
$$\cos\alpha\, \ddot{x}_1 + l\ddot{\alpha} = -g \sin\alpha , \tag{2}$$

representing a model of the investigated system. In accordance with general properties of Lagrangian formalism, the equations (2) are solved relative to $\ddot{x}_1$, $\ddot{\alpha}$ and at known initial values of generalized coordinates and velocities we are enabled to find the coordinates and velocities of points at any moment of time.

The nonlinear system of fourth order (2) is easily reduced to a second order equation, for example, by elimination of $\ddot{x}_1$

$$l(m_1 + m_2 \sin^2\alpha)\, \ddot{\alpha} = -\sin\alpha[m_2 l \cos\alpha \cdot \dot{\alpha}^2 + (m_1 + m_2)g] . \tag{3}$$

This outcome is the consequence of the invariance of the Lagrangian (1) relative to two one-parameter families of transformations. The first of them is given by the formula $x_1^* = x_1 + \beta$ (L does not vary at a shift of coordinate x_1), and the second by formula $\alpha^* = \alpha + m \,\mathrm{sign}\, \beta\, 2\pi$, where $m = 1, 2, \ldots$; β is the the parameter of transformation (L does not vary when the system of coordinates is rotated at an angle multiple to 2π). According to Noether's theorem (subsection 4, section 1) a system possesses two first integrals defined via formula (23), section 1, and consequently its order can be reduced by two units. One more integral of the system is obvious: the Lagrangian (1) does not depend explicitly on time (conservatism), and the total energy $H = T + V$ is conserved. This property ensures the possibility of lowering of the order of (2) by one more unit and reducing (3) to a first order equation (exercise 1).

The given example well illustrates the difference between the Lagrangian and Newtonian approaches describing the motion of mechanical systems. Newton's equations for the suspension and pendulum in coordinate form are as follows

$$m_1\ddot{x}_1 = R_1(x_2 - x_1)/l ,$$
$$m_2\ddot{x}_2 = -R_1(x_2 - x_1)/l , \tag{4}$$
$$m_2\ddot{y}_2 = -R_2 y_2/l - m_2 g ,$$

where $R_1 = R_2 = R$ is the module of a vector of reaction of the rod acting on the masses m_1 and m_2 (see Fig. 39), so that it is obvious that $\vec{R}_1 = -\vec{R}_2$. The reaction is formed by the tension of the rod, which in the ideal case is considered to be absolutely rigid, and its deformation is neglected.

Three equations (4) contain four unknown variables: x_1, x_2, y_2, R. The system (4) can be closed using the connections $(x_2 - x_1)^2 + y_2^2 = l^2$, and one will have a second order nonlinear equation (see also exercise 2). However, in the case of more complex systems this cumbersome procedure becomes actually impracticable. While deducing the Lagrangian equations this procedure is not required (this fact was the initial reason for the development of Lagrangian formalism). Besides, the invariant properties of the Lagrangian clearly indicate the existence of first integrals of motion, which essentially simplifies the research.

The equation following from (3) is of the first order and can be studied relatively easily in a plane of functions α, $d\alpha/dt$ (phase plane) and all characteristics of movement depending on the initial data can be determined. We shall confine ourselves to a consideration of small oscillations of the system, when $\alpha \ll 1$. Omitting in (3) the higher order terms, we come to the equation

$$\ddot{\alpha} = -\frac{g}{l}\,\frac{m_1 + m_2}{m_1}\,\alpha\,,$$

which obviously has a general solution

$$\alpha(t) = A\,\sin\omega t + B\,\cos\omega t\,,$$

where the constants A and B are determined from the initial data, and the frequency of oscillations is given by the formula

$$\omega = \sqrt{\frac{g}{l}\left(1 + \frac{m_2}{m_1}\right)}\,.$$

In comparison with the rigidly fixed pendulum, for which $\omega_0 = \sqrt{g/l}$, the frequency is increased, depending on m_1, m_2 and increases more, the bigger is the ratio m_2/m_1, which is connected with the free motion of the attaching point. This explains one more difference, which is the following. Let in an initial moment $t = 0$ the shift of the pendulum be $\alpha(0) > 0$, and its velocity, as well as the velocity of the suspension, equal zero, i.e. the energy of the system is concentrated on the potential energy of the pendulum. It will be completely transformed to kinetic energy while passing the lowest point. In that moment the velocity of pendulum is $v_{2x} = \dot{x}_1 + l\alpha$. In estimating it we have to take into account that in this case $\alpha(t) = \alpha(0)\,\cos\omega t$ and that $\ddot{x}_1 = -l\ddot{\alpha} - g\alpha$ (the latter equality follows from the linearized system (2)). Thus,

$$\dot{x}_1 = -l\dot{\alpha} - \int_0^1 g\alpha(t)\,dt\,,$$

or

$$v_{2x} = -g \int_0^1 \alpha(0) \cos \omega t \, dt \, .$$

In the moment $t = \pi/(2\omega)$

$$v_{2x}(\alpha = 0) = -ga(0) \int_0^{\pi/(2\omega)} \cos \omega t \, dt = -\frac{g\alpha(0)}{\omega} \, .$$

This value is ω/ω_0 times less than the maximal velocity of the pendulum on a rigid suspension – the initially reserved energy partially transforms into kinetic energy of suspension.

If $m_1 \to \infty$ (massive suspension), then naturally, both small and finite oscillations of the system coincide with the motion of the rigidly fixed pendulum.

2. Non-potential oscillations. We now take into account the action of forces of friction on a pendulum and suspension, considering them proportional to velocities:

$$\vec{F}_1 = -\mu_1 \vec{v}_1 \, , \qquad \vec{F}_2 = -\mu_2 \vec{v}_2 \, , \qquad \mu_1 > 0 \, , \qquad \mu_2 > 0 \, .$$

In so far as the forces of friction depend on velocities, the movement is not potential, and it is necessary to use Lagrangian equations (9), section 1, with generalized forces on the right hand side. Choosing, as before, $q_1 = x_1$, $q_2 = \alpha$, we obtain from the general formula (10), section 1,

$$\Phi_1 = \Im_{1x} \frac{\partial x_1}{\partial q_1} + \Im_{1y} \frac{\partial x_2}{\partial q_1} + \Im_{2x} \frac{\partial x_2}{\partial q_1} + \Im_{2y} \frac{\partial y_2}{\partial q_1} \, ,$$

$$\Phi_2 = \Im_{1x} \frac{\partial x_1}{\partial q_2} + \Im_{1y} \frac{\partial x_2}{\partial q_2} + \Im_{2x} \frac{\partial x_2}{\partial q_2} + \Im_{2y} \frac{\partial y_2}{\partial q_2} \, ,$$

where $\Im_{1x}$, $\Im_{1y}$, $\Im_{2x}$, $\Im_{2y}$ are component resultant forces acting on the masses m_1, m_2 (see Fig. 39). By taking into account that $\Im_{1y} = 0$, and equalities $\partial x_1/\partial q_1 = \partial x_2/\partial q_1 = 1$, $\partial y_1/\partial q_1 = \partial x_1/\partial q_2 = 0$, $\partial x_2/\partial q_2 = l \cos q_2$, $\partial y_2/\partial q_2 = l \sin q_2$, we simplify the expressions for Φ_1, Φ_2:

$$\begin{aligned} \Phi_1 &= \Im_{1x} + \Im_{2x} \, , \\ \Phi_2 &= \Im_{2x} l \cos q_2 + \Im_{2y} l \sin q_2 \, . \end{aligned} \tag{5}$$

We now express the components of acting forces in coordinates $q_1 = x_1$, $q_2 = \alpha$:

$$\Im_{1x} = F_{1x} + R_{1x} = -\mu_1 v_{1x} + R_{1x} = -\mu_1 \dot{x}_1 + R\,\sin\alpha\,,$$

$$\Im_{2x} = F_{2x} + R_{2x} = -\mu_2 v_{2x} + R_{2x} = -\mu_2(\dot{x}_1 + l\dot{\alpha}\,\cos\alpha) - R\,\sin\alpha\,,$$

$$\Im_{2y} = F_{2y} + R_{2y} - mg = -\mu_2 v_{2y} + R_{2y} - \mu_2 g =$$

$$= -\mu_2 l\dot{\alpha}\,\sin\alpha + R\,\cos\alpha - m_2 g\,,$$

where R is the force of reaction of the rod. Substituting them into (5), we obtain

$$\begin{aligned}\Phi_1 &= -(\mu_1 + \mu_2)\,\dot{x}_1 - \mu_2 l\dot{\alpha}\,\cos\alpha\,,\\ \Phi_2 &= \mu_2 l\dot{x}_1\,\cos\alpha - \mu_2 l^2\dot{\alpha} - m_2 gl\,\sin\alpha\,.\end{aligned} \tag{6}$$

According to the general property of Lagrangian formalism, the reactions of connection, as it is seen from (6), have not entered into the final expression for Φ_1, Φ_2. The kinetic energy of system T is found in subsection 1:

$$T = \frac{m_1 + m_2}{2}\,\dot{x}_1^2 + \frac{m_2 l}{2}(l\dot{\alpha}^2 + 2\dot{x}_1\dot{\alpha}\,\cos\alpha)\,.$$

Calculating analogously to subsection 1 the derivatives $dT/dq_{1,2}$ and $dT/d\dot{q}_{1,2}$ and differentiating by t, we arrive at Lagrangian equations with respect to the considered system

$$\begin{aligned}(m_1 + m_2)\ddot{x}_1 + m_2 l\,\cos\alpha\cdot\ddot{\alpha} &= m_2 l\,\sin\alpha\cdot\dot{\alpha}^2 + \Phi_1\,,\\ m_2 l\,\cos\alpha\cdot\ddot{x}_1 + m_2 l^2\ddot{\alpha} &= \Phi_2\,.\end{aligned} \tag{7}$$

As for the potential motion, the nonlinear system of fourth order (7) is solved relative to higher derivatives, and at given initial values $x_1(0)$, $\dot{x}_1(0)$, $\alpha(0)$, $\dot{\alpha}(0)$ from it the positions and velocities of masses m_1, m_2 are determined at any moment of time.

However, as distinct from the system (2), the considered motion has no three first integrals (Nether's theorem is valid for potential motions), and its order can be lowered only by one (exercise 3). One more difference is in the form of the energy balance relation

$$E(0) = E(t) + A(t)\,, \tag{8}$$

where $E(0)$ is the total initial energy of system, $E(t) = T(t) + V(t)$ is the current total energy, $A(t)$ is the work of forces of friction up to the moment t.

The mechanical content of (8) is such that the amount of dissipated energy of system is equal to the work performed by the non-potential forces of friction.

We obtain the equality (8) for simplicity in a case $\mu_2 = 0$ (friction is acting on the suspension only). From (6) we have

$$\Phi_1 = -\mu_1 \dot{x}_1\,, \qquad \Phi_2 = -m_2 lg \,\sin\alpha\,.$$

Substituting these expressions into (7), multiplying the first equation on $\dot{x}_1$, the second – on $\dot{\alpha}$, and summing both equations, we come to an equality

$$\begin{aligned}\ddot{x}_1((m_1+m_2)\dot{x}_1 + m_2 l\,\cos\alpha\cdot\dot{\alpha}) + \ddot{\alpha}(m_2 l\dot{x}_1\,\cos\alpha + m_2 l^2\,\dot{\alpha}) = \\ = m_2 l\,\sin\alpha\cdot\dot{\alpha}\cdot\dot{x}_1 - m_2 lg\,\sin\alpha\cdot\dot{\alpha} - \mu_1\dot{x}_1^2\,.\end{aligned} \tag{9}$$

The relation (9) coincides with equality (8) differentiated by time, where the total energy at the moment t is

$$E(t) = \frac{m_1+m_2}{2}\,\dot{x}_1^2 + \frac{m_2 l}{2}(l\dot{\alpha}^2 + 2\dot{x}_1\dot{\alpha}\,\cos\alpha) - m_2 lg\,\cos\alpha\,,$$

and the work of the force of friction is given by the formula

$$A(t) = -\int_0^1 F_1\,dx_1 = \int_0^1 \mu_1 v_1\,dx_1 = \int_0^1 \mu_1\frac{dx_1}{dt}\,dx_1 = \int_0^1 \mu_1\left(\frac{dx_1}{dt}\right)^2 dt\,.$$

Thus, (9) is equivalent to equality

$$\frac{d}{dt}(E(t)+A(t)) = \frac{d}{dt}\left(E(t) + \int_0^1 \mu_1\left(\frac{dx_1}{dt}\right)^2 dt\right) = 0\,,$$

and hence, to equality (8). In so far as

$$\frac{dE(t)}{dt} = -\mu_1\left(\frac{dx_1}{dt}\right)^2 \le 0\,,$$

then in comparison with a conservative system of subsection 1, the total energy in this case is not constant, but is decreased in time.

For small oscillations of the rigidly fixed pendulum from (7) we obtain the equation

$$\ddot{\alpha} = -\frac{\mu_2}{m_2}\,\dot{\alpha} - \frac{g}{l}\alpha\,,$$

which is equivalent to the equation of motion in viscous medium of a ball with a spring (subsection 3, section 4, Chapter 1) and with a simple general solution.

3. Small oscillations of a string. The applicability of the Hamiltonian principle is not limited to systems of material points. It can also be used for systems which are not, strictly speaking, systems of point masses. An example can be the elastic thread or string – a continuous medium, which nevertheless can be considered as a system of closely associated material points. Assume a string with thickness much less than its length l, and constant linear density ρ_0. The string of tension F_0 at balance is motionless and represents a straight line. While shifted from equilibrium, for example, as a result of an impact, the string is bent and its parts begin to move (Fig. 40). The oscillations are considered flat and small – their amplitude is considerably less than the length of a string. This assumption enables one to neglect the longitudinal displacement and velocities of parts of the string, considering only their transversal movement.

Consider a string as a system of N material points of equal masses $m_i = \rho_0 l/N = \rho_0 \Delta x$, $i = 1, \ldots, N$. The length of piece of a string of mass m_i is Δx and is considered small at equilibrium, and it does not vary strongly in view of the size of oscillations. Therefore the position of i-th "material point" at any moment in time can be characterized by quantities $x_i(t)$ (longitudinal coordinate of center of i-th piece) and $y_i(t)$ (transversal shift of the center of a piece from the state of equilibrium).

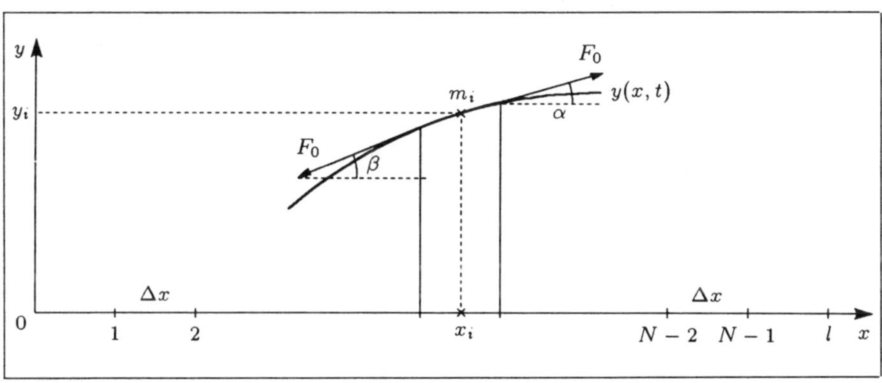

Fig.40.

The introduced generalized coordinates (in this case – Cartesian) completely describe a flat motion of the considered system. In view of the small size of the shifts, as already mentioned, $dx_i(t)/dt = v_{ix} = 0$, i.e. the coordinates x_i, do not depend on time.

The "material points" forming the string are connected among themselves and at $N \to \infty$, $\Delta x \to 0$ represent at any moment of time in a plane

x, y a curve

$$y = y(x,\, t)\,. \tag{10}$$

As distinct from the mechanical connections considered above which are always considered as given, "the connection" (10) is unknown, and the function $y(x,\, t)$ is a subject of finding. If it is found, then the coordinates $y_i = y(x_i, t)$ are known, and since the coordinates x_i do not vary by time, the motion of the system is known completely.

The kinetic energy of i-th mass is determined via the formula

$$T_i = \frac{1}{2}\, m_i v_{iy}^2 = \frac{1}{2} m_i \left(\frac{dy_i}{dt}\right)^2 = \frac{1}{2}\, m_i \left(\frac{\partial y_i}{\partial t}\right)^2 .$$

When deducing this formula the equality $dy_i/dt = \partial y_i/\partial t$ has been used, valid in view of equality $dx_i/dt = 0$. The total kinetic energy of system is equal

$$T = \frac{1}{2} \sum_{i=1}^{N} m_i \left(\frac{\partial y_i}{\partial t}\right)^2 = \frac{1}{2} \sum_{i=1}^{N} \rho_0 \left(\frac{\partial y}{\partial t}\right)_i^2 \Delta x\,. \tag{11}$$

Now we calculate the forces acting onto i-th mass. In accordance with the assumption about small oscillations $\Im_{ix} = 0$. The vertical component $\Im_{iy}$ is determined as the sum of the tension of a string on the right and left edges of i-th piece. Its lengthening at departure from the state of equilibrium is small, therefore the tension of the string can be considered constant and equal to F_0. Then the vertical components of forces acting to the right and left edges respectively are equal to

$$\Im_- = -F_0\, \sin\alpha = -F_0\, \frac{\partial y}{\partial x}\left(x_i + \frac{\Delta x}{2}\right),$$

$$\Im_+ = F_0\, \sin\beta = F_0\, \frac{\partial y}{\partial x}\left(x_i - \frac{\Delta x}{2}\right)$$

(see Fig. 40; recall, that the force of tension is directed tangentially to the string). As a result we obtain

$$\Im_{iy} = \Im_+ + \Im_- = F_0[y_x(x - \Delta x/2) - y_x(x + \Delta x/2)]\,,$$

or taking into account the small size of a piece Δx,

$$\Im_{iy} = -F_0(y_{xx})_i\, \Delta x\,.$$

Taking into account that $\partial y = y_x \cdot \partial x$ follows from (10), we rewrite the latter expression as

$$\Im_{iy} = -F_0\, \Delta x \left(\frac{\partial}{\partial x} y_x\right)_i = -F_0 \Delta x \left(y_x \frac{\partial}{\partial y} y_x\right)_i = -\frac{1}{2}\, F_0\, \Delta x \left(\frac{\partial}{\partial y} y_x^2\right)_i .$$

From here it is clear that all forces $\Im_{iy}$, $i = 1, \ldots, N$ are potential, and the potential energy of i-th mass is given by the formula

$$V_i = \frac{1}{2}\, F_0\, (y_x^2)_i\, \Delta x\,,$$

and the total potential energy of the string is

$$V = \frac{1}{2}\, F_0 \sum_{i=1}^{N} (y_x^2)_i\, \Delta x\,. \tag{12}$$

In so far as the motion of the system is potential, then one can apply the Hamilton principle from section 1. From (11) and (12) we obtain the Lagrangian

$$L(y,\, \dot{y},\, t) = T - V = \frac{1}{2} \sum_{i=1}^{2} \left[\rho_0 \left(\frac{\partial y}{\partial t}\right)^2 - F_0 \left(\frac{\partial y}{\partial x}\right)^2\right]_i \Delta x\,, \qquad \dot{y} = \frac{\partial y}{\partial t}\,.$$

The action is obtained via the Hamiltonian by the formula

$$Q(y,\, \dot{y},\, t) = \int_{t_0}^{t_1} L\, dt = \int_{t_0}^{t_1} \frac{1}{2} \sum_{i=1}^{N} \left[\rho_0 \left(\frac{\partial y}{\partial t}\right)^2 - F_0 \left(\frac{\partial y}{\partial x}\right)^2\right]_i \Delta x\, dt\,, \tag{13}$$

where t_0, t_1 are two arbitrary moments in time, when the system has coordinates $y_i(t_0)$ $y_i(t_1)$. There are several ways of moving from the state with $y_i(t_0)$ to the state with $y_i(t_1)$. The Hamiltonian principle from this set selects the "true" direct path, for which the variation of action δQ is equal to zero. Calculate with the help of (13) the variation δQ, at the variation δy_i of coordinates and $\delta \dot{y}_i$ of velocities of i-th point

$$\begin{aligned} \delta Q &= \delta \int_{t_0}^{t_1} L\, dt = \\ &= \frac{1}{2} \int_{t_0}^{t_1} \sum_{i=1}^{N} \left\{ 2\rho_0\, \frac{\partial y}{\partial t}\, \delta \left(\frac{\partial y}{\partial t}\right) - F_0\, \frac{\partial}{\partial y} \left[\left(\frac{\partial y}{\partial x}\right)^2\right] \cdot \delta y \right\}_i \Delta x\, dt\,. \end{aligned} \tag{14}$$

The first member on the right hand side of (14) results from variation of speed $\dot{y}_i$, the second results from variation of coordinate y_i. For transformation of the first member we integrate it in parts, using the commutativity of operations of variation and differentiation by t, and considering that $\delta y(t_0) = \delta y(t_1) = 0$, we obtain

$$\frac{1}{2} \int_{t_0}^{t_1} \sum_{i=1}^{N} \left[2\rho_0\, \frac{\partial y}{\partial t}\, \delta \left(\frac{\partial y}{\partial t}\right)\right]_i \Delta x\, dt = \int_{t_0}^{t_1} \sum_{i=1}^{N} \rho_0 \left(\frac{\partial^2 y}{\partial t^2}\right)_i \delta y_i\, \Delta x\, dt\,.$$

The second member in view of equality $\partial y = y_x \partial x$ corresponds to

$$-\frac{1}{2}\int_{t_0}^{t_1}\sum_{i=1}^{N} F_0 \frac{\partial}{\partial y}\left[\left(\frac{\partial y}{\partial x}\right)_i^2\right] \delta y_i = -\int_{t_0}^{t_1} F_0 \left(\frac{\partial^2 y}{\partial x^2}\right)_i \delta y_i \,\Delta x \, dt .$$

After the substitution of these expressions into (14) we obtain

$$\delta Q = \int_{t_0}^{t_1}\sum_{i=1}^{N}\left(\rho_0 \frac{\partial^2 y}{\partial t^2} - F_0 \frac{\partial^2 y}{\partial x^2}\right)_i \delta y_i \,\Delta x \, dt .$$

We now move from a discrete system of "material points" assumed to describe the string, to a continuous medium. Then, in the last equality we tend $N \to 0$ replacing Δx by dx, and lower the script i:

$$\delta Q = \int_{t_0}^{t_1}\int_0^1 \left(\rho_0 \frac{\partial^2 y}{\partial t^2} - F_0 \frac{\partial^2 y}{\partial x^2}\right) \delta y \, dx \, dt .$$

On a straight path $\delta Q = 0$, which is possible only if in the latter equality the integrand is equal to zero at all x and t, that is

$$\frac{\partial^2 y}{\partial t^2} = a_0^2 \frac{\partial^2 y}{\partial x^2}, \qquad a_0^2 = \frac{F_0}{\rho_0}, \qquad 0 < x < l, \quad t > 0. \tag{15}$$

Hence, for small fluctuations of a string, its deviation satisfies equation (15), from which under the appropriate boundary conditions, the function $y = y(x, t)$ is determined.

With respect to the considered situation, it is possible to treat the Hamiltonian principle as a way of deriving the equation of "connection" (10). In view of the properties of the Lagrangian the total energy $H = T + V$ of string is preserved by time (the conservatism of motion can also be easily established directly from equation (15); see exercise 4).

The equation of small oscillations of a string (15) ($\omega = 2\pi a_0/\lambda$ is the frequency of oscillations with wavelength λ) is a second order linear equation in partial derivatives of hyperbolic type. The principle of superposition allows us to obtain its general solution as a sum of partial solutions, using the appropriate methods of the theory of equations of mathematical physics.

The basic boundary problem for (15) is the first boundary problem on interval $[0, l]$, when to determine uniquely the solution the initial deviations $y(x, 0) = y_0(x)$, $0 < x < l$, and velocities $\dot{y}(x, 0) = \dot{y}_0(x)$, $0 < x < l$, and boundary conditions for function $y(0, t) = y_1(t)$, $t > 0$ and $y(0, l) = y_2(t)$, $t > 0$ have to be given. The basic problem admits various modifications; the simplest among them – the Cauchy problem – is solved for $-\infty < x < \infty$. Such idealization is justified when the motion of the central part of a string during a short time is considered, and the influence of boundaries can be neglected. For the solution of the Cauchy problem it is enough to know the initial velocities and coordinates of the string, i.e. the functions $y_0(x)$, $\dot{y}_0(x))$ at $\infty < x < \infty$.

For a special type of motions possessing the property $\partial y/\partial t = c\,\partial y/\partial x$, $c = \text{const}$ (simple waves), the equation (15) turns to the hyperbolic equations of the first order, which were investigated in section 1, Chapter II, or transfer equation (see also exercise 5).

Note that equation (15) is usually obtained with the help of direct application of Newton's second law and Hook's law for an element of a string. Then, the assumptions about the small size of fluctuations and the uniformity of the string, etc. are the same in both approaches. Therefore the mathematical models of motion of a string in both cases are identical.

4. Electromechanical analogy. The application of the Hamiltonian principle is possible not only for motion of continuous media, but also for certain non-mechanical objects. Consider the oscillatory circuit already studied in section 5, Chapter I, consisting of a capacitor of capacity C_0 and induction coils L_0. At the initial moment of time the circuit is open, the charge is concentrated on plates of the capacitor. When the circuit is closed the capacitor begins to discharge, and a current appears in the circuit.

The electromechanical analogy is as follows. The generalized coordinate corresponds to the charge on plates of the capacitor, i.e. is an unknown function of time $q = q(t)$. The magnitude of the electrical current $\dot{q}(t) = dq(t)/dt = i(t)$ plays the role of a generalized velocity. To obtain a correct definition of analogies of kinetic (energy of motion) and potential (energy of the capacitor) energies, we shall be guided by the following considerations. The energy of charges moving in the conductor (energy of current) is proportional to the square of the velocity v of their directed motion. On the other hand, the charge passing through the cross section S of a conductor in a unit of time (current) is equal $i = q_0 nSv$, where q_0, n are the elementary charge and concentration of carriers of current, respectively. Hence, the energy of motion of particles, $T \sim v^2 \sim i^2$, i.e. is proportional to a square of current $\dot{q}(t) = i(t)$. The coefficient of proportionality (analog of mass) is taken to be L_0, that is

$$T = T(\dot{q}) = \frac{1}{2}\,L_0\,\dot{q}^2\,.$$

The potential energy of a circuit is contained in the capacitor. To charge it one has to separate opposite charges. According to Coulomb's law the opposing force as a function of charges q_1, q_2 is proportional to their product q_1q_2 (if $q_1 = q_2 = q$, then the force is proportional to q^2). Thus, the work on separating the charges, i.e. the potential energy of system V, is proportional to the square of "generalized coordinate"

$$V = V(q) = \frac{1}{2C_0}\, q^2 \,,$$

where $1/C_0$ is the analog of elastic force coefficient in Hook's law (system ball-spring), or of quantity $\sqrt{g/l}$ in the case of oscillations of a pendulum.

Now we take into account that the forces acting in a circuit have purely electrostatic origin (resistance of conductors is neglected, i.e. "the friction" is absent, as there are no also losses of energy on emission of electromagnetic waves). In accordance with Coulomb's law these forces are determined by "generalized coordinate" q and do not depend on $\dot{q}$. In this sense the forces are "potential", and hence, is also "potential" the considered system. Therefore it has a "Lagrangian" $L = T - V$ and an analog of the Hamiltonian principle is valid: for a "true" path of the system the variation of "action"

$$Q = \int\limits_{t_0}^{t_1} L\, dt$$

is equal to zero (here as usual t_0, t_1 are arbitrary moments in time). Let the function $q(t,\, 0) = q^0(t)$ correspond to a direct path of the system in an interval $t_0 < t < t_1$. The variation of coordinate $q(t,\, \alpha)$, $\alpha \neq 0$ is equal to $\delta q = q(t,\, \alpha) - q^0(t)$, where $q(t,\, \alpha)$ are all possible "trajectories", with identical coordinates $q(t_0,\, \alpha)$, $q(t_1,\, \alpha)$. For variation of "action" we have

$$\delta Q = \delta \int\limits_{t_0}^{t_1} L\, dt = \delta \int\limits_{t_0}^{t_1} \frac{1}{2} \left(L_0 \dot{q}^2 - \frac{1}{C_0}\, q \right) dt = \frac{1}{2} \int\limits_{t_0}^{t_1} [L(q) - L(q^0)]\, dt \,.$$

In so far as $q = q^0 + \delta q$, the integrand can be represented as

$$L(q) = -L(q^0) = L_0[(\dot{q}^0)^2 + 2\dot{q}^0\, \delta\dot{q} + \delta\dot{q}^2] -$$

$$- \frac{1}{C_0}[(q_0)^2 + 2q^0\, \delta q + \delta q^2] - L_0(\dot{q}^0)^2 + \frac{1}{C_0}(q^0)^2 \,.$$

Neglecting the second order members we obtain

$$\delta Q = \int\limits_{t_0}^{t_1} \left(L_0 \dot{q}^0\, \delta\dot{q} - \frac{1}{C_0}\, q^0\, \delta q \right) dt \,.$$

Integrating by parts the term $L_0 q^0 \, \delta\dot{q}$, where $\delta\dot{q} = d(\delta q)/dt$ and taking into account that $\delta q(t_0) = \delta q(t_1) = 0$, we come to the final expression for δQ

$$\delta Q = \int\limits_{t_0}^{t_1} \left(L_0 \ddot{q}^0 + \frac{1}{C_0} q^0 \right) \delta q \, dt = 0 \, .$$

From this for the charge $q(t)$ (the upper script at q^0 is omitted) the equation yields

$$L_0 \ddot{q} = -\frac{1}{C_0} q \, ,$$

describing the oscillations in an electrical circuit (derived in another way in section 5, Chapter I.) Note that the total energy of oscillations $H = T + V$ is conserved in time, which is in agreement with the invariance of the "Lagrangian" relative to a shift of time.

The considered analogy is also applicable to electrical circuits whit much more complex configurations, and on the basis of this mathematical models of corresponding processes are constructed. The given example is by no means the only one illustrating the wide applicability of Hamiltonian and other variational principles. They are often used to construct to mathematical models, not only of mechanical or physical phenomena, but also of many chemical, biological and other phenomena.

EXERCISES

1. Using replacement $d\alpha/dt = v$, reduce equation (3) to a form $dv/d\alpha = f(v, \alpha)$.

2. By transition in equations (4) to coordinates $x_2 = x_1 + l \sin\alpha$, $y_2 = -l \cos\alpha$ deduce equation (3).

3. The quantity x_1 does not enter system (7) explicitly. Using this property, show that (7) can be reduced to a system of equations of third order relative to functions $X(t) = dx_1/dt$, $Y(t) = d\alpha/dt$, $Z(t) = \alpha(t)$.

4. Multiply both sides of equation (15) by $\partial y/\partial t$ and, integrating the obtained equality from $t = 0$ up to $t > 0$ and from $x = 0$ up to $x = 1$, prove that $H(t) = H(0)$ for any $t > 0$.

5. Derive the solution of equation (15), describing the motion of a string for which $\partial y/\partial t = c\, \partial y/\partial x$.

3 The Boltzmann Equation and its Derivative Equations

We now construct a hierarchical chain of models describing the dynamics of a large number of material particles. The kinetic Boltzmann equation for distribution function will serve as a basis of the initial model. Then, by the principle "from above downwards" by means of successive use of simplifying assumptions we shall obtain models for viscous thermal conductive gas, Euler equations and equations of acoustics.

1. The description of a set of particles with the help of the distribution function. In space x, y, z one has a "liquid" or "gas" – a large number of material particles (molecules, atoms, electrons, ions), freely moving within intervals between "collisions". If the coordinates and velocities of any particle as functions of time are known, the basic characteristics of the considered ensemble are thus known completely. They are determined by the initial conditions of each particle, the character of interaction between them (for example, electrons and ions are interacting by Coulomb's law), and also by acting external forces.

Setting the positions and velocities of particles at the moment $t = 0$ and knowing all acting forces, in principle it is possible to solve the basic problem of mechanics (section 1) for the considered system (this means that they satisfy the laws of classical mechanics). Such an approach (with respect to usual gas it is called *molecular dynamics*) gives exhaustive information and has a supreme place in the hierarchy of mathematical descriptions of gas. However, for a sufficiently large number of particles, it is practically impossible to realize, at sufficiently least due to the fact that the initial positions and velocities are never known with accuracy. Besides, as a rule there is no need to trace the behavior of every particle, in so far as only the average characteristics of system are of interest; they include the density, velocity, temperature, etc.

Therefore, the description with the help of *first principles*, i.e. by applying of Newton's second law to each of the particles, is used only in special cases. The transition to the following stage in the hierarchy of models is based on a refusal to study the fate of individual particles. A statistical probability description of their ensemble is given with the help of *distribution function* $f(\vec{r}, \vec{v}, t)$, where $\vec{r}$, $\vec{v}$ are the radius-vector and velocity, respectively. The function f depends on "coordinates" of six-dimensional *phase space* x, y, z, v_x, v_y, v_z, (space of states) and time (compare with distribution function of photons in section 3, Chapter II). The quantity

$$f(\vec{r}, \vec{v}, t)\, d\vec{r}\, d\vec{v}$$

by definition is equal to the number of particles in the moment t (more precisely, to their average value in a short time interval dt) in element $d\vec{r}\, d\vec{v}$ of phase space, with coordinates in an interval from $\vec{r}$ to $\vec{r} + d\vec{r}$ and velocity from $\vec{v}$ to $\vec{v} + d\vec{v}$. The element $d\vec{r} d\vec{v}$ is considered small in comparison with the characteristic sizes of the system, but contains a sufficiently large number of particles. For simplicity, a gas consisting of particles of one type is considered.

Through the distribution function the average quantities describing the condition of gas in space and in time are calculated. For example the expression

$$n(\vec{r},\, t) = \int f(\vec{r},\, \vec{v},\, t)\, d\vec{v}\,,$$

where the integration is conducted over all velocities, is clearly nothing other than the number of particles in a unit of physical volume with coordinate $\vec{r}$ at the moment in time t.

In the general case, the average values are obtained as follows. Let $\Phi(\vec{v})$ be an arbitrary function of velocity of a particle (kinetic energy, velocity and so on). Denote through $\sum \Phi$ the average during time scale dt the sum of values of function Φ over all particles in elementary physical volume $d\vec{r}$. Then the average (i.e. per particle) value of Φ is obtained by dividing $\sum \Phi$ by the number of particles in volume $d\vec{r}$, which is equal to $n\, d\vec{r}$

$$(\Phi) = \frac{\sum \Phi(\vec{v})}{n\, d\vec{r}}\,.$$

On the other hand, the number of particles in elementary phase volume $d\vec{r}\, d\vec{v}$ is equal to $f\, d\vec{r}\, d\vec{v}$, and each of them contributes in $\sum \Phi$ by amount $\Phi(\vec{v})$, and their total elementary contribution is equal to $\Phi\, f\, d\vec{r}\, d\vec{v}$. Now we take into account the fact, that the particles in physical volume $d\vec{r}$ can have any velocities $\vec{v}$. Therefore to obtain their total contribution it is necessary to sum the elementary contributions over all velocities

$$\sum \Phi(\vec{v}) = d\vec{r} \int \Phi(\vec{v})\, f(\vec{r},\, \vec{v},\, t)\, d\vec{v}\,.$$

Comparing the two latter formulae, we obtain for the average of Φ

$$(\Phi) = \frac{\int \Phi\, f\, d\vec{v}}{\int f\, d\vec{v}} = \frac{\int \Phi\, f\, d\vec{v}}{n}\,. \tag{1}$$

This is the relation between an arbitrary averaged function describing gas, with distribution function. For example, if $\Phi(\vec{v}) = \vec{v}$, for the velocity of gas we obtain

$$\vec{V}(\vec{r},\, t) = \int \vec{v}\, f\,(\vec{r},\, \vec{v},\, t)\, d\vec{v} \cdot n^{-1}\,.$$

Analogously, using the known distribution function it is possible to calculate as functions of $\vec{r}$ and t all other macroscopic quantities describing the state of the gas.

2. Boltzmann equation for distribution function. We will give a non-strict deduction of this equation based on the following assumptions:

1) the time scale of collision (direct effective interaction) of particles is much less than the time between their collisions;

2) it is possible to neglect the influence of external forces acting on the particles (gravitational, electrical, etc);

3) the particles do not decay and merge.

Consider particles located with moment t in phase volume $d\vec{r}\,d\vec{v}$ (with coordinates and velocities within $\vec{r}$ up to $\vec{r}+d\vec{r}$ and $\vec{v}$ up to $\vec{v}+d\vec{v}$, respectively). Let collisions between them be absent. Then, at the moment $t^* = t + dt$ the velocities of particles do not change, while their coordinates will be changed in accordance with initial velocities

$$\vec{v}^* = \vec{v}\,, \qquad \vec{r}^* = \vec{r} + \vec{v}\,dt\,,$$

where the quantities $\vec{v}$ and $\vec{r}$ lie within the ranges mentioned above. We calculate the phase volume of particles $d\vec{r}^*\,d\vec{v}^*$ at the moment t^*. From two latter equalities we have

$$d\vec{v}^* = d\vec{v}\,, \qquad d\vec{r}^* = d\vec{r} + d\vec{v}\,dt = d\vec{r} + \frac{d\vec{v}}{dt}(dt)^2 + \ldots\,,$$

i.e. $d\vec{r}\,d\vec{v} = d\vec{r}^*\,d\vec{v}^*$, and the phase volume of particles is conserved with accuracy of members of order $(dt)^2$. Thus the number of particles $f(\vec{r},\,\vec{v},\,t)\,d\vec{r}\,d\vec{v}$ located within the volume $d\vec{r}\,d\vec{v}$ is equal to $f(\vec{r}^*,\,\vec{v}^*,\,t^*)\,d\vec{r}^*\,d\vec{v}^*$ in volume $d\vec{r}^*\,d\vec{v}^*$. In other words, in the absence of collisions $f = f^*$, the distribution function does not vary in time, the particles only change their phase volume (in this case they change only their coordinates).

Take into account the role of collisions via the introduction of the concept of *integral of collisions* $S(f)$. By its content the quantity $S(f)\,d\vec{r}\,d\vec{v}\,dt$ is the difference between the number of particles having left the volume $d\vec{r}\,d\vec{v}$ due to collisions during dt (and not entered into the volume $d\vec{r}^*\,d\vec{v}^*$), and the number of particles having entered volume $d\vec{r}^*\,d\vec{v}^*$ because of collisions during dt (and not being in the initial volume $d\vec{r}\,d\vec{v}$). Then the equation of conservation of the number of particles during dt at transition from volume $d\vec{r}\,d\vec{v}$ to volume $d\vec{r}^*\,d\vec{v}^*$ is rewritten as follows

$$f(\vec{r}^*,\,\vec{v}^*,\,t^*)\,d\vec{r}^*\,d\vec{r}^* - f(\vec{r},\,\vec{v},\,t)\,d\vec{r}\,d\vec{v} = -S(f)\,d\vec{r}\,d\vec{v}\,dt\,.$$

Obviously, when the collisions are taken into account (as distinct from processes without collisions) generally speaking, $f \neq f^*$.

Expand the left hand side of the equation of balance via the Taylor series, omitting the members of the order $(dt)^2$ and higher

$$f(\vec{r},\,\vec{v},\,t)\,d\vec{r}^*\,d\vec{v}^* + \frac{\partial f}{\partial t}\,d\vec{r}^*\,d\vec{v}^*\,dt + \frac{\partial f}{\partial \vec{r}}\,\frac{d\vec{r}}{dt}\,d\vec{r}^*\,d\vec{v}^*\,dt +$$

$$+\frac{\partial f}{\partial \vec{v}}\frac{d\vec{v}}{dt}\,d\vec{r}^*\,d\vec{v}^* - f(\vec{r},\,\vec{v},\,t)\,d\vec{r}\,d\vec{v} = -S(f)\,d\vec{r}\,d\vec{v}\,dt\,.$$

In the equality obtained $d\vec{r}/dt = \vec{v}$ and $m\,d\vec{v}/dt = \vec{F}$, where $\vec{F}$ is the external force equal by assumption to zero, m is the mass of a particle. We choose the interval dt to be small, so that during t and $t^* = t + dt$ particles do not collide. Then the phase volume is conserved, and $d\vec{r}\,d\vec{v} = d\vec{r}^*\,d\vec{v}^*$. Dividing both parts of the equality on $d\vec{r}\,d\vec{v}\,dt$ and tending dt to zero, we come to *the Boltzmann equation*

$$\frac{\partial f}{\partial t} + \vec{v}\,\frac{\partial f}{\partial \vec{r}} + S(f) = 0\,, \tag{2}$$

which is the next order of complexity (after models derived from first principles), in the hierarchy of mathematical descriptions of gas. It represents a nonlinear integro-differential equation. Its particular form depends on the character of collisions of particles, i.e. from the form of function $S(f)$. Note that (2) is easily generalized for a gas consisting of particles of different types, and for a case in which the external force $\vec{F}$ is different from zero (this can be caused, for example, by the presence of electromagnetic fields). In the absence of collisions, i.e. at $S(f) = 0$, (2) turns to the transfer equation of section 1, Chapter II.

3. Maxwell distribution and the H-theorem. One of the widely used integrals of collisions is given as follows:

$$S(f) = \int\!\!\int\!\!\int (f'\,f_1' - f\,f_1)\,gb\,db\,d\psi\,d\vec{v}_1\,. \tag{3}$$

Without representing the tedious derivation of (3), we shall explain the content of variables in it. Through $f = f(\vec{r},\,\vec{v},\,t)$, $f_1 = f_1(\vec{r},\vec{v}_1,\,t)$ we denote the distribution functions before collision, through $f' = f'(\vec{r},\,\vec{v}')$, $f_1' = f_1',(\vec{r},\,\vec{v}_1,\,t)$ we denote the distribution functions after collision, where $\vec{v}$ and $\vec{v}_1$ is the velocity of two colliding particles. The quantity g is the module of their relative velocity, b is the so-called target parameter (distance of the least rapprochement between particles), ψ is the angular characteristic of their interaction. The integration is performed over all possible values of b, ψ, $\vec{v}_1$.

The integral of collisions in Boltzmann form (3) is deduced via summation of the elementary acts of mechanical interaction between particles. Thus the following are assumed: the collisions are elastic (total mass, momentum, rotational moment and energy of particles are conserved, which implies, in particular, the fulfilling of equalities $g = g'$, $b = b'$); the force of interaction of particles depends only on the distance between them and more strongly than external forces; the number of collisions involving more than two particles is negligible (gas is not too dense). For the known particular laws of interaction of particles from (3) the concrete expressions for integral of collisions can be obtained, for example, by studying phenomena in plasma, where electrons and ions are interacting by Coulomb's law.

Using properties of integral (3), we shall obtain a simple, but rather important solution of the Boltzmann equation. Consider a gas in state of a *thermodynamic equilibrium*, i.e. in a situation where all its macroscopic characteristics are constant in space and do not depend on time. Then, obviously, the function $f(\vec{r}, \vec{v}, t)$ also does not depend on $\vec{r}$ and t, i.e. $f = f(\vec{v})$. From (2) we readily have the equation

$$\int\int\int (f' f_1' - f f_1)\, gb\, db\, d\psi\, d\vec{v}_1 = 0\,,$$

which is satisfied only at condition

$$f f_1 = f' f_1'\,,$$

or

$$\ln f(\vec{v}) + \ln f_1(\vec{v}_1) = \ln f'(\vec{v}\,') + \ln f_1'(\vec{v}_1')\,.$$

In other words, the sum of logarithms of distribution functions is conserved, being an invariant of collision. But for elastic collisions invariant are also the total energy of particles

$$\frac{mv^2}{2} + \frac{mv_1^2}{2} = \frac{mv'^2}{2} + \frac{mv_1'^2}{2}\,,$$

and their total momentum

$$m\vec{v} + m\vec{v}_1 = m\vec{v}\,' + m\vec{v}_1'$$

are also invariant. These properties provide a way of obtaining of functional dependence of f on $\vec{v}$. For the sake of simplicity we shall establish this dependence for the "one-dimensional" case, where the vectors of all velocities are parallel to each other. Then, multiplying the latter two equalities by constants a_1, a_2 and subtracting them from the previous one, we obtain

$$\Im(v) + \Im(v_1) = \Im(v') + \Im(v_1')\,,$$

where the notation $\Im(x) = \ln f(x) - a_1 m/2 \cdot x^2 - a_2 m x$ is used. In this equality the arguments are arbitrary, so that it can be fulfilled only at $\Im(x) \equiv a_3$ (a_3 is constant). Thus, for the distribution function the following is valid

$$\ln f(v) = \frac{a_1 m}{2} v^2 + a_2 m v + a_3\,.$$

Allocating on the right hand side of this expression a complete square,

$$\ln f(v) = \frac{a_1 m}{2}\left(v + \frac{a_2}{a_1}\right)^2 - \frac{a_2^2 m}{2a_1} + a_3\,,$$

and getting rid of the logarithm, we come to the formula

$$f(v) = \exp\left\{\frac{2a_1 a_3 - a_2^2 m}{2a_1}\right\}\exp\left\{\frac{a_1 m}{2}\left(v + \frac{a_2}{a_1}\right)^2\right\}, \qquad (4)$$

expressing the dependence of f on v. In view of properties of function $f(v)$ the following equalities are valid (see (1))

$$n = \int f\,dv\,, \qquad nV = \int vf\,dv\,, \qquad \frac{3}{2}nkT = \int \frac{m(v-V)^2}{2} f\,dv\,,$$

where n is the concentration of particles; V is their average (macroscopic) velocity; T is the average temperature of particles (by definition), expressed through average kinetic energy of their chaotic motion with thermal velocity $c = v - V$ and measured in Kelvin degrees (see also subsection 1, section 2, Chapter II); k is the Boltzmann constant. Using these equalities and notations, we shall exclude from (4) the quantities a_1, a_2, a_3

$$f(v) = n\left(\frac{m}{2\pi kT}\right)^{3/2}\exp\left\{-\frac{m}{2kT}(v-V)^2\right\}. \qquad (5)$$

A similar expression is also valid (with replacement of v on $\vec{v}$ and V on $\vec{V}$) in the "many-dimensional" case. Formula (5) represents one of the solutions of the Boltzmann equation, describing the distribution of particles by velocities in a gas in thermodynamic equilibrium (*Maxwell distribution*). The same distribution is fair not only at complete, but also at *local thermodynamic equilibrium* (LTE), when macroscopic characteristics n, T, V are slowly varying functions of $\vec{r}$ and t. More precisely, at LTE the functions $n(\vec{r}, t)$, $T(\vec{r}, t)$, $V(\vec{r}, t)$ vary weakly on distances of the order of the length of free path l and during the free run τ. Therefore in LTE conditions the considerations preceding the deduction of the formula (5), are applicable to any point $\vec{r}$ at any moment in time t (therefore the quantities appearing in (5) (n, T, V) are different at different points in space and moments of time).

With the help of the Maxwell distribution we can establish the *Boltzmann's H-theorem* which is of fundamental importance;, according to this theorem, the entropy of gas (described by equation (2)) increases in time. Assume that the distribution of particles in space is homogeneous (but time-dependent). Then the Boltzmann equation (2) has a form

$$\frac{\partial f}{\partial t} = \int (f' f_1' - f f_1)\, gb\, db\, d\psi\, d\vec{v}_1\,. \qquad (6)$$

Boltzmann's H-function is introduced via the formula

$$H(t) = \int f \ln f \, d\vec{v}$$

and by definition represents the entropy of a unit volume of gas taken with opposite sign, $S(t) = -H(t)$. Its derivative by time is

$$\frac{dH}{dt} = \int (1 + \ln f) \frac{\partial f}{\partial t} \, d\vec{v} \, ,$$

or, in view of equation (6),

$$\frac{dH}{dt} = \int (1 + \ln f) \, (f' \, f_1' - f \, f_1) \, gb \, db \, d\psi \, d\vec{v} \, d\vec{v}_1 \, .$$

Similarly, in view of the symmetry of equation (6) relative functions f and f_1, we have

$$\frac{dH}{dt} = \int (1 + \ln f_1)(f' \, f_1' - f \, f_1) \, gb \, db \, d\psi \, d\vec{v} \, d\vec{v}_1 \, ,$$

that is

$$\frac{dH}{dt} = \frac{1}{2} \int (f' \, f_1' - f \, f_1)(2 + \ln f f_1) \, gb \, db \, d\psi \, d\vec{v} \, d\vec{v}_1 \, .$$

Using the symmetry (reflecting the reversibility of the process of elastic collisions) of equation (6) one more time, it is easy to prove that for "backward" collisions the same formula is fair

$$\frac{dH}{dt} = \frac{1}{2} \int (f \, f_1 - f' \, f_1')(2 + \ln f' f_1') \, g' b' \, db' \, d\psi' \, d\vec{v}' \, d\vec{v}_1' \, ,$$

but the variables f', f_1' and f, f_1 have exchanged places. As mentioned above, at elastic collisions $g = g'$, $b = b'$ (the same is true for differentials: $db \, d\psi = db' \, d\psi'$, $d\vec{v} \, d\vec{v}_1 = d\vec{v}' \, d\vec{v}_1'$). Therefore from two latter equations we finally obtain

$$\frac{dH}{dt} = -\frac{1}{4} \int (f' \, f_1' - f \, f_1) \ln \frac{f' f_1'}{f \, f_1} \, gb \, db \, d\psi \, d\vec{v} \, d\vec{v}_1 \, . \tag{7}$$

The sign of derivative dH/dt is determined by the sign of function $\Im(x, y) = (x - y) \ln(x/y)$ under the integral. It is easily established that $\Im(x, y) \geq 0$ (the equality $\Im(x, y) = 0$ corresponds to an equilibrium condition, when the distribution function is not changed after collision). Therefore, $-dH/dt = dS/dt \geq 0$, i.e. the entropy of gas in nonequilibrium state is increased, up to its maximal value. It is reached in a state thermodynamic equilibrium, when the particles are distributed by Maxwell's law. Thus, Boltzmann's kinetic equation, as distinct to the equations of classical mechanics, describes irreversible processes. It is a consequence of the transition from models based on first principles, to models using an averaged statistical description of system of particles with the help of distribution function.

4. Equations for the moments of distribution function. The next level of equation in the hierarchy of mathematical description of particle systems are the hydrodynamical models of gas. For their construction we shall first establish some properties of integral of collisions, valid not only for Boltzmann gas, but also in the general case. The only requirement is that the collisions are elastic. Then from the invariance of the number of particles before collision, one has

$$\int S(f(\vec{v}))\, d\vec{v} = 0\,, \tag{8}$$

while (exercise 3) from the laws of conservation of momentum and energy follows the relation

$$\int m\vec{v}S(f(\vec{v}))\, d\vec{v} = 0\,, \qquad \int \frac{mv^2}{2}\, S(f(\vec{v}))\, d\vec{v} = 0\,. \tag{9}$$

Below the following formulae will be necessary as well

$$\begin{aligned} &\int \Phi(\vec{v})\, \frac{\partial f}{\partial t}\, d\vec{v} = \frac{\partial}{\partial t} \int \Phi(\vec{v})\, f\, d\vec{v} = \frac{\partial}{\partial t}\, (n\langle \Phi \rangle)\,, \\ &\int \Phi(\vec{v})\, v_i\, \frac{\partial f}{\partial x_i}\, d\vec{v} = \frac{\partial}{\partial x_i} \int \Phi(\vec{v})\, v_i\, f\, d\vec{v} = \frac{\partial}{\partial x_i} (n\langle \Phi[\vec{v}] v_i \rangle)\,. \end{aligned} \tag{10}$$

These are easily deduced from the formula (1) for average functions describing the state of gas.

Now from Boltzmann's equation we deduce the equations for the moments of distribution function, i.e. for quantities of the form $\int \Phi(\vec{v})\, f\, d\vec{v}$. For the sake of simplicity, we shall consider the one-dimensional case, when $\vec{r} = x$, $\vec{v} = v$. Multiplying (2) by $\Phi(\vec{v}) = l$ and integrating the obtained expression by velocities, we have

$$\int \frac{\partial f}{\partial t}\, dv + \int \frac{\partial f}{\partial x}\, dv + \int S(f)\, dv = 0\,.$$

Taking into account that due to (8) the third member on the left hand side of the latter equation is equal to zero, and using formulae (10), we rewrite the latter equation as follows

$$\frac{\partial n}{\partial t} + \frac{\partial}{\partial x}\, (n\langle v \rangle)\,.$$

So far as $\langle v \rangle = V$, we have

$$\frac{\partial n}{\partial t} + \frac{\partial}{\partial x}(nV) = 0 \,,$$

or, multiplying the given equation by m, we come to the continuity equation

$$\frac{\partial p}{\partial t} + \frac{\partial}{\partial x}(\rho V) = 0 \,, \tag{11}$$

where $\rho = nm$ is the gas density, V is its macroscopic velocity. In the many-dimensional case the analog of (11) is

$$\frac{\partial \rho}{\partial t} + \operatorname{div}(\rho \vec{V}) = 0$$

or equivalently

$$\frac{d\rho}{dt} = -\rho \operatorname{div} \vec{V} \,. \tag{12}$$

Let the function $\Phi(v)$ equals the momentum, i.e. $\Phi(v) = mv$. Repeating the previous procedure, we shall obtain from (2)

$$\int \frac{\partial f}{\partial t}\, mv\, dv + \int v\, \frac{\partial f}{\partial x}\, mv\, dv + \int S(f)\, mv\, dv = 0 \,.$$

In view of the first formula in (9), the third member on the left hand side of this equation is equal to zero. Using (1) and (10), we come to the equation

$$\frac{\partial}{\partial t}\,(nmV) + \frac{\partial}{\partial x}\,(nm\langle v^2 \rangle) = 0 \,, \tag{13}$$

where $\langle v^2 \rangle$ is the average square of the velocity of particles. To calculate this we shall introduce the chaotic velocity of particles c (thermal velocity), representing the difference between own velocity v and the velocity of macroscopic motion of gas V as a whole, $c = v - V$. For an average of v^2 we have

$$\langle v^2 \rangle = \langle (V+c)^2 \rangle = V^2 + 2\langle cV \rangle + \langle c^2 \rangle = V^2 + \langle c^2 \rangle \,.$$

In the latter equality we have taken into account that $\langle cV \rangle = V\langle c \rangle = 0$ in view of the chaotic nature of thermal motion. Using the derived expression for $\langle v^2 \rangle$, from (13) consequently we have

$$\frac{\partial}{\partial t}(\rho V) + \frac{\partial}{\partial x}(\rho V^2) + \frac{\partial}{\partial t}(\rho \langle c^2 \rangle) = 0 \,,$$

$$\frac{\partial \rho}{\partial t} + \rho\, \frac{\partial V}{\partial t} + V\, \frac{\partial}{\partial V} + \rho V\, \frac{\partial V}{\partial x} + \frac{\partial}{\partial x}(\rho \langle c^2 \rangle) = 0 \,.$$

The first and third terms on the left hand side in the latter equation in view of (11) are mutually cancelled. Therefore it can be rewritten as

$$\frac{\partial V}{\partial t} + V\frac{\partial V}{\partial x} = \frac{1}{\rho}\frac{\partial P}{\partial x}. \tag{14}$$

In the many-dimensional case the analog of the one-dimensional equation (11) has a form

$$\frac{\partial \vec{V}}{\partial t} + \vec{V}\,(\nabla\vec{V}) = \frac{1}{\rho}\,\mathrm{grad}\,P\,,$$

or equivalently,

$$\frac{d\vec{V}}{dt} = \frac{1}{\rho}\,\mathrm{grad}\,P\,. \tag{15}$$

In equations of gas motion (14), (15) the following notations are used: $P_{ij} = -\rho\langle c_i c_j\rangle$, $i,\ j = 1,\ 2,\ 3$; $x_i = x,\ y,\ z$ for $i = 1,\ 2,\ 3$.

Finally, we shall deduce the equation for change of energy, by taking instead of $\Phi(\vec{v})$ the quantity $mv^2/2$, multiplying it by the equation (2) and integrating over the velocities

$$\int \frac{\partial f}{\partial t}\frac{mv^2}{2}\,dv + \int v\frac{\partial f}{\partial x}\frac{mv^2}{2}\,dv + \int S(f)\,\frac{mv^2}{2}\,dv = 0\,.$$

The third term on the left hand side of this expression is equal to zero (see the second formula in (9)), and the first two in view of formula (1) can be rewritten through the averages:

$$\frac{\partial}{\partial t}\left(n\left\langle\frac{mv^2}{2}\right\rangle\right) + \frac{\partial}{\partial x}\left(n\left\langle\frac{mv^2}{2}v\right\rangle\right) = 0\,.$$

Again we shall introduce a thermal velocity $c = v - V$ and calculate the quantity $\langle v^3\rangle = \langle (V+c)^3\rangle = \langle V^3 + 3V^2c + 3Vc^2 + c^3\rangle = V^3 + 3V\langle c^2\rangle + \langle c^3\rangle$ (exercise 4; it is taken into account that $\langle 3V^2c\rangle = 3V^2\langle c\rangle = 0$ due to the chaotic nature of thermal motion). Using the expressions for $\langle v^2\rangle$ and $\langle v^3\rangle$ and equality $\rho = nm$, we have

$$\frac{1}{2}\frac{\partial}{\partial t}(\rho V^2) + \frac{1}{2}\frac{\partial}{\partial t}(\rho\langle c^2\rangle) + \frac{1}{2}\frac{\partial}{\partial x}(\rho V^3) +$$

$$+\frac{3}{2}\frac{\partial}{\partial x}(\rho V\langle c^2\rangle) + \frac{1}{2}\frac{\partial}{\partial x}(\rho\langle c^3\rangle) = 0\,,$$

or differentiating the first and third terms on the left hand side of this equation

$$\frac{1}{2}\frac{\partial}{\partial t}(\rho V^2) = \frac{V^2}{2}\frac{\partial \rho}{\partial t} + \rho V\frac{\partial V}{\partial t}\,, \quad \frac{1}{2}\frac{\partial}{\partial x}(\rho V^3) = \frac{V^2}{2}\frac{\partial \rho V}{\partial x} + \rho V^2\frac{\partial V}{\partial x}\,,$$

and considering the equations (11), (14) multiplied by V^2 and ρV, respectively, we come to the equation

$$-V\frac{\partial}{\partial x}(\rho\langle c^2\rangle)+\frac{1}{2}\frac{\partial}{\partial t}(\rho\langle c^2\rangle)+\frac{3}{2}\frac{\partial}{\partial x}(\rho V\langle c^2\rangle)+\frac{1}{2}\frac{\partial}{\partial x}(\rho\langle c^3\rangle)=0\,.$$

Representing its third term on the left hand side as

$$\frac{3}{2}\frac{\partial}{\partial x}(\rho V\langle c^2\rangle)=\frac{3}{2}V\frac{\partial}{\partial x}(\rho\langle c^2\rangle)+\frac{3}{2}\rho\langle c^2\rangle\frac{\partial V}{\partial x}\,,$$

we obtain

$$\frac{1}{2}V\frac{\partial}{\partial x}(\rho\langle c^2\rangle)+\frac{1}{2}\frac{\partial}{\partial t}(\rho\langle c^2\rangle)+\frac{3}{2}\rho\langle c^2\rangle\frac{\partial V}{\partial x}+\frac{1}{2}\frac{\partial}{\partial x}(\rho\langle c^3\rangle)=0\,. \qquad (16)$$

The quantity $\rho\langle c^2\rangle/2=nm\langle c^2\rangle/2$ in (16) represents the energy of chaotic motion of particles in a unit volume of gas, or its internal energy (for simplicity, we consider that it is determined only by the forward motion of particles). Obviously, the internal energy ε referred to unit mass, is $\varepsilon=\langle c^2\rangle/2$. Then, the first and second terms on the left hand side of (16) can be represented as

$$\frac{1}{2}V\frac{\partial}{\partial x}(\rho\langle c^2\rangle)=\rho V\frac{\partial\varepsilon}{\partial x}+\frac{1}{2}V\langle c^2\rangle\frac{\partial\rho}{\partial x}\,,$$

$$\frac{1}{2}\frac{\partial}{\partial t}(\rho\langle c^2\rangle)=\rho\frac{\partial\varepsilon}{\partial t}+\frac{\langle c^2\rangle}{2}\frac{\partial\rho}{\partial t}=\rho\frac{\partial\varepsilon}{\partial t}-\frac{\langle c^2\rangle}{2}\frac{\partial}{\partial x}(\rho V)=$$

$$=\rho\frac{\partial\varepsilon}{\partial t}-\frac{1}{2}\rho\langle c^2\rangle\frac{\partial V}{\partial x}-\frac{1}{2}V\langle c^2\rangle\frac{\partial\rho}{\partial x}\,,$$

where the continuity equation (11) is used to derive the second expression. By substituting these expressions into (16), we obtain

$$\rho\frac{\partial\varepsilon}{\partial t}+\rho V\frac{\partial\varepsilon}{\partial x}+\rho\langle c^2\rangle\frac{\partial V}{\partial x}+\frac{1}{2}\frac{\partial}{\partial x}(\rho\langle c^3\rangle)=0\,,$$

or finally,

$$\rho\frac{d\varepsilon}{dt}=P\frac{\partial V}{\partial x}-\frac{\partial W}{\partial x}\,. \qquad (17)$$

In the many-dimensional case the equation of energy has a form

$$\rho\frac{d\varepsilon}{dt}=-P\operatorname{div}\vec{V}+\sum_{i,j}\Pi_{i,j}\frac{\partial V_i}{\partial x_j}\operatorname{div}\vec{W}\,. \qquad (18)$$

In (17) and (18) the quantity $\vec{W}$ is defined as

$$\vec{W} = \frac{\rho\langle c^2\vec{c}\rangle}{2} \tag{19}$$

and is called *a vector of heat flux* (the reason for this name will become clear later).

The stress tensor P, introduced in equation (15), has a more complex structure:

$$P_{ij} = -\rho\langle c_i c_j\rangle = -p\delta_{ij} + \Pi_{ij}\,,$$

where δ_{ij} is the Kronecker symbol, and p, Π_{ij} are given by the formulae

$$p = \rho\frac{\langle c^2\rangle}{3}\,, \qquad \Pi_{ij} = \rho\left(\frac{\langle c^2\rangle}{3}\delta_{ij} - \langle c_i c_j\rangle\right)\,. \tag{20}$$

Here, as it is easy to see, p is the pressure of gas particles connected with the average energy of their chaotic motion, i.e. with temperature $T = [m/(3k)]\langle c^2\rangle$, by relation $p = nkT = \rho kT/m$ (strictly speaking, this is the definition temperature for states close to equilibrium). The second term in the formula for P is *the viscous stress tensor* Π_{ij}. The viscous forces are caused by "friction" between parts of gas with different macroscopic velocities. Its origin is easy to understand assuming, for example, two plane gas layers moving with respect to each other. The particles passing through the boundary between the "fast" and "slow" layers, will increase the average velocity of the latter, and vice versa. Thus, the velocities of layers tend to become equal, implying the presence of a " friction force" between them.

The equations (11), (12), (14), (15), (17), (18) for the moments are rewritten relative to the average (hydrodynamical) quantities describing the gas: density, velocity, pressure, internal energy. When deriving them, no essential additional assumptions have been involved, and in this sense they belong the same hierarchical level as the Boltzmann equation. However, as distinct from equation (2), they still cannot be considered as a model of gas, as they are not close: except ρ, the components of velocity $\vec{V}$, p, ε they include also Π_{ij}, $\vec{W}$, i.e. the number of equations is less than the number of unknown variables.

The attempt to deduce the analogous equations for the moments of higher order with the aim of finding Π_{ij}, W, leads to appearance of new unknown variables in these equations. Therefore in order to construct a hydrodynamical model of gas one has to express Π_{ij}, $\vec{W}$ through the sought hydrodynamical parameters ρ, $\vec{V}$, p, ε. Then, five unknown variables will appear in five equations (recall that the internal energy is considered as known function of density and pressure: $\varepsilon = \varepsilon(\rho, p)$, or density and temperature: $\varepsilon = \varepsilon(\rho,\, T)$).

To obtain these expressions for Π_{ij}, $\vec{W}$ one can proceed in various ways, for example, setting some semi-empirical dependence. More strict is the way of finding an approximate "solution" of the kinetic equation (2). Then, an essential assumption is made that the length of the free path l (time of free motion is τ) of particles is much less than the characteristic size L (of characteristic time t) of the system. In other words, the system is close to LTE state, and the distribution function f is not so different from the local Maxwellian one. Thus, expanding the solution as a form of infinite series $f = f^{(0)} + f^{(1)} + \ldots$ (where $f^{(0)}$ is Maxwellian (5), and $f^{(1)}$ is a small deviation from it), considering other terms of series to be small relative $f^{(1)}$ and substituting the expression $f = f^{(0)} + f^{(1)}$ into (2), one can find an explicit form of the perturbation $f^{(1)}$. Then, via the already known function $f = f^{(0)} + f^{(1)}$ using the formula (1), the thermal velocity $\vec{c} = \vec{v} + \vec{V}$ is obtained, and by means of formulae (19) and (20) the quantities Π_{ij}, $\vec{W}$ are determined. For example, $\vec{W}$ is calculated as follows

$$\vec{W} = \int \frac{m|\vec{c}|^2}{2}\, \vec{c}\,(f^{(0)} + f^{(1)})\, d\vec{c}.$$

Note that at $f = f^{(0)}$ the calculations give $\Pi_{ij} \equiv 0$, $\vec{W} \equiv 0$: in gas with Maxwellian distribution of particles viscous tensions and heat fluxes are absent.

We shall not describe the complex and tedious process of finding function $f^{(0)}$ by means of solving (2) (the assumption $l \ll L$ ($\tau \ll t$) is used in that procedure). We present the final result

$$f^{(1)} =$$

$$= f^{(0)} \left[-A(\xi)\, \vec{c} \sqrt{\frac{m}{2kT}}\, \frac{\partial \ln T}{\partial \vec{r}} - \frac{m}{2kT}\, B(\xi) \sum_{i,j} \left(c_i c_j - \frac{1}{3}\, \delta_{ij}\, |\, \vec{c}\,|^2 \right) \frac{\partial V_i}{\partial x_j} \right],$$

where $A(\xi)$ and $B(\xi)$ are scalar functions of the argument $\xi = m|\,\vec{c}\,|^2/(2kT)$; their form depends, in particular, on the form of the integral of collisions between particles.

The expression for $f^{(1)}$ contains the unknown macroscopic quantity $T(\vec{r},\, t)$, $V(\vec{r},\, t)$. Therefore one cannot call the procedure of obtaining $f^{(1)}$ as a solution of the Boltzmann equation in a complete sense, i.e. finding an unknown function $f(\vec{r},\, \vec{v},\, t)$ in the whole area of its definition. This solution should be understood as establishing certain connections between hydrodynamical parameters of medium using (2).

The use of two latter formulae gives for the heat flux

$$\vec{W} = -\kappa\, \mathrm{grad}\, T\,. \tag{21}$$

For the components of the viscous stress tensor the following expressions are valid

$$\Pi_{ij} = \lambda \operatorname{div} \vec{V} + 2\mu \frac{\partial V_i}{\partial x_i}\,; \qquad \Pi_{ij} = \mu \left(\frac{\partial V_i}{\partial x_j} + \frac{\partial V_j}{\partial x_i} \right) \qquad i \neq j\,. \tag{22}$$

The quantity $\kappa > 0$ is called *the coefficient of thermal conductivity*, $\lambda > 0$, $\mu > 0$ – *viscosity coefficient. The transfer coefficients* κ, λ, μ can, generally speaking, depend on ρ and T, i.e. on the state of gas. Their functional dependence on ρ, T and numerical value are determined, as follows from the form of $f^{(0)}$, $f^{(1)}$, by the properties of particles (mass, character of their interaction, etc.).

The formula (21) represents Fourier's law for thermal conducting media (see also section 2, Chapter II), the formula (22) represents the Navier-Stokes law for viscous liquids and gases, connecting the components of viscous stress tensor with *deformation rates* (a generalization of Newton's law of viscosity).

5. Chain of hydrodynamical gas models. The laws (21) and (22) enable one at given transfer coefficients to close the equations deduced in subsection 4, for the moments, and to obtain models of gas in hydrodynamical approximation. The substitution of (21) and (22) in (15) and (18) (with the account of (12)) leads to a system of five equations, consisting of the continuity equation

$$\frac{d\rho}{dt} = -\rho \operatorname{div} \vec{V}\,, \tag{23}$$

three equations of motion (rewritten in vector form; the coefficients λ, μ are considered constant)

$$\frac{d\vec{V}}{dt} = -\frac{1}{\rho} \operatorname{grad} p + \frac{\lambda + \mu}{\rho} \operatorname{grad} \operatorname{div} \vec{V} + \nu \Delta \vec{V} \tag{24}$$

and the equations of energy

$$\rho \frac{d\varepsilon}{dt} = -p \operatorname{div} \vec{V} + \lambda (\operatorname{div} \vec{V})^2 + 2\mu \sum_i \left(\frac{\partial V_i}{\partial x_i} \right)^2 +$$

$$+\mu \sum_{i \neq j} \left(\frac{\partial V_i}{\partial x_j} + \frac{\partial V_j}{\partial x_i} \right)^2 + \operatorname{div} \kappa \operatorname{grad} T\,, \tag{25}$$

where $\nu = \mu/\rho$, $df/dt = \partial f/\partial t + (\vec{V} \operatorname{grad} f)$ – is the total (substantive) derivative by time.

At known boundary conditions from these equations the five sought functions can be obtained: ρ, p and three components of velocity $\vec{V}$. The system

(23)–(25) represents the following after the Boltzmann equation level in the hierarchy of mathematical descriptions of large numbers of interacting particles – the model of compressible viscous thermal conducting gas. Instead of finding the function f depending on six "coordinates" $\vec{r}$, $\vec{v}$ (and also on time t), and the finding through it the average quantities the problem is reduced to a direct search of hydrodynamical parameters, being functions only of three Cartesian coordinates and time. Note that with the help of similar constructions from equation (2) there follow hydrodynamical models which are more complex than (23)–(25), and which are taking into account, for example, the anisotropy, presence of external forces, chemical reactions in medium, charged particles and electromagnetic fields and so on.

Obtained from the initial kinetic equation (2), the models (23)–(25) have a clear interpretation in hydrodynamical terms. Equation (23) represents the law of conservation of mass of a liquid particle, absolutely coinciding with the continuity equation introduced in section 4, Chapter II, in another way. Equation (24) expresses Newton's second law applied to a fixed liquid particle: its acceleration is determined by the sum of forces of pressure and viscous pressure. Finally, (25) describes the variation in time of internal energy of a liquid particle as a result of forces of pressure (first term on the right hand side), of viscous friction (the following three terms) and heat transfer (the last term). As distinct from thermal conductivity, the viscosity, as it is readily seen from (25), always leads to the increase of internal energy of the gas (the same is also true for entropy of the medium).

Models of lower hierarchical levels follow from (23)–(25) with the appropriate simplifications and concrete definitions of the considered object determined by the physical process taking place in it, its geometry, etc. The number of such hierarchical chains can be rather big. We shall demonstrate some of them.

In the absence of thermal conductivity, i.e. at $\kappa = 0$, the model (23)–(25) corresponds to a viscous, compressible non-thermal conductive gas (to *Navier-Stokes equations*, widely used to describe of various processes in natural science and technology). For an important particular case of incompressible liquid, when $\rho(\vec{r}, t) = \text{const}$, it follows from (23) that $\operatorname{div}\vec{V} = 0$, and the Navier-Stokes model consists of four equations

$$\frac{d\vec{V}}{dt} = -\frac{1}{\rho}\operatorname{grad} p + \frac{\mu}{\rho}\Delta\vec{V}\,, \qquad \operatorname{div}\vec{V} = 0\,, \tag{26}$$

and the equation of energy (25) is replaced by the given dependence $\varepsilon = \varepsilon(\rho, p) = \varepsilon(p)$.

When the forces of pressure greatly exceed the forces of viscous stress, it is possible to put $\mu = 0$, and from (26) follow *Euler's equations of motion for an incompressible liquid*, and together with the remaining unchanged equation of continuity, we obtain

$$\frac{d\vec{V}}{dt} = -\operatorname{grad}\frac{p}{\rho}, \qquad \operatorname{div}\vec{V} = 0. \tag{27}$$

If the motion is also potential, i.e. a scalar function $\phi(x, y, z, t)$ exists such that $\vec{V} = \operatorname{grad}\phi$ (in this case, as follows from the well-known formula of vector calculus, $\operatorname{rot}\vec{V} = 0$), the system (27) is reduced to *the Laplace equation* for a potential ϕ

$$\Delta\phi = 0,$$

and the first of equations (27) is satisfied automatically (see exercise 5).

In the absence of viscosity, the system (23)–(25) describes the flow of *compressible thermal conductive gas*, often studied in various phenomena. In this case the equation of energy (one-dimensional geometry)

$$\rho\frac{d\varepsilon}{dt} = -p\frac{\partial V}{\partial x} - \frac{\partial W}{\partial x}, \qquad W = -\kappa\frac{\partial T}{\partial x},$$

coincides with the need to account for the transition from Eulerian coordinates to Lagrangian ones, with equation (22) of section 4, Chapter II, obtained by the direct application of by the energy conservation law to a liquid particle. For a motionless medium ($\vec{V} \equiv 0$) the model represents *the heat transfer equation* (5) of section 2, Chapter II

$$C\frac{\partial T}{\partial t} = \operatorname{div}\kappa\,(\operatorname{grad}T), \qquad C = \rho\frac{\partial\varepsilon(T)}{\partial T},$$

which at constant C, κ for stationary process turns to the Laplace equation for temperature T

$$\Delta T = 0.$$

Now consider a medium without dissipative processes of viscosity and thermal conductivity ($\lambda = 0$, $\mu = 0$, $\kappa = 0$). Then from (23)–(25) *the Eulerian equations of motion for a compressible liquid* readily follow

$$\frac{\partial\vec{V}}{\partial t} + (\vec{V}\operatorname{grad})\,\vec{V} = -\frac{1}{\rho}\operatorname{grad}p,$$

complemented by the equations of continuity and energy

$$\frac{d\rho}{dt} = -\rho\operatorname{div}\vec{V}, \qquad \rho\frac{d\varepsilon}{dt} = -p\operatorname{div}\vec{V},$$

i.e. we come to a model coinciding with systems (4), (10), (14) of section 4, Chapter II, and at $\rho =$ const coinciding with system (27). One of the hierarchical sequences produced by the given model can be formed, for example, in the following way.

First, assume that the process is one-dimensional, and we turn to mass coordinate m. Then we obtain the system (18)–(20) of section 4, Chapter II

$$\frac{\partial}{\partial t}\left(\frac{1}{\rho}\right) = \frac{\partial V}{\partial m}, \qquad \frac{\partial V}{\partial t} = -\frac{\partial p}{\partial m}, \qquad \frac{\partial \varepsilon}{\partial t} = -p\,\frac{\partial V}{\partial m},$$

where $\partial/\partial t$ denotes the total time derivative. This system consisting of three equations, in a case of iso-entropic motion of ideal gas is reduced to *a second order quasi-linear equation* (equation (30) of section 4, Chapter II):

$$\frac{\partial^2 \rho}{\partial t^2} = \frac{\partial}{\partial m}\left(a_0 \rho^{\gamma+1}\,\frac{\partial \rho}{\partial m}\right).$$

Then, for the motion representing small variation from constant flow, follows *the equation of acoustics*

$$\frac{\partial^2 \rho}{\partial t^2} = c_0\,\frac{\partial^2 \rho}{\partial m^2}$$

(similar to *the equation of string oscillations* from section 2, Chapter III; $c_0 = \sqrt{\gamma p_0/\rho_0}$ is the speed of sound), or if the current is a simple wave, *the Hopf equation*

$$\frac{\partial \rho}{\partial t} + \sqrt{a_0}\,\rho^{\frac{\gamma+1}{2}}\,\frac{\partial \rho}{\partial m} = 0,$$

considered in subsection 7, section 4, Chapter II. Finally, from the equation of acoustics either for a flow of a simple wave type or from the Hopf equation for small perturbations follows the *linear transfer equation* derived in subsection 1, section 1, Chapter II,

$$\frac{\partial \rho}{\partial t} + \left(a_0 \rho_0^{\gamma+1}\right)^{1/2}\,\frac{\partial \rho}{\partial m} = 0,$$

which can be considered as the simplest model of gas in the class of equations in partial derivatives.

In summary we recall the logic of constructing a hierarchy of mathematical descriptions of large numbers of interacting particles:

1) the most complex models are based on first principles, i.e. on applying the laws of classical mechanics to each particle of a medium;

2) at transition to the following level of hierarchy – the Boltzmann kinetic equation (2) – the probabilistic description of gas with the help of distribution functions was used, and for concrete definition of integral of collisions,

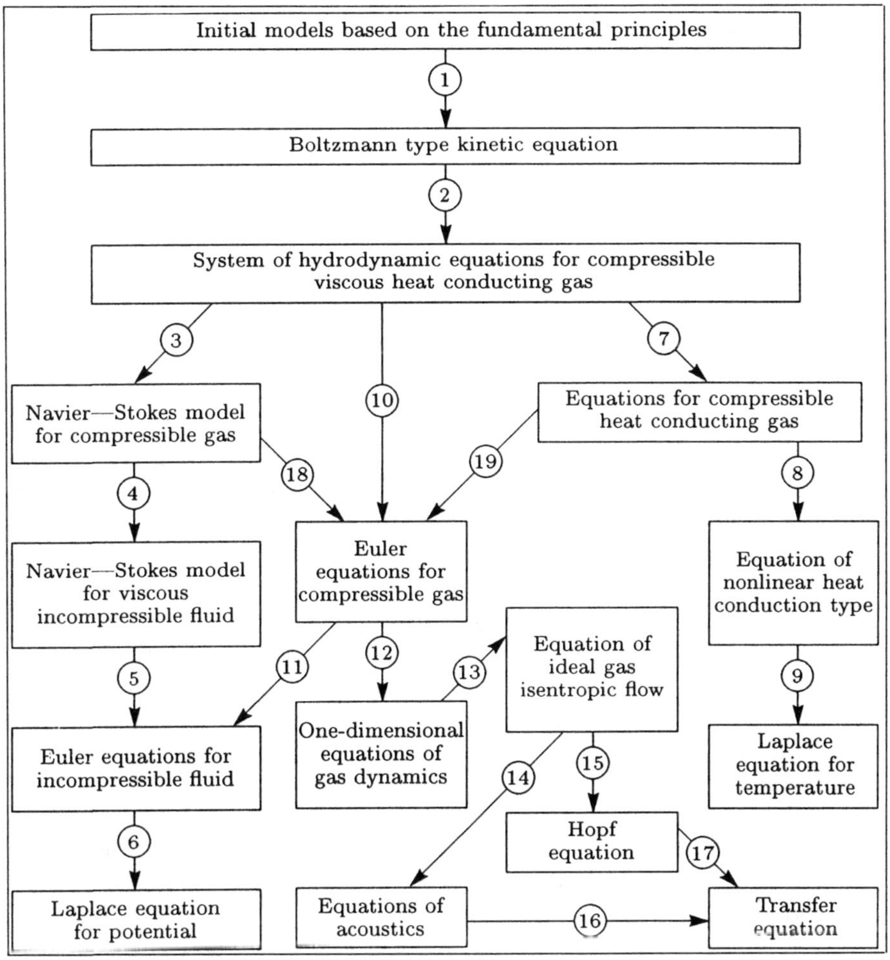

Fig.41. *Hierarchical chains of models of gas. Arrows denote transition to a model of a lower hierarchical level, numbers denote corresponding assumptions about the object: 1 – possibilities of the statistical description of gas with the help of distribution function, elastic collisions; 2 – closeness of processes to local thermodynamic equilibrium; 3, 19 – absence of thermal conductivity; 4,11 – incompressibility of a liquid; 5,7,18 – absence of viscosity; 6 – ideal liquid, potential currents; 8 – absence of gas motions; 9 - stationary process, constance of coefficient of thermal conductivity; 10 – absence of viscosity and thermal conductivity; 12 – one-dimensional flows; 13 – ideal gas, entropy is constant; 14, 17 – small perturbations of gas; 15, 16 – simple wave type flows*

assumptions on their elastic character, on absence of triple collisions, etc., have been made;

3) an assumption of local thermodynamic equilibrium is the basic one for transition to the description of gas in hydrodynamical approximation, i.e. to equations (23)–(25);

4) the basic hydrodynamical model (23)–(25) creates, depending on the character of the considered processes, various hierarchical chains, partly represented above (Fig. 41).

The hierarchy of gas models contains a wide spectrum of equations which are essentially different from each other also from a purely mathematical point of view. This includes the systems of equations of classical mechanics of higher dimensions, the kinetic equations, equations of the mechanics of continuous media. The latter can in their turn be divided into linear and nonlinear equations of hyperbolic (Euler and Hopf equations), parabolic (equation of thermal conductivity) and elliptic (Laplace equation) types, as well as of mixed types, stationary and non-stationary, many-dimensional and one-dimensional equations, etc. The additional variations of the constructed models are connected with various versions of boundary conditions and other input data.

The method of constructing models by the principle "from above downwards", which has been demonstrated in the present section, is more universal than the one based on the principle "from below upwards". Thus, for example, the Boltzmann equation cannot be derived from any models of a lower hierarchical level.

In the construction and analysis of any model, it is always useful to know its place in the general hierarchy of models of the investigated object. It enables us to correctly evaluate the area of its applicability and to clearly realize its connections with models of other levels, thus supporting the deeper understanding of the studied phenomena.

EXERCISES

1. Using the properties of distribution function $f(v)$, given by the formula (1), deduce (5) from (4).
2. Check whether the function $\Im(x, y)$ appearing in (7) is non-negative.
3. Deduce the formulae (8), (9), proceeding from the definition of integral of collisions $S(f(\vec{v}))$.
4. Using simple particular examples, prove that the quantity $\langle c^3 \rangle$ is not equal identically to zero (as distinct of average thermal velocity $\langle c \rangle$).
5. Using the formula of the vector calculus

$$\frac{1}{2}\,\mathrm{grad}\,|\vec{V}|^2 = \vec{V}\,(\mathrm{rot}\,\vec{V}) + (\vec{V}\nabla)\,\vec{V}$$

and applying operation rot to both hand sides of the first equation (27), obtain the equation $\partial\,(\mathrm{rot}\vec{V})/\partial t = \mathrm{rot}\,[\vec{V}\,\mathrm{rot}\,\vec{V}]$, identically fulfilled in the case of potential flows.

Bibliography for Chapter III: [3, 11, 20, 42, 43, 45, 69, 72].

Chapter IV

MODELS OF SOME HARDLY FORMALIZABLE OBJECTS

1 Universality of Mathematical Models

We will now construct mathematical models of the dynamics of clusters of amoebas and of random Markov process. We shall describe the behavior of living matter (amoeba) and "non-material" quantities (probability density) by the same parabolic equations, as for the phenomena of "dead" nature investigated in section II. We shall also consider the analogies between some mechanical or physical objects and economic processes.

1. Dynamics of a cluster of amoebas. Amoeba is a one-cell object of about ten micron (10^{-3} cm) dimensions, inhabiting in a soil and moving with the help of false-feet, i.e. parts of the body. Amoebas feed mainly on bacteria, absorbing them together with the soil (if the food is enough, amoebas multiply via division into two parts).

From observations and experiment it is known that the dynamics of development of their community – that is, of a sufficiently large population of amoebas in small distance from each other – can occasionally be quite complex. For example, depending on external conditions amoebas gather in huge (up to hundreds of thousands) clusters, which start to move as a single unit, though the individuality of each amoeba is conserved. It is noticed that this macroscopic "organized" motion occurs towards a higher concentration of some chemical substance, developed by the amoebas. The mathematical model of dynamics of a cluster is based on the following assumptions:

1) the distance between amoebas is small as compared with the sizes of the clusters (hundreds of microns), the latter can be considered as a "continuous medium" and one can introduce a concentration $N(x, y, z, t)$ – the number of amoebas in unit volume;

2) the process is one-dimensional, i.e. the concentration of amoebas and other quantities are functions only of coordinate x and time t;

3) amoebas are not born and do not die in a process of macroscopic motion, i.e. the characteristic time of motion (several hours) is small relative to the characteristic times of multiplication and life duration of amoebas;

4) the individual motion of amoebas in the absence of stimulating external influences (food, heat, etc.) is random, chaotic; there are no preferred directions and each amoeba with equal probability can move both to the right and to the left;

5) if there is an "attracting" chemical substance in the medium, then to the own disordered motion of amoebas a directed movement towards the area of higher density of this substance is added.

We deduce the equation of balance of amoebas in volume element dx during dt, using "the conservation law" of their total number (assumption 3). In this case the total number of amoebas in volume dx (the area of cross-section is unit) varies only due to the difference of flows of amoebas $W(x, t)$ on the left and right boundaries of the element. The quantity $W(x, t)$ is understood in the usual sense: the number of amoebas crossing the unit area per unit of time. The sought equation looks as follows (compare with the equation of heat balance in section 2 Chapter II)

$$[\bar{N}(x, t + dt) - \bar{N}(x, t)]\, dx = [\bar{W}(x, t) - \bar{W}(x + dx, t)]\, dt\,,$$

where $\bar{N}$, $\bar{W}$ is some average of quantities in small intervals dx, dt. When dx and dt tend to zero, we come to a differential equation of balance of the number of amoebas

$$\frac{\partial N}{\partial t} = -\frac{\partial W}{\partial x}\,.$$

The quantity $W = W_c + W_d$ is composed of two components, W_c and W_d. The part W_c of general flow is formed due to chaotic motion of amoebas, and consequently by analogy to Fourier's law for heat diffusion (section 2, Chapter II), it can be rewritten through the gradient of their concentration

$$W_c = -\mu \frac{\partial N}{\partial x}\,,$$

where $\mu > 0$ is a coefficient describing the "medium". This formula for W_c and the quantity μ can be easily obtained from a more detailed analysis of the process at microlevel, using the same considerations as those applied to the phenomena of heat transfer.

When deriving the expression for the component W_d, describing the directed flow of amoebas, we consider that the greater is W_d, the larger is the gradient of density of the "attracting" substance:

$$W_d = \eta N \frac{\partial \rho}{\partial x}.$$

Here $\eta > 0$ is a constant, $\rho(x, t)$ is the density of substance, and the multiplier N before the gradient means that with the given gradient of ρ the component of flow W_d is proportional to concentration of amoebas in the given point. Unifying the expressions for W_c, W_d and substituting them into the equation of balance, we obtain

$$\frac{\partial N}{\partial t} = \frac{\partial}{\partial x}\left(\mu \frac{\partial N}{\partial x} - \eta N \frac{\partial \rho}{\partial x}\right). \qquad (1)$$

There are two unknown functions – N and ρ in equation (1). Therefore it is necessary to obtain the equation of balance for ρ using the conservation law of substance. Thus, one has to take into account that the rate of creation of the chemical substance is proportional to the concentration of amoebas. Take into account also the disintegration of substance, with a rate naturally proportional to its concentration (similar to the process of radioactive decay; see section 1, Chapter I). Thus, in a unit of time in a unit of volume there appears and disappears an amount of substance, equal to

$$f = \alpha N - \beta \rho,$$

where $\alpha > 0$, $\beta > 0$ are constants describing the rates of its creation by amoebas and of its decay, respectively (this is the difference of model for the substance from the model for amoebas, when they do not die and are not born). The change of density of substance in the elementary volume of the medium also occurs due to the difference of its flows on the left and right boundaries of the element. It diffuses through the medium from areas with a higher concentration to those with a smaller concentration, just as the heat propagates from hotter areas of thermal conductive medium to cooler areas. According to Fick's law, this motion creates a flow W_ρ

$$W_\rho = -D \frac{\partial \rho}{\partial x},$$

where $D > 0$ is the coefficient of diffusion (the derivation of Fick's law is similar to that of Fourier's law).

Thus, the equation of balance of substance has a form

$$\frac{\partial \rho}{\partial t} = -\frac{\partial W_\rho}{\partial x} + f\,,$$

or taking into account expressions for W_ρ and f,

$$\frac{\partial \rho}{\partial t} = D\,\frac{\partial^2 \rho}{\partial x^2} + \alpha N - \beta \rho\,. \tag{2}$$

The equations (1) and (2) together with the input data μ, η, α, β, D are the model of the dynamics of a cluster of amoebas under the above made assumptions. For an unique determination of the solution, i.e. the functions $N(x,\,t)$, $\rho(x,\,t)$, it is necessary, as usual, to know the corresponding initial and boundary conditions. They look most simple in the case of the Cauchy problem, when the process is considered within an infinite space $-\infty < x < \infty$; then it is enough to set in the moment $t = 0$ the initial concentration of amoebas $N(x,\,0) = N_0(x)$ and the density of substance $\rho(x,\,0) = \rho_0(x)$.

Equations (1) and (2) are interconnected: ρ enters into the first of them, and N appears in the second. The system (1), (2) is non-linear because of the presence of the member $\eta N\,\partial\rho/\partial x$ in brackets in the right hand side of (1). Considered with respect to the concentration of amoebas N and the density ρ, the equations (1) and (2) clearly belong to parabolic type.

If amoebas do not create the "attracting" substance and $\rho(x,\,t) \equiv 0$, (1) turns to the equation of thermal conductivity (or diffusion)

$$\frac{\partial N}{\partial t} = \mu\,\frac{\partial^2 N}{\partial x^2}\,,$$

which can be explained easily, since in a flow W only the component W_c remains corresponding to the chaotic, not directed, motion of amoebas. When amoebas for any reasons cease to create the substance, the coefficient α in (2) turns to zero, and the equation (2) turns to (diffusion with disintegration)

$$\frac{\partial \rho}{\partial t} = D\,\frac{\partial^2 \rho}{\partial x^2} - \beta \rho \tag{3}$$

and by simple replacement (exercise 1) is again reduced to the equation of thermal conductivity

$$\frac{\partial u}{\partial t} = D\,\frac{\partial^2 u}{\partial x^2}\,.$$

Because of the nonlinearity of system (1), (2) one cannot deduce its general solution, and therefore the determination of spatial-temporal dynamics of clusters of amoebas is a rather complicated problem. However it is considerably simplified, if the small variations from the solution $N \equiv N_0$, $\rho \equiv \rho_0$, which is constant in time and space, are being studied, i.e. when the nonlinear problem becomes linear. Such a solution exists only at a ratio

$$\alpha N_0 = \beta \rho_0 ,$$

implying that the creation and the disintegration of the substance counterbalance each other.

Linearized in a vicinity of the constant solution the system (1), (2) (compare with the equation of acoustic oscillations in section 4, Chapter II) has a form

$$\begin{aligned} \frac{\partial \tilde{N}}{\partial t} &= \frac{\partial}{\partial x} \left(\mu \frac{\partial \tilde{N}}{\partial x} - \eta N_0 \frac{\partial \tilde{\rho}}{\partial x} \right) , \\ \frac{\partial \tilde{\rho}}{\partial t} &= D \frac{\partial^2 \tilde{\rho}}{\partial x^2} + \alpha \tilde{N} - \beta \tilde{\rho} , \end{aligned} \tag{4}$$

where $\tilde{N}$ and $\tilde{\rho}$ are small perturbations ($\tilde{N} \ll N_0$, $\tilde{\rho} \ll \rho_0$). Its general solution for the infinite space $-\infty < x < \infty$ (in this case there is no necessity to satisfy the boundary conditions) is possible to construct as the sum of partial solutions (harmonics)

$$\tilde{N} = C_1 \sin kx \, e^{\gamma t} , \qquad \tilde{\rho} = C_2 \sin kx \, e^{\gamma t} ,$$

where $k > 0$ is the wave number, C_1, C_2 are constants. For partial solutions the following expressions have to be fulfilled

$$\begin{aligned} C_1 (\gamma + \mu k^2) &= C_2 \, \eta \, N_0 \, k^2 , \\ C_2 (\gamma + \beta + D k^2) &= C_1 \, \alpha , \end{aligned} \tag{5}$$

connecting wavelength of a harmonic $\lambda = 2\pi/k$ with γ – its increment (or decrement), describing the increase or damping of perturbation in time. Excluding C_1 and C_2 from (5), we obtain the following square equation for γ

$$\gamma^2 + b\gamma + c = 0 , \tag{6}$$

where $b = \beta + k^2(\mu + D)$, $c = \mu k^2(\beta + Dk^2) - \eta \alpha N_0 k^2$. Both roots of equation (6) are negative only in the case when $c > 0$, i.e. at the fulfillment of the inequality

$$\mu(\beta + Dk^2) > \eta \alpha N_0 . \tag{7}$$

If (7) is fulfilled, than for any k the amplitude of perturbations of any wavelength decreases in time, and the constant solution is stable. The inequality (7) is obviously fulfilled at

$$\mu\beta > \eta\alpha N_0 \qquad (\text{or}\, \mu > \eta\rho_0)\,,$$

i.e. for the fixed parameters of the problem with a sufficiently small concentration of amoebas (and sufficient density of "attracting" substance). Otherwise the instability of the constant solution is possible (if the initial perturbation spectrum will contain long wavelength harmonic of small k, growing in time), inducing a more complex picture of evolution of clusters of amoebas. Certainly, the linearized model does not describe exhaustively the process, but it enables us to reveal a number of the features useful for more complete study.

2. Random Markov process. A typical example of such a process can be the motion of a small rigid particle in a liquid, performing a chaotic motion under the action of random collisions with molecules of the liquid (Brownian motion). Its position at any moment in time $t > t_0$ is given via coordinates x, y, z of three-dimensional space $\Re^3$. To simplify the calculations we shall consider below a one-dimensional motion, i.e. the random Brownian wandering of a particle along the axis x, $x \in \Re^1$.

The random process is called *Markovian*, if by the position of a point x at an moment of time t the probability of its presence at an arbitrary moment $t' > t$ in a certain region of space $\Re^1$ (in any measurable subset E) is determined uniquely. In other words, it is a process without memory, when the events happened within a time interval t and t' do not influence the position of a point in the moment t'.

The Markovian process is completely characterized by the function

$$p\,(t,\, x,\, t',\, x')\,, \qquad x \in \Re^1$$

called *probability density* in a point x'; knowing it, it is easy to calculate the probability of the presence of a particle in certain vicinity $E(x')$ of a point x' at the moment of time t'

$$p\,(t,\, x,\, t',\, E) = \int\limits_{E(x')} p\,(x,\, t,\, x',\, t')\, dx'\,.$$

Obviously, for function p the normalization condition is fulfilled

$$\int\limits_{\Re^1} p\,(t,\, x,\, t',\, x') = 1\,, \tag{8}$$

i.e. at any moment the particle has to be located at certain point in space $\Re^1$.

To construct a model of Markovian process the assumption of *a strong continuity* is essential. It is considered that there is a small probability that the particle in small time intervals Δt can undergo a remarkable increase of coordinate $\Delta x \geq \delta$. It means that for any $\delta > 0$

$$\int\limits_{|x'-x|\geq\delta} p\,(t-\Delta t,\, x,\, t,\, x')\, dx' = o(\Delta t)$$

or equivalently,

$$\lim_{\Delta t\to 0} \frac{1}{\Delta t} \int\limits_{|x'-x|\geq\delta} p\,(t-\Delta t,\, x,\, t,\, x')\, dx' = 0\,. \tag{9}$$

It is also assumed that for any $\delta > 0$ there are limits which are uniform by x

$$\lim_{\Delta t\to 0} \frac{1}{\Delta t} \int\limits_{|x'-x|<\delta} (x'-x)\, p\,(t-\Delta t,\, x,\, t,\, x')\, dx' = b > 0\,, \tag{10}$$

$$\lim_{\Delta t\to 0} \frac{1}{\Delta t} \int\limits_{|x'-x|<\delta} (x'-x)^2\, p\,(t-\Delta t,\, x,\, t,\, x')\, dx' = 2a > 0\,. \tag{11}$$

These assumptions have the following interpretation: the probability of a particle being at moment t within an interval $|x'-x|<0$ is proportional to Δt (decreases with reduction of time interval, after the initial moment $t-\Delta t$, which is natural) and in inverse proportion to certain "average" interval $|x'-x|$ (and to its square, which is also natural). The quantities a and b depend, generally speaking, on the point x and the moment t, $a = a(x,\, t)$, $b = b(x,\, t)$. However, for the sake of simplicity we shall consider the particular case, when a and b are constants.

Finally, the last assumption is that for any t, x, t', x' there are continuous partial derivative functions of p on x

$$\frac{\partial p}{\partial x}\,, \qquad \frac{\partial^2 p}{\partial x^2}\,. \tag{12}$$

The basic property of the considered process is expressed by *Markov identity*

$$p\,(t,\, x,\, t',\, x') = \int\limits_{\Re^1} p\,(x,\, t,\, \bar{t},\, \bar{x})\, p\,(\bar{t},\, \bar{x},\, t',\, x')\, d\bar{x}\,, \tag{13}$$

where $t < \bar{t} < t'$ is some moment in time within interval t and t', and $\bar{x}$ is the coordinate of a particle in the moment t, $t < \bar{t} < t'$ is an intermediate point in motion from x to x'. The content of identity (13) is revealed by considering the passage from point x to x' via a sequence of two transitions – first from x to $\bar{x}$, and then from x to x' (in Fig. 42 possible combinations of these transitions are shown, and the intermediate motions are represented by dashed lines, while the main transition is represented via a continuous line). The probability of an event consisting of two successive independent events is equal to the product of probabilities of each event, therefore under the integral in (13) one will have a product of the corresponding quantities. The integral is taken over all possible intermediate points $\bar{x} \in \Re^1$.

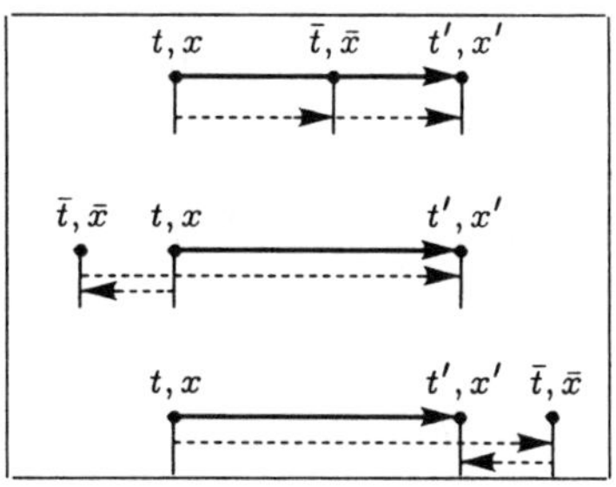

Fig.42.

For the considered process the Markov identity plays the role of a so-called "fundamental law", connecting in certain way the values of function p in points t, x and t', x'. With its help we shall first calculate the difference of values of p in the moments $t - \Delta t$ and t

$$p\,(t - \Delta t,\, x,\, t',\, x') - p\,(t,\, x,\, t',\, x') =$$

$$\int\limits_{\Re^1} p(t - \Delta t,\, x,\, t,\, \bar{x})p(t,\, \bar{x},\, t',\, x')\, d\bar{x} \quad p(t,\, x,\, t',\, x') \int\limits_{\Re^1} p(t - \Delta t,\, x,\, t,\, \bar{x})\, dx.$$

The first term on the left hand side of this equality corresponds to the first integral in its right hand side (see (13)) while the second corresponds to the term identical to it, multiplied by an integral, which is equal to one in accordance with the normalization (8). The multiplier $p(t,\, x,\, t',\, x')$ does not depend on $\bar{x}$, and consequently, inserting it under the sign of integral, we rewrite the equality as

$$p\,(t - \Delta t,\, x,\, t',\, x') - p\,(t,\, x,\, t',\, x') =$$

$$\int\limits_{\Re^1} [p\,(t,\, \bar{x},\, t',\, x') - p\,(t,\, x,\, t',\, x')]\, p\,(t - \Delta t,\, x,\, t,\, \bar{x})\, d\bar{x}\,.$$

Now we divide both parts of this equality by Δt and split the integral into two parts – over areas $|\bar{x} - x| \geq \delta$ and $|\bar{x} - x| < \delta$:

$$\frac{p\,(t - \Delta t,\, x,\, t',\, x') - p\,(t,\, x,\, t',\, x')}{\Delta t} = I_1 + I_2\,, \tag{14}$$

$$I_1 = \frac{1}{\Delta t} \int\limits_{|\bar{x}-x|\geq\delta} [p\,(t,\, \bar{x},\, t',\, x') - p\,(t,\, x,\, t',\, x')]\, p\,(t - \Delta t,\, x,\, t,\, \bar{x})\, d\bar{x}\,,$$

$$I_2 = \frac{1}{\Delta t} \int\limits_{|\bar{x}-x|<\delta} [p\,(t,\, \bar{x},\, t',\, x') - p\,(t,\, x,\, t',\, x')]\, p\,(t - \Delta t,\, x,\, t,\, \bar{x})\, d\bar{x}\,.$$

The integral I_1 by virtue of the property of strong continuity (9) tends to zero at $\Delta t \to 0$ (the first multiplier in the integrand does not depend on Δt and does not influence the behavior of I_1 at $\Delta t \to 0$).

The integral I_2 can be transformed, expanding in view of (12) the first multiplier over the degrees of $(\bar{x} - x)$

$$\begin{aligned} I_2 = {}& \frac{1}{\Delta t} \int\limits_{|\bar{x}-x|<\delta} \frac{\partial p(t,\, x,\, t',\, x')}{\partial x}\, (\bar{x} - x)\, p(t - \Delta t,\, x,\, t,\, \bar{x})\, d\bar{x} + \\ &+ \frac{1}{\Delta t} \int\limits_{|\bar{x}-x|<\delta} \frac{1}{2} \frac{\partial^2 p(t,\, x,\, t',\, x')}{\partial x^2}\, (\bar{x} - x)^2\, p(t - \Delta t,\, x,\, t,\, \bar{x})\, d\bar{x} + \\ &+ \frac{1}{\Delta t} \int\limits_{|\bar{x}-x|<\delta} o[(\bar{x} - x)^2] \frac{1}{2} \frac{\partial^2 p(t,\, x,\, t',\, x')}{\partial x^2}\, p(t - \Delta t,\, x,\, t,\, \bar{x})\, d\bar{x}\,. \end{aligned}$$

When Δt tends to zero in this equality, we note that the partial derivative functions p under the sign of the integral do not depend on $\bar{x}$. Removing them from the integral, for the first two terms we obtain by virtue of the assumptions (10), (11), that their limits are equal, respectively

$$b\, \frac{\partial p(t,\, x,\, t',\, x')}{\partial x}\,, \qquad a\, \frac{\partial^2 p(t,\, x,\, t',\, x')}{\partial x^2}\,,$$

and the third term we represent in a form

$$\bar{\varepsilon}(\bar{x} - x)\, \frac{1}{2}\, \frac{\partial^2 p(t,\, x,\, t',\, x')}{\partial x^2}\, \frac{1}{\Delta t} \int\limits_{|\bar{x}-x|<\delta} (\bar{x} - x)^2\, p(t - \Delta t,\, x,\, t,\, \bar{x})\, d\bar{x}\,,$$

where $\bar{\varepsilon}(\bar{x} - x)$ is the average value of function $\varepsilon(\bar{x} - x)$, so that by definition of $o[(\bar{x} - x)^2]$ we have $\bar{\varepsilon}(\bar{x} - x) \to 0$ at $\delta \to 0$. In the latter expression we consider first the limit when $\delta \to 0$. This limit, obviously, is equal to zero at any $\Delta t > 0$. Therefore its limit at $\Delta t \to 0$ is also equal to zero. Finally, the limit of the left hand side (14) at $\Delta t \to 0$ is equal to the derivative of function p by time. Summarizing these results, we obtain from (14) *the Kolmogorov equation* for probability density, valid for all $t > t_0$, $-\infty < x < \infty$

$$\frac{\partial p(t,\, x,\, t',\, x')}{\partial t} = a\,\frac{\partial^2 p(t,\, x,\, t',\, x')}{\partial x^2} + b\,\frac{\partial p(t,\, x,\, x,\, t',\, x')}{\partial x}\,. \tag{15}$$

Equation (15) is a linear parabolic equation. Its generalizations also have the same properties. For example, in a case, when b and a in (10), (11) depend on t, x, the analog of equation (15) has a form

$$\frac{\partial p}{\partial t} = a(t,\, x)\,\frac{\partial^2 p}{\partial x^2} + b(t,\, x)\,\frac{\partial p}{\partial x}\,. \tag{16}$$

If the point x belongs to n-dimensional space, i.e. $x = (x_1,\, x_2,\, \ldots,\, x_n)$, then for function p the following generalization of (15) and (16) is valid

$$\frac{\partial p}{\partial t} = \sum_{i,j=1}^{n} a_{ij}(t,\, x)\,\frac{\partial^2 p}{\partial x_i\, \partial x_j} + \sum_{i=1}^{n} b_i(t,\, x)\,\frac{\partial p}{\partial x_i}\,, \tag{17}$$

where $b_i(t,\, x)$ and $a_{ij}(t,\, x)$ are calculated via the formulae (10), (11), but in (10) the multiplier $(x' - x)$ is replaced by $(x' - x)_i$, and in (11) instead of $(x' - x)^2$ there appears the expression $(x' - x)_i\,(x' - x)_j$, $i,\, j = 1,\, 2,\, \ldots,\, n$.

Note that (17) is by no means a formal generalization of (15), (16). Random Markov processes can proceed not only in real physical space (Brownian motion), but also in so-called phase space. They are peculiar to a number of technical and other systems, when a state is described by set phase variables x_1, x_2, ..., x_n, with number greatly exceeding three.

The simplest version of Kolmogorov equation follows from (15) at $b = 0$

$$\frac{\partial p}{\partial t} = a\,\frac{\partial^2 p}{\partial x^2}\,, \tag{18}$$

and represents *the equation of thermal conductivity* (or diffusion).

However there is an essential difference between mathematical models of heat transfer and a random Markov process. When deriving of equation (15), for the function $p(t,\, x)$ (as distinct from the derivation of the equation of thermal conductivity) the condition of strong continuity (9)–(11) was used. Hence, the function $p(t,\, x)$ cannot be an arbitrary solution of the Kolmogorov equation. It appears to be the so-called *fundamental solution of equations* (15)–(18).

It is easiest to explain this property of function $p(t, x)$ in the case of elementary equation (18). Let function $u(t, x)$ be a solution of equation (18) determined at $t > t_0$, $-\infty < x < \infty$ and satisfying the given initial condition

$$u(t, x) \to u_0(x) \geq 0 \qquad \text{by } t \to t_0 \,. \tag{19}$$

Then, if $p(x, t, t', x')$ is a fundamental solution of (18), the function $u(t, x)$ is obtained via the formula

$$u(t, x) = \int_{\Re^1} p(t, x, t', x')\, u_0(x')\, dx' \,. \tag{20}$$

Whether or not $u(t, x)$ given by (20) is a solution of (18), can be easily established by differentiation (exercise 4). The property (19) is proved by splitting the integral in (20)

$$u(t, x) = \int_{|x'-x|<\delta} p(t, x, t', x')\, u_0(x')\, dx' + \int_{|x'-x|\geq\delta} p(t, x, t', x')\, u_0(x')\, dx' \,.$$

At $t = t_0$ the point has a coordinate x, and therefore due to the property of strong continuity, the probability of its presence at $t' \to t \to t_0$ within the area $|x' - x| \geq \delta$ is zero, i.e. $p(t, x, t', x') \to 0$ at $t' \to t_0$. Hence, from the latter formula we have

$$\lim_{t' \to t_0} \int_{\Re^1} p(t, x, t', x')\, u_0(x')\, dx' = \lim_{t' \to t_0} \int_{|x-y|<\delta} p(t, x. t', x')\, u_0(x')\, dx' \,.$$

Using the condition of normalization (8) at $t' \to t_0$, $x' \to x$ and taking into account the independence of the left limit in the latter formula on δ, we obtain

$$\lim_{t' \to t_0} \int_{\Re^1} p(t, x, t', x')\, u_0(x')\, dx' = u_0(x) \,,$$

i.e. the formula (19).

The fact that the function $p(t, x)$ is not arbitrary but is a fundamental solution of Kolmogorov equation (for equations (15)–(17) the proof is similar is not a defect of the considered model, and reflects the natural property of random Markov processes. Indeed, at the initial moment $t = t_0$ the wandering point has some coordinate x_0, and hence, $p(x, t_0) \equiv 0$ with $x \neq x_0$. At the same time, from the condition (8) at $t' = t = t_0$ follows

$$\int_{\Re^1} p(x, t_0)\, dx = 1 \,,$$

i.e. the initial data for equation (18) (and equations (15)–(17)) are given by δ-function. The solutions of Cauchy problem for linear parabolic equations with such initial condition are their fundamental solutions. The simplest example is the function of an instant point heat source for the equation of thermal conductivity (11) from section 2, Chapter II, given by the simple formula (20) from section 2, Chapter II, and possessing the same property as function $p(t, x)$. For more general equations (15)–(17) there are no such simple representations of their fundamental solutions. However the circumstance, that $p(t, x)$ satisfies the Kolmogorov equation and is its fundamental solution, is essentially used in the research of objects with random Markov processes, in particular in control problems of those objects.

Table 1

Universality of mathematical models. The parabolic equations.

Object (process)	**Basic assumptions and laws**
Motion of underground waters.	Conservation of mass, Darsi law.
Heat transfer; diffusion of substance.	Conservation of energy, Fourier law; conservation of mass, Fick's law.
Motion of a cluster of amoebas.	Conservation of number of amoebas, randomness of a movement of amoebas in absence of "attracting" substances.
Random Markov process.	Markov identity, strong continuity of the process.
Dynamics of distribution of power in a hierarchy.	Following the rules, postulate on the mechanisms of redistribution of power in a hierarchy.

Thus, the parabolic equations are an example of the universality of mathematical models (Table 1). They describe a wide range of processes of quite different natures (the latter example in Table 1 is considered in section 4). Note that the parabolic equations are frequently associated with chaotic disordered phenomena (heat transfer, diffusion, etc.). However they are applicable also to many processes considered as determined (motion of underground waters, filtration of gas in porous medium, etc.).

The universality of mathematical models is a reflection of the unity of the world surrounding us and ways of its description. Therefore the methods and results developed by relatively simple analogy in the mathematical modeling of one phenomenon can be transferred to a wide class of completely different processes.

3. Examples of analogies between mechanical, thermodynamic and economic objects. "The method of analogies" has a special meaning in the mathematical modeling of hardly formalizable objects, for which the fundamental laws, variational principles and other general and mathematically strict statements are either unknown, or do not exist at all. Systems involving remarkable intervention of people and, in particular, economic systems concern such objects.

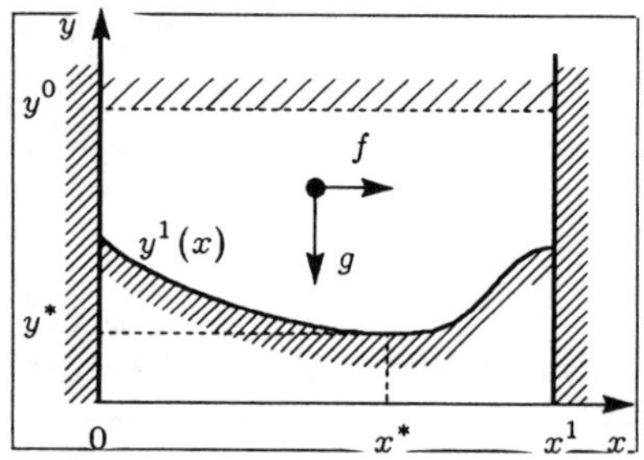

Fig.43.

One of the most important mechanical-economic analogies is the analogy between the equilibrium of a material particle in a potential field of external forces and the choice of an optimal plan of manufacturing. Consider for the sake of simplicity a particular case. Let a unit mass (the point in Fig. 43) be in a gravity field with any position in a space limited from below by a rigid ideal surface. The equation of the surface is given by dependence $y^1(x) \geq 0$, where x, y are, respectively, the horizontal and vertical coordinates of the particle, $y^1(x)$ is a smooth function. Obviously, y – the distance of the particle from an abscissa axis, satisfies the inequality

$$y \geq y^1(x), \qquad 0 \leq x \leq x^1 . \tag{21}$$

The gravity field is potential, i.e. there is a function (potential) $P(x, y)$ such, that the components of its gradient determine the external force acting on a material particle in a given point of the field. In the considered case

$$\operatorname{grad} P = (0, -g)\,,$$

where g is the free-fall acceleration (the horizontal component, naturally, is equal to zero). Hence, the potential is given by the formula

$$P = -gy\,, \tag{22}$$

where an insignificant additive constant is omitted.

The variation of potential in the motion of the particle within the gravity field is equal to the work A, performed by this force, and is determined only by the initial and final positions of a particle (in this case only via coordinate y):

$$A = \int_{x_1,\, y_1}^{x_2,\, y_2} dP = P\,(x_2,\, y_2) - P\,(x_1,\, y_1) = P\,(y_2) - P\,(y_1)\,. \qquad (23)$$

The point x^*, y^* is called *the equilibrium position*, if the particle placed in it and having zero velocity, remains in it any time (at the posed connections (21) and external force grad $P(x^*,\, y^*)$). The potential in the equilibrium point reaches its extreme value. This property is well illustrated in Fig. 43, where the equilibrium points are points of minimum of function $y^1(x)$ (where the gravity force is balanced by the opposite force of reaction of support). Any *virtual* (not contradicting connections (21)) small displacement of a particle from these points will result in a negative work, as seen from (23), and the potential will decrease (if the motion is completely or partially performed over the surfaces $y^1(x)$, $x = 0$, $x = x^1$, then since they are considered ideal and the forces of friction are absent and their reaction is perpendicular to the motion, it has no influence on the work). At all other points in area $y \geq y^1(x)$, $0 \leq x \leq x^1$ the potential (22) does not reach its maximal value.

Thus, the search for a stable equilibrium is reduced to the solution of the problem

$$P\,(x,\, y) \to \max \quad \text{by} \quad y \geq y^1(x), \quad 0 \leq x \leq x^1\,. \qquad (24)$$

Maupertuis law of rest (24) is similarly formulated for general mechanical systems, when the equilibrium points are obtained not so simply as in the given case. For example, if other potential forces as well as gravity are acting on the particle, then the equilibrium points are not necessary to coincide with points of minimum of function $y^1(x)$.

In an economic interpretation the problem (24) is called *the problem of nonlinear programming* and often arises in industrial planning.

Let a certain enterprise produces some goods (bricks) of an amount x, $0 \leq x \leq x^1$. To produce this it is necessary to expend certain resources (clay); we denote by y the resources remaining after the necessary production (the initial resource is y^0 and is considered independent from the amount of plan x, $y^0 \leq y \leq 0$). It is known that production occurs with some resource restrictions from above, that is

$$y^0 \leq y \leq y^1(x)\,, \qquad 0 \leq x \leq x^1\,, \qquad (25)$$

where $y^1(x)$ is the minimum amount of unused resources, which the enterprise (for technological, financial or other reasons) is obliged to have after performance of the plan x ($y^1(x)$ is considered as given smooth function x).

In the simplified formulation the profit $P(x,\, y)$ is equal to the difference between the cost of the final production fx and the cost of the resources used $g(y - y^0)$

$$P(x, y) = fx + g(y^0 - y), \tag{26}$$

where f, g are the prices of unit production and resource, respectively (other expenses are considered insignificant).

The problem in planning the production is to choose the plan x^* with the maximal profit (26) under the resource restrictions (25)

$$P(x, y) \to \max \qquad \text{by} \qquad y^0 \leq y \leq y^1(x), \quad 0 \leq x \leq x^1. \tag{27}$$

Formulations similar to (27) are valid for rather general problems of planning.

To reach an absolute analogy we compound the problem about the equilibrium of a material particle. In addition to gravity force we introduce a potential external force (see Fig. 43), acting on the particle in the direction of axis x and equal to f. For example, for a charged particle this force appears in the presence of an appropriate electrical field. Then the potential is

$$P(x, y) = fx + gy,$$

i.e. within an insignificant additive constant, coincides with (26). The analogy is completed via the introduction of an ideal rigid surface $y = y^0$, restricting the movement of the particle from above; then (21) takes the form of (25) (taking into account that the quantities are always considered negative).

Thus, the problems (24) and (27) are absolutely similar and have coinciding solutions x^*, y^*. Note, that the mechanical-economic apply not only for general formulations of problems, but also for many relevant concepts (force – limiting profit, reaction of connections – limiting costs, etc.).

In the plan x^*, y^* (*the optimal plan*) the maximal value of the economic analog of potential (profit) is reached, but there is also a maximum of one more quantity, corresponding not to a mechanical but a thermodynamic concept – the entropy. It is known that an isolated thermodynamic system, for example, gas in an isolated vessel (see subsection 3, section 3, Chapter III), with highest probability will evolve to a state with least ordering of parameters describing the particle of the system (atoms, molecules). In this state the system is in equilibrium, its parameters are the same in all its points. Therefore there is no way to distinguish its parts from each other: the highest disorder (in comparison with other possible states) – "chaos" is reached. The measure of this disorder is the entropy, being a function of the state of a system and reaching its maximal value when the system is in equilibrium.

For an arbitrary non-optimal plan $(x, y) \neq (x^*, y^*)$ of the problem (27), consider ε-vicinity of all close plans $(\bar{x}, \bar{y})$ (that is the conditions $|\bar{x} - x| < \varepsilon$, $|\bar{y} - y| < \varepsilon$ are fulfilled). In so far as the plan (x, y) is non-optimal, then in

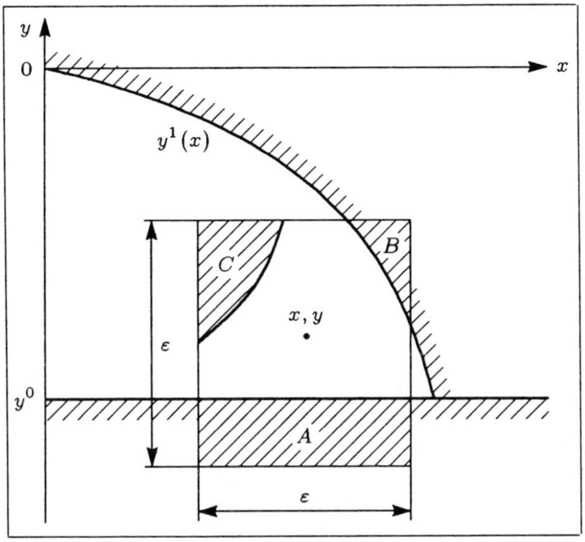

Fig.44.

its ε-vicinity there is always a non-empty set of plans, which either do not satisfy resource restrictions (areas A and B in Fig. 44) or have no advantages of profit in comparison with the plan (x, y) (area C in Fig. 44), or both. It means that in the vicinity of (x, y) it is possible to carry out a partial ordering of plans relative to each other, so far as the procedure of preference of one plan with respect to others is known. The volume of the set of these "bad" plans V depends on x, y, ε and, obviously, is always smaller the volume of ε-vicinity

$$V(x, y, \varepsilon) < V_\varepsilon = \varepsilon^2 .$$

Consider now a ε-vicinity of the optimal plan (x^*, y^*). As distinct from the case of the plan (x, y) all plans from this area are less preferable relative to (x^*, y^*) (either due to resource restrictions, or the size of profit having a maximum in (x^*, y^*)). In this case, the volume of "bad" plans is equal to the volume of all ε-vicinity

$$V(x^*, y^*, \varepsilon) = V_\varepsilon .$$

Introduce a function

$$E(x, y, \varepsilon) = \frac{V(x, y, \varepsilon)}{V_\varepsilon} ,$$

which is the measure of the ordering of various states of the given system. The "closer" the point (x, y) to "the equilibrium" optimal state (x^*, y^*), the greater is the value of function E. Obviously, $E < 1$ at all $(x, y) \neq (x^*, y^*)$, and reaches its maximum $E = 1$ at (x^*, y^*). It is just an economic analog of entropy in the considered problem.

As with the mechanical analogies, the thermodynamic ones are valid for many rather general economic systems and are widely applied in their studies.

EXERCISES

1. Using a replacement $\rho(x, t) = \varphi(t) \cdot u(x, t)$ reduce (3) to the equation of thermal conductivity.

2. Prove that the square equation (6) always has real roots, and that a necessary and sufficient condition of negativity of both roots is the condition $c > 0$.

3. Deduce equations (16) and (17) using the same assumptions as in the deduction of (15).

4. Differentiating the integral on the right hand side of (20) by t, x, and using equation (18), prove that the function $u(t, x)$ is a solution of equation (18).

2 Some Models of Financial and Economic Processes

We shall consider the models of advertising campaigns, procedures of repaying mutual duties of enterprises, simple macromodels of equilibrium and the growth of an economic system. We shall discuss the role of analogies used in the construction of models and some conclusions following from their analysis.

1. Organization of an advertising campaign. A firm starts to advertise a new product or service. Certainly, the profit from the future sales should cover the costs of the expensive campaign. Clearly, in the beginning the outlay can exceed profit, as only a small fraction of the potential customers will be informed of the novelty. Then, with the increase in number of sales, it is already possible to expect a wealthy profit and finally, a time should be reached when the market will be saturated, and it will become senseless to advertise the goods further.

The model of an advertising campaign is based on the following basic assumptions. It is considered that dN/dt is the rate of change in time of numbers of consumers informed about the goods and ready to buy them (t is the time since the beginning of advertising campaign, $N(t)$ is the number of already informed consumers) and is the proportional to the number of customers not yet aware of it, i.e. to $\alpha_1(t)(N_0 - N(t))$, where N_0 is the total number of potential customers, $\alpha_1(t) > 0$ characterizes the intensity of the advertising campaign (actually determined by the publicity expenses at the given moment in time). It is also assumed that the consumers who are aware of the goods somehow distribute that information among those who are unaware, acting as a kind of additional advertising agents. Their contribution is equal to $\alpha_2(t)N(t)(N_0 - N(t))$ and is larger, the more agents there are. The quantity $\alpha_2(t) > 0$ characterizes a degree of interaction of the consumers among themselves (this can be established, for example, with the help of questionnaires).

As a result we have the equation

$$\frac{dN}{dt} = [\alpha_1(t) + \alpha_2(t)\, N(t)](N_0 - N)\,. \tag{1}$$

At $\alpha_1(t) \gg \alpha_2 N(t)$ from (1), the model similar to the Malthus model (10) of section 1, Chapter 1 is obtained, with an opposite inequality – the equation of logistic curve (see (12) in section 1, Chapter 1)

$$\frac{dN}{d\tau} = N\,(N_0 - N)\,, \qquad d\tau = \alpha_2(t)\,dt\,.$$

Its solution is investigated in section 1, Chapter I (see also Fig. 7).

The analogy obtained is quite understandable, in so far as in the construction of the present model and the model of population growth the same idea of "saturation" was used: the growth rate in time of a quantity is proportional to the product of its current value $N(t)$ and the difference $N_0 - N(t)$ between its equilibrium (population) or limiting (consumers) and current values.

The analogy between both processes fails, if at any moment in time the quantity $\alpha_1 + \alpha_2 N$ turns to zero or even becomes negative (for this it is necessary that one or both coefficients $\alpha_1(t)$, $\alpha_2(t)$ become negative). A similar negative effect rather frequently occurs in advertising campaigns of various kinds and should prompt the organizers of campaigns either to change the character of the advertising or cancel further advertising. The actions aimed at increasing the popularity of goods, depending on the values of quantities $\alpha_1(t)$, $\alpha_2(t)$, $N(t)$ can be directed towards the improvement of results both of direct (parameter α_1) and indirect (parameter α_2) advertising.

The model (1) avoids the obvious shortage peculiar to the logistics equation. Indeed, it has no solutions turning to zero in a finite time scale (from the corresponding formula for $N(t)$ in subsection 5, section 1, Chapter I, it follows that $N(t) \to 0$ with $t \to -\infty$). Concerning the advertising, this would mean that part of customers were already aware about the new product even prior to the beginning of campaign. Consider the model (1) in a vicinity of the point $N(t = 0) = N(0) = 0$ ($t = 0$ is the moment at which the campaign is begun), assuming that $N \ll N_0$, $\alpha_2(t)N \ll \alpha_1(t)$, then the equation (1) takes the form

$$\frac{dN}{dt} = \alpha_1(t)\, N_0$$

and has a solution

$$N(t) = N_0 \int_0^t \alpha_1(t)\, dt\,, \tag{2}$$

satisfying the natural initial condition at $t = 0$.

From (2) it is relatively easy to estimate the ratio between the advertising costs and the profit at the beginning of campaign. Denote through p the amount of profit from a single sale, if expenses are absent. For simplicity, it is considered that each customer gets only a single production unit. The coefficient $\alpha_1(t)$ by its content is the number of equivalent advertising actions in a unit of time (for example, sticking up identical posters). Through s we denote the cost of an elementary piece of advertising. Then the total profit is

$$P = pN(t) = pN_0 \int_0^t \alpha_1(t)\, dt\,, \tag{3}$$

and the expenses are

$$S = s \int_0^t \alpha_1(t)\, dt\,.$$

The profit overwhelms the costs at $pN_0 > s$, and if the advertising is effective and inexpensive, and the market is capacious enough the profit is achieved from the very beginning of campaign (in reality between the payment of advertising, the advertising action and subsequent purchase a so-called *lag* or temporary delay does exist, which can be taken into account in more complete models). With not particularly effective or expensive advertising the firm has losses from the outset. However this circumstance, generally speaking, cannot become a reason to stop advertising. Indeed, the expression (3) and the condition obtained with its help, $pN_0 > s$ are valid only for small values of $N(t)$, when the functions P and S grow in time by identical laws. At the increase of $N(t)$ the terms omitted from (1) become essential, in particular, the role of indirect advertising increases. Therefore $N(t)$ can become a more "rapid" function of time than in formula (3). This nonlinear effect in the variation of $N(t)$ at a constant rate of growth of costs enables us to compensate for the financial failure of an initial stage of a campaign.

We shall explain the given statement in a particular case of equation (1) with constant coefficients α_1, α_2. By replacement

$$\bar{N} = \alpha_1/\alpha_2 + N$$

it is reduced to a logistic equation

$$\frac{d\bar{N}}{dt} = \alpha_2 \bar{N}\,(\bar{N}_0 - \bar{N})\,, \qquad \bar{N}_0 = \frac{\alpha_1}{\alpha_2} + N_0\,, \tag{4}$$

with a solution,

$$\bar{N}(t) = \bar{N}_0[1 + (\bar{N}_0\alpha_2/\alpha_1 - 1)\,\exp(-\bar{N}_0\alpha_2 t)]^{-1}\,. \tag{5}$$

Thus $\bar{N}(0) = \alpha_1/\alpha_2$, so that $N(0) = 0$, and the initial condition is fulfilled. From (4) it is seen that the derivative of function $N(t)$ and, hence, the function $N(t)$ at $t > 0$ can be more than its initial value (at condition $\bar{N}_0 > 2\alpha_1/\alpha_2$ or $N_0 > \alpha_1/\alpha_2$). The maximum of the derivative is reached at $\bar{N} = \bar{N}_0/2$, $N = (\alpha_1/\alpha_2 + N_0)/2$:

$$\left(\frac{d\bar{N}}{dt}\right)_m = \left(\frac{dN}{dt}\right)_m = \alpha_2\,\frac{\bar{N}_0^2}{4} = \alpha_2\,\frac{(\alpha_1/\alpha_2 + N_0)^2}{4}\,.$$

In this current period, i.e. for the profit received in a unit of time, we have

$$P_m = p\,\frac{dN}{dt} = p\,\alpha_2\,\frac{(\alpha_1/\alpha_2 + N_0)^2}{4}\,.$$

Subtracting from P_m the initial current profit $P_0 = p\,(dN/dt)_{t=0} = \alpha_1 N_0$ (see (2)), we obtain

$$P_m - P_0 = p\,\frac{(\alpha_1/\sqrt{\alpha_2} - \sqrt{\alpha_2}\,N_0)^2}{4}\,,$$

i.e. the difference between the initial and maximal current profit can be rather significant (see also exercise 1). The total economic benefit of the campaign (its necessary condition is, obviously, the fulfilling of an inequality $P_m = p\,(\alpha_1/\sqrt{\alpha_2} + \sqrt{\alpha_2}\,N_0)^2/4 > \alpha_1 s$) is determined by all its processes, the characteristics of which are calculated from (4), (5) with the help of quadrature (see also exercise 2).

As follows from (4), at some point the continuation of advertising becomes unprofitable. Indeed, at $\bar{N}(t)$, close to $\bar{N}_0$, the equation (4) has a form

$$\frac{d\bar{N}}{dt} = \alpha_2\bar{N}_0\,(\bar{N}_0 - \bar{N})\,. \tag{6}$$

Its solution tends at $t \to \infty$ to limiting value $\bar{N}_0$ (and the function $N(t)$ – to N_0) via slow exponential law (exercise 2). A negligible number of new consumers appear in a unit of time, and the profit obtained in any case cannot cover the continuing expenses.

Similar characteristics are calculated for equation (1) and its various generalizations are also widely used to describe applications of technological and other innovations.

2. Mutual offset of debts of enterprises. Any economic system significant by its scales includes tens of thousands of enterprises (firms, companies and so on), exchanging goods and services among themselves. Even a small firm with a relatively small number of direct partners is indirectly connected (through direct and secondary partners) with a huge number of enterprises included in the system, and its success directly depends on their situation. This statement is moreover true for large companies with partnerships with hundreds and thousands of economic agents.

The interconnection of all parts of an economic system is well displayed when estimating expenses between the enterprises for the product exchanged. Indeed, upon receiving the money from the customers for the product, the enterprise spends on a raw materials and machines from other companies, on salaries, on advertising and other actions necessary for normal functioning. Thus a large number of partners in the given enterprise are involved in the economic process. In turn the customers, having received goods from the enterprise, are forwarding them for a further resale, using their for own products and so on, again increasing the number of agents participating in the economic activity.

If the deliveries and payments are carried out in time (in an ideal world this would take place instantly), from the financial point of view nothing threatens the economic system. To continue their activity the enterprises are not obliged to use the significant fraction of their financial resources in their bank accounts, moreover to sell the basic capital (lands, buildings, equipment, technologies). In reality, between the delivery of goods and payment (or prepayment for the goods and the ensuing delivery) there is always a time delay. The minimal delay is determined by purely technical reasons, so far as a time always is required for transportation and packaging of goods, to realization bank transfers, etc. With small lags of small amounts of goods (or small money sums) the enterprises involve for short periods a minor fraction of their free finances and then quickly compensate them from payments received from the partners.

However, certain situations are possible, when for economic, financial, internal or external political, social, psychological and other reasons the time by which payment is delayed becomes comparable to the time taken to circulate finances, and the absolute value (volume) of delayed payments or deliveries is comparable with the volume of free resources of the enterprises.

In this case there occurs a so-called *crisis of nonpayment*, capable of leading to a serious crisis in all economic systems.

Indeed, the enterprise which has not received money for the delivered product (or which has paid for the goods, but has not received them), cannot pay the suppliers (in so far as the volume of debts to the enterprise is comparable with the amount of its free resources, using these cannot essentially improve the situation). In turn, the suppliers do not pay the clients, who in turn do not pay their clients, etc. Long chains of nonpayment appear, penetrating the whole system. They can obviously consist of N parts, and their total number can reach up to $N!$ (N is the total number of enterprises). The sum of absolute amounts of debts on all chains can not only exceed the free resources of the enterprises, but can also become comparable with the cost of their basic capital (we mean the sum of absolute debts, in so far as any enterprise can simultaneously be both debtor and creditor of its partners). The system is in crisis – the enterprises either should stop production or ask again for credits from the partners, increasing their total mutual debt.

In principle, the situation can be saved if an authorized establishment (for example, the main national bank) provides all enterprises when a lump sum credit equal to the sum of all the debts. Then they pay among themselves and return the credit. However such credit can provoke high inflation (the production of the goods has not increased, while the amount of circulating money has increased significantly), with all its negative consequences.

In any non-payment crisis a certain role is always played by a purely "technical" component connected with the shortages in the procedure of accounts. Below we shall consider the crises induced just by these factors, distracting from economic, political and other reasons for their occurrence.

First we shall explain the essence of the problem on by a simple numerical example for a system of three enterprises, each with free resources equal conditionally to one financial unit, and basic capital equal to 10 units. Let the first enterprise owes 100 units to the second, the second owes 100 units to third, and, finally, the third also owes 100 units to the first enterprise. The total absolute debt of the enterprises is equal to 600 units and is huge in comparison with their funds (30 units), not to mention the free resources (3 units). At the same time the financial situation of the system is actually safe, in so far as the total "debt" of each enterprise separately (i.e. the sum of money which the enterprise owes to others, and others owe to it) is equal to zero. The obvious procedure of mutual offset consists in the simultaneous cancelling (repayment) of all debts: it is announced that nobody owes anything to anybody, and the partners continue the work, being free from the burden of debt. A centralized credit naturally, is thus not required at all.

It is certainly impossible to realize a similar operation "manually" for a large number of enterprises with numerous financial obligations. Deeper approaches are required, first of all, with necessary formalization of the problem.

Let an economic system consist of N enterprises able to have mutual debts. Denote the debts of the n-th enterprise to the m-th one through x_{nm}, where $1 \leq n,\ m \leq N$ ($x_{nm} < 0$, if the first enterprise owes to the second, and $x_{nm} > 0$ in opposite case). It is clear that

$$x_{nm} = -x_{mn}, \qquad x_{nn} = 0\,,$$

i.e. the set of debts is described by a skew-symmetrical matrix $N \times N$ with zero diagonal ($x_{nn} = 0$, as the enterprise cannot owe to itself).

The sum of all mutual debts is calculated through the individual debts via the simple formula

$$X = \sum_{n=1}^{N} \sum_{m=1}^{N} |x_{nm}|\,. \tag{7}$$

The quantity (7) serves one of the integral quantitative characteristics of a financial situation of a system: if it is comparable to the sum of all free resources of the enterprises X_0, i.e.

$$X \geq X_0 = \sum_{n=1}^{N} x_n\,, \tag{8}$$

then the situations described by inequality (8) just imply a non-payment crisis (here $x_n \geq 0$ are the individual free money of cnterprises).

One more important characteristic is the balance of the credits and debts of each enterprise

$$S_n = \sum_{m=1}^{N} x_{nm}\,, \tag{9}$$

and, as is obvious from (9), versions $S_n > 0$, $S_n < 0$, $S_n = 0$ are possible. At $S_n > 0$ the enterprise in some sense is the creditor of the debtors, i.e. those whose $S_n < 0$ (at $S_n = 0$ the enterprise is "neutral" with respect to debts). At $|S_n| < x_n$ the individual financial situation of the enterprise is in fact normal, as its real total debts (or the credits "given" to others) are less than its free resources.

Similarly, the total absolute balance of system

$$S = \sum_{n=1}^{N} |S_n| \tag{10}$$

serves as a macroindex of its possible financial "health". If $S < X_0$, then the free resources in the system are greater than the actual debts, and potentially it can operate successfully (similarly to the system of three enterprises from the above mentioned example).

Between X and S there always exists certain correspondence. For an arbitrary matrix of debts the following inequality is fulfilled

$$X \geq S\,, \tag{11}$$

i.e. the total debt in any case cannot be less than the total balance.

The problem of repayment of mutual debts is as follows: knowing the matrix x_{nm} to find the matrix x'_{nm} "new" debts, so that the inequality $X' < X$ is fulfilled. It is obvious that ideal solution would be $X' = S$, when the inequality (11) turns to an equality. Note that then for a secure system with $S \leq X_0$ the relation $X' = S \leq X_0$ would be reached, and after mutual offset it could operate normally (though the reduction of quantity X in any case is useful).

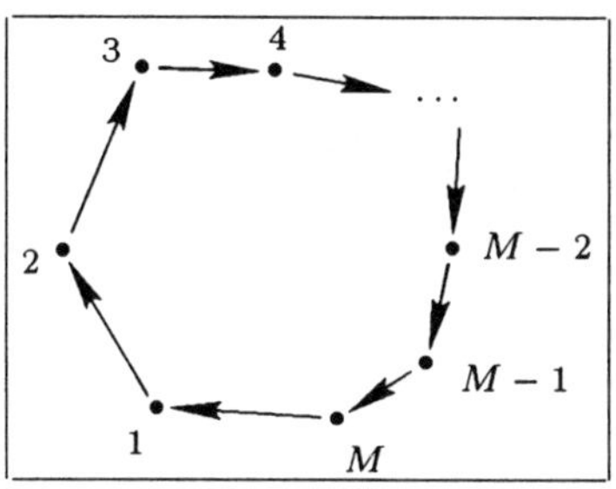

Fig.45.

In the construction of a mathematical model of the procedure of mutual offset of debts, a number of actions are used which are similar to those applied in the study of natural sciences objects. The first is in the refusal at certain stage of the detailed consideration of the set of individual debts and corresponding connections between the enterprises. The transition from microlevel to macrolevel is similar to the refusal to trace the trajectory of each particle and to introduce certain average characteristics in the description of a large number of particles of gas; the knowledge of the new quantities, however, is quite enough to map the detailed behavior of the object (see, for example, the deduction of Boltzmann equation in section 3, Chapter III). The procedure of tracing chains of nonpayment, applied above for the three enterprises, is not only hardly realizable for N enterprises, but has also a basic inadequacy. Indeed, consider first, a chain, where each enterprise from the first to the M-th ($M \leq N$) owes an identical sum to the other, and the same sum owes the M-th enterprise to the first one (Fig. 45). The chain is closed, and the solution is obvious – all debts in the chain are repaid. Let now the M-th enterprise owes nothing to the first (Fig. 46). Then the chain is open, and this method is inapplicable. At the same time the simple solution is that the debts of the enterprises from the second to the $(M-1)$-th are cancelled, and the debt of the first is readdressed to the M-th (Fig. 47). The economic content of readdressing corresponds to a bill manipulation, when the liability changes the owners, and as a result the debtor (the first enterprise) gets a new creditor (the M-th enterprise).

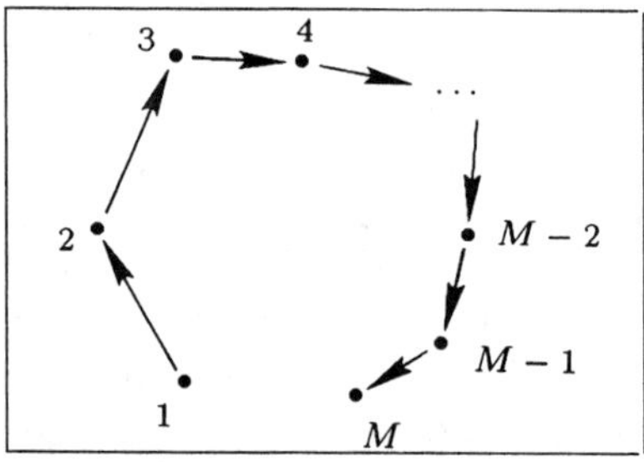

Fig.46.

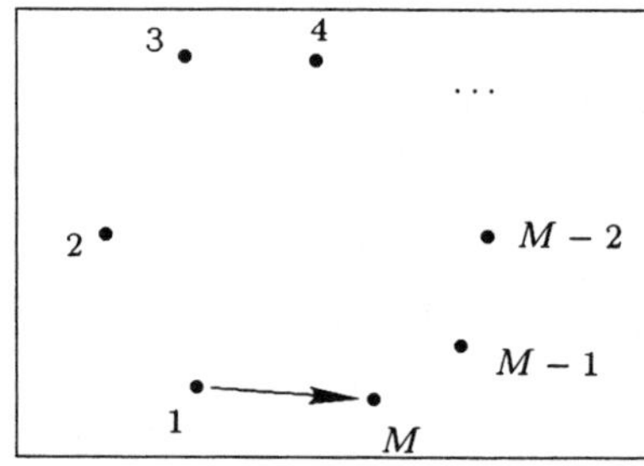

Fig.47.

As distinct from the situation with debts in chains the complete system of debts in all chains is closed, in so far as only the mutual debts are considered. Indeed, from the property $x_{nm} = -x_{mn}$ it follows that

$$\sum_{n=1}^{N} \sum_{m=1}^{N} x_{nm} = 0$$

for any set nonpayment. Taking into account that $S_n = \sum_{m=1}^{N} x_{nm}$, from the last equality we obtain

$$\sum_{n=1}^{N} S_n = 0\,, \tag{12}$$

or

$$\sum_{S_n > 0} S_n = - \sum_{S_n < 0} S_n = \frac{S}{2}\,, \tag{13}$$

i.e. the sum of positive balances of enterprises is equal in absolute amount to the sum of negative balances. Considered at the macrolevel the system of mutual debts has a property of "symmetrical conservatism" (13), while "the conservation law" (12) is an analog of usual conservation laws (mass, energy, etc.) concerning the investigated situation.

The equality (13) reveals the construction of a mathematical model of ideal mutual offset for the following natural conditions:

1) all debts x_{nm} are known and are admitted by the enterprises;

2) at realization of mutual offset the balances of enterprises S_n remains constant: $S'_n = S_n$, i.e. the individual financial position of each enterprise does not change in this sense;

3) a part of debts x_{nm} is withdrawn, another part is readdressed, enterprises can obtain new debtors and creditors while part of the old ones can disappear.

The essence of macroprocedure of mutual offset is that instead of x_{nm} the quantities S_n are considered. The enterprises with $S_n < 0$ are announced as debtors (within their balances), the enterprises with $S_n > 0$ are announced as creditors (within the same amounts). Then the debts of enterprises with $S_n < 0$ are somehow distributed between the creditors, i.e. a new system of debts x'_{nm} is found. In the meantime the conservation law (12) and condition 2) hold, and the equality $X' = S$ is fulfilled, hence the solution of the problem is optimal.

Generally speaking a lot of such optimal solutions may exist in so far as the distribution of debts between the creditors can be performed in different ways. We represent the two most simple and evident ones. The first of them is given by a simple formula, in which the new debts are calculated through the old ones

$$x'_{nm} = \frac{S_n|S_m| - S_m|S_n|}{S}\,. \tag{14}$$

According to algorithm (14), debts of any enterprise (equal to S_n, if $S_n < 0$) are assigned to the enterprises-creditors in fractions proportional to the amount of their balances (equal to S_m, if $S_m > 0$). The large part of the debts is assigned to the enterprises with a larger positive balance, and in total they equal S_m (exercise 5). For enterprises with zero balances the mutual offset is reduced to repayment of all their debts and all debts to them.

Note that in solution (14) for new debts we have $x'_{nm} = 0$ at $S_n < 0$, $S_m < 0$ or at $S_n > 0$, $S_m > 0$ (after mutual offset the debtors do not owe to the debtors, and creditors do not owe to the creditors). This means that the number of financial connections formed between the enterprises is significantly less than the largest one possible, when each enterprise is a debtor or creditor of any other, and the matrix of debts has no zero elements (except, certainly, diagonal ones).

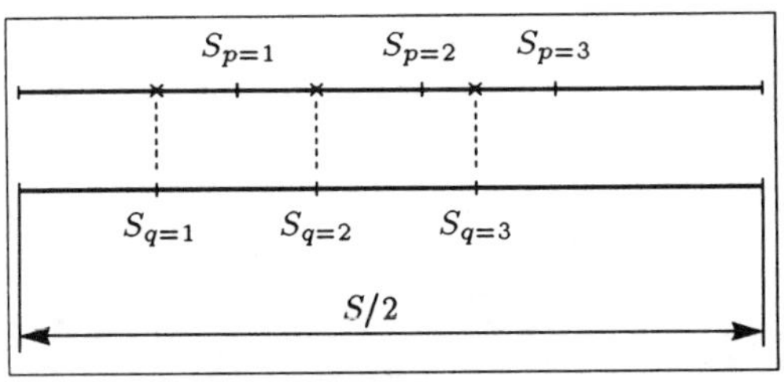

Fig.48.

The number of connections can be considerably reduced if a preliminary ordering of enterprises in accordance with the absolute values of their balances can be performed and direct connections between the debtors and creditors of same scale can be established (big debtors with big creditors, small debtors with small creditors, etc.). This procedure allows a simple geometrical interpretation. In Fig. 48 on the upper straight line, distribution of balances of the creditors (in decreasing order) is described. The length of intervals is equal to the amount of balance of each enterprise, $S_p > 0$, $1 < p < N$, and its total length, is obviously S/2. On the lower straight line the distribution of balances of the debtors $S_q < 0$, $l < q < N$ is described, $p + q \leq N$ (the balances are taken with opposite sign) again in decreasing order. Its length in accordance with (13) is also equal to S/2. The shaped lines passing through the nodes of the bottom line, divide "the line of creditors" into q pieces equal to the amount of the debt of each enterprise. This debt is either assigned to one creditor, or is divided between several of them according to the arrangement of the nodes of the upper line relative to the given interval.

Table 2

An example of mutual offset in a system with $N = 10$ and initial matrix of debts with 90 non-zero non-diagonal elements

	1	2	3	4	5	6	7	8	9
Initial matrix ($X = 3729$)									
2	-25								
3	-1	-20							
4	4	25	-2						
5	25	-450	25	30					
6	-15	150	-30	20	-928				
7	3	-40	3	3	5	25			
8	1	-22	-2	-2	4	-15	5		
9	10	322	-15	-25	498	-800	-10	20	
10	1	-25	-2	1	-20	15	-1	-3	30
Final matrix ($X' = S = 62$)									
2	2								
3	0	0							
4	0	0	0						
5	0	0	0	0					
6	0	0	0	0	-28				
7	1	0	0	0	0	0			
8	0	-7	0	0	0	0	0		
9	0	-18	0	0	-2	0	0	0	
10	0	0	0	0	0	4	0	0	0

The algorithm described is optimal by criterion $X' = S$ and might be the best according to the number of connections remaining after the mutual offset.

An example of similar mutual offset in a system with $N = 10$ and an initial matrix of debts with 90 non-zero non-diagonal elements is given in Table 2. The final matrix contains only 14 non-zero elements. In special cases one debtor still has one creditor, and vice versa (exercise 6).

Note that these and other procedures of mutual offset only make sense if conditions 1)–3) are fulfilled, i.e. in the presence of certain agreement between the enterprises. The reasons which are not allowing them to adhere the given agreement, can be quite different – from unwillingness to pay the debts since it is favorable to the debtor, up to consequences of sanctions by international or other organizations, when the financial resources of the enterprises are frozen. These circumstances determine the frameworks of applicability of the model of mutual offset, when analogies with models of some natural sciences objects were essentially used.

3. Macromodel of equilibrium of a market economy. Any participant in a market economic process acts according to its own individual interests (extraction of profit, improvement of working conditions, minimization of risk, economy of resources, etc.). The elementary version of such a system is the economy with perfect competition, when each subject is economically insignificant and has no direct influence on the level of production, prices, salary and other macroindices. At the same time, separate actions of economic agents can develop due to the relations existing in the system on the sale and purchase in the cumulative coordinated actions of the employers and hired workers, financiers and investors and so on.

If as a result of such collective interaction the common production of the goods and services in a system is coordinated with the common demand for them, this state of economy is called *equilibrium*, and the installed prices are called *equilibrium market prices.* The balance between supply and demand occurs, as it is intended, not at arbitrary but just at these market prices which implies in particular the payability of the demand.

One important problem of economic science is the definition of conditions of balance in economy, including the equilibrium market prices. The most simple mathematical models of economic balance are constructed under the following assumptions:

1) perfect market competition implying the absence of both large industrial corporations (and, especially, monopolies), and workers' associations, able to dictate the conditions for the whole system;

2) stationary industrial possibilities of the system: equipment, industrial areas and technologies do not change in time;

3) economic interests of the partners constant in time, which are the employers do not try to increase the profit, workers do not push for more the salary, the investors are satisfied with the percentages received for stocks,

etc.

Models satisfying this assumptions describe the rather special case of an ideal market economy "frozen" in time. However, they answer to a question about the possibility that there can exist an economic balance formed from market "chaos", and, besides, mutually connects the basic macroindices of the economic system.

One such macromodel – *Keynes' model* – considers agents of the employers and employees, consumers and savers, producers and investors working in the labor markets, products and money markets, distributing and exchanging these goods (labor, products, money) among themselves.

The first macroindex of a system is *the national income* Y, being for the sake of simplicity the only product made in a unit of time. This product is developed by the industrial sector of economy, and its amount is given by the function F, depending on the quantity and quality of resources, the structure of basic capital and *the number of engaged workers* R (second macroindex). According to the assumption 2) in equilibrium conditions the production function R, and also the product Y, are determined only by the employment, that is

$$Y = F(R). \tag{15}$$

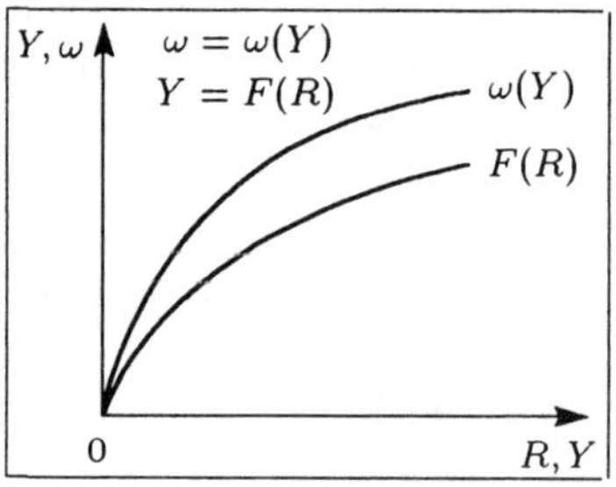

Fig.49.

With respect to $F(R)$ it is usually considered that $F(0) = 0$, $F'(R) > 0$, $R > 0$ and $F''(R) < 0$ with $R > 0$ (Fig. 49). The function $F(R)$ has a property of "saturation": as R grows the outcome grows slower. Such an approach is quite justified, since if many people are engaged in the production, there simply can be not enough work. One or several workers, having discovered a gold mine, will reach maximal productivity quickly and without problems; with a greater number of workers they will begin to disturb each other, and their individual productivity will decrease; finally, with a very large number of workers the production of gold will stop growing, since the newly arriving ones cannot reach the site.

Eq. (15) gives the connection between the labor markets (R) and product (Y) ones. The additional relation is determined with the help of one of basic postulates of classical political economy:

4) *the salary* s of a worker is equal to the cost of a product, which would be lost if employment was reduced by one unit (the salary is equal to a limiting labor product).

Note that in a postulate 4) other expenses are not taken into account, which would disappear as a result of reducing the work force by one (expenses for resources, equipment, etc.). Thus, from this postulate we obtain

$$\Delta Y^{(1)} \cdot p = s ,$$

where $\Delta Y^{(1)}$ is the amount of product lost with the reduction of employment per unit, p is *the price of a product* (thus, on the left hand side of this equality the amount of lost expenses is given). If the employment has changed on ΔR, from the latter equality we have

$$\Delta Y \cdot p = s \cdot \Delta R ,$$

where $\Delta Y = \Delta Y^{(1)} \Delta R$ are the expenses lost or gained with change in number of workers on ΔR. Considering ΔR and ΔY to be small in comparison with R and Y, we rewrite the latter equality in a differential form

$$\frac{dY}{dR} = \frac{s}{p} ,$$

or, considering (15),

$$F'(R) = \frac{s}{p} . \tag{16}$$

As $F(R)$ is given (and with it the function $F'(R)$), then at known macroindices s and p from (16) it is possible to find the employment level R, and from (15) an amount of a product Y. Recall: this level corresponds to the number of workers, who have agreed to work for the given salary and given prices and other characteristics of system, and not to the generally possible number of hired workers. It is assumed that to maintain an equilibrium occupation level there will be always enough people wishing to work in existing conditions, that is

5) the offer of labor does not constrain the production process, the number of people engaged is determined by demand for labor by employers.

Two equations – (15) and (16)– contain four variables. Concerning one of them, we assume that:

6) the salary s in the model is assumed given.

It is determined as a result of compromise between the employers and employees (the real salary depends also on the price level).

Obviously, to construct a closed model, the further study of the market for a product and of the financial market is necessary. The product is partially spent on consumption and is partially saved up:

$$Y = S + \omega ,$$

where ω is the *consumed part* (does not return into the economy), and S is the saved part which returns into the economic system (or a fund-forming product).

The ratio between quantities S and ω is determined from the following considerations. Concerning ω it is assumed that

7) the consumed part depends on the amount of production, i.e. $\omega = \omega(Y)$.

Thus the function $\omega(Y)$ has a property of "saturation", the same as function $F(R)$: the greater the production, the smaller fraction of additional production ΔY is spent for consumption (Fig. 49), and the greater is the fraction saved. The quantity $d\omega/dY = c(Y)$ is called *inclination to consumption* and is within limits $0 < c < 1$, otherwise at small production more product would be consumed than would be made ($d = 1 - c$ – inclination to accumulation).

Forming fund product

$$S = Y - \omega(Y) \tag{17}$$

is put by the investors into the economy with the purpose of gaining an income in future from these investments. It is considered that the investments are equivalent to the postponed consumption and consequently are determined by one more financial macroindex of system – by *norm of the bank percentage rate* r. Indeed, by making *investments* in an amount of A and by receiving in one year an income $D = Ar$, the investor loses nothing (in the given example he does not win either) in comparison with keeping these means in a bank with percentage rate r. In both cases the present consumption is postponed for the sake of an opportunity of greater consumption in the next year. The demand on the investment is given by function $A(r)$, such that $A'(r) < 0$ at $0 < r < r_1$ and $A(r) = 0$ at $r \geq r_1$: at large norm of percentage the investments are absent (Fig. 50).

In conditions of equilibrium the offer of a forming fund product $S(Y)$ is balanced with the demand on the investments $A(r)$

$$S(Y) = A(r),$$

or, taking into account (17),

$$Y - \omega(Y) = A(r). \tag{18}$$

To finally close the model we consider the financial market. The money is necessary to economic agents for purchase of the fund forming product, for consumption, and also as one of the means of accumulation. It is considered that the money is provided by the state, and their quantity (*offer*) Z is the given control parameter of the system. Concerning the demand for money the following assumption is made:

8) the demand for money represents the sum of operational and speculative demands.

Operational demand is determined by the amount of money required for the purchase of the goods Y (both for fund forming and for consumption). If the price of a product is p, and *the circulation time* is τ, then the operational demand is obviously equal to $\tau p Y$.

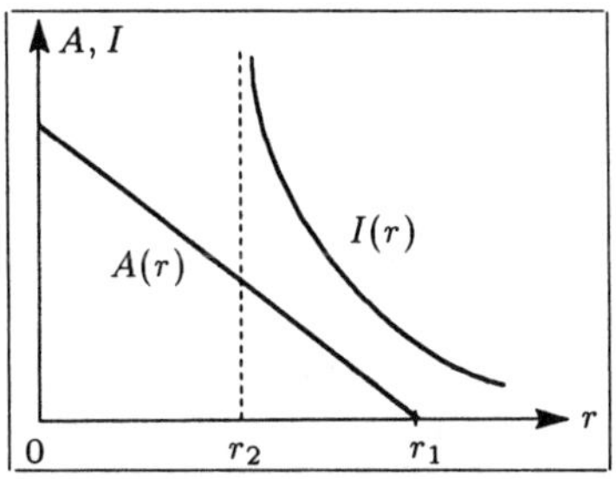

Fig.50.

The speculative demand is connected with the norm of percentage r. If the norm of percentage is high, the owners prefer to keep a large part of money in a bank, expecting a good income and refusing from higher liquid banknotes (in comparison with the bank obligations). At low interest rates, the speculative demand increases: the owners wish to keep ever more banknotes, accumulating their savings. Therefore the speculative demand is given by function $I(r)$ (Fig. 50), so that $I'(r) < 0$ at $r > r_2$ and $I(r)$ grows sharply at $r \to r_2$ ($\lim I(r) = \infty$, $r \to r_2$; the investors do not buy the obligations of bank). It is natural to consider $r_2 < r_1$, otherwise either the investments are equal to zero, and one cannot speak of an economic balance, or the function $I(r)$ is not determined, and the considerations are not meaningful.

In so far as the financial market is in equilibrium, the balance ("the conservation law") of money in the system is given by the equation

$$Z = \tau p Y + I(r). \tag{19}$$

Unifying the equations (15), (16), (18), (19), we come to the *mathematical model of market balance*, obtained through assumptions 1) –8)

$$\begin{aligned} &Y = F(R), \\ &F'(R) = s/p, \\ &Y - \omega(Y) = A(r), \\ &Z = \tau p Y + I(r). \end{aligned} \tag{20}$$

In model (20) the parameters of system s (salary unit), Z (offer of money) and a technical parameter τ are given. Functions F, F', ω, A, I are known functions of the arguments with the properties described above. By this input data the four unknown variables are determined: Y (production of a product), R (employment), p (price of a product) and r (income norm).

Excluding p, r, Y from (20), the equations (20) are easily reduced to a single equation for R

$$-\frac{\tau s F(R)}{F'(R)} + Z = I\left\{A^{-1}[F(R) - \omega(F(R))]\right\}, \tag{21}$$

where A^{-1} is an inverse function to A. Obtaining the values of R from (21), it is easy to determine all other required quantities from (20).

Now with the help of non-strict, but simple constructions we prove the existence of a solution to (21), based on the analysis of diagrams of functions which are included on its left and right hand sides.

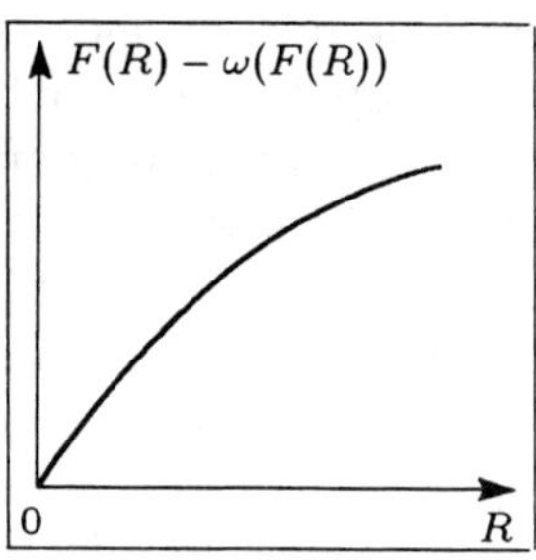

Fig.51.

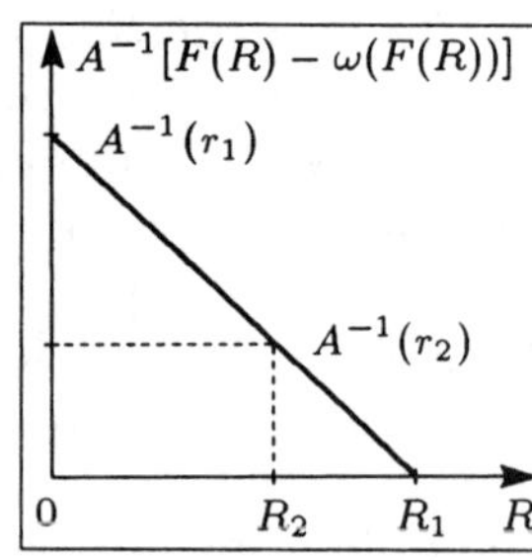

Fig.52.

The function $F(R) - \omega(F(R))$ is a monotonously growing function R, equal to zero at $R = 0$ (Fig. 51). Its monotony follows from a condition $d\omega(F(R))/d(F(R)) = c < 1$, and the growth via increase of R – from a condition $dF(R)/dR > 0$. The given function is an argument for monotonous function A^{-1}, and from the properties of function A (Fig. 50) it is easy to establish a qualitative dependence of A^{-1} on R (Fig. 52), and $A^{-1} \equiv 0$ at $R > R_1$ (R_1 is certain value of R, $0 < R_1 < \infty$). In turn A^{-1} serves as an argument for monotonous function I, with properties (Fig. 50), such that as function R has a form represented in Fig. 53 (for values $R > R_2$ the function I is not determined).

Consider now the left hand side of equation (21). The function $-\tau s F(R)/F'(R)$ is equal to zero at $R = 0$ (it is considered that $F'(0) \neq 0$); (see Fig. 49). Its first derivative by R, as follows from properties of functions $F'(R) > 0$, $F''(R) < 0$, is negative, i.e. it is monotonously decreasing (Fig. 54).

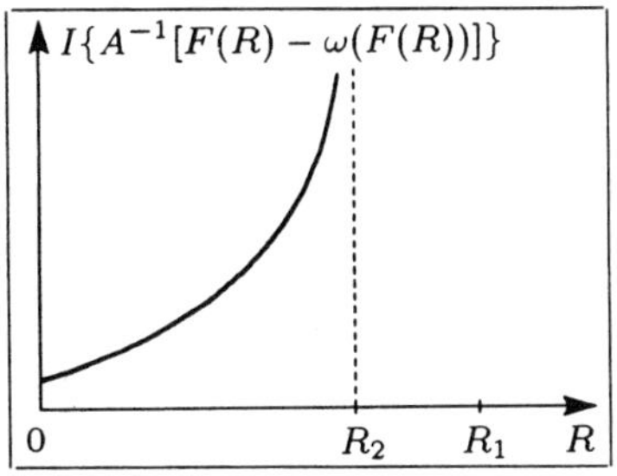

Fig.53.

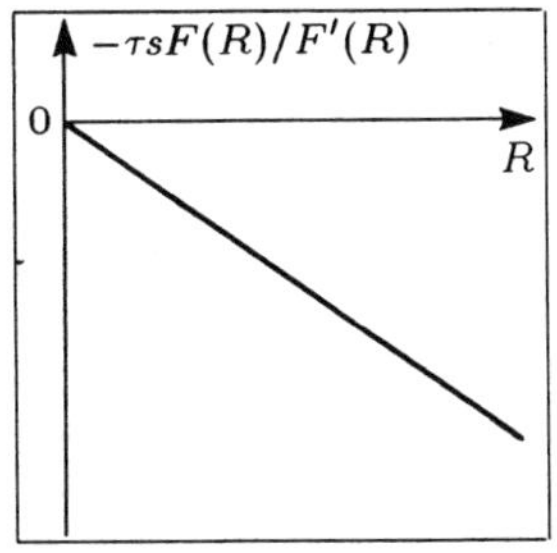

Fig.54.

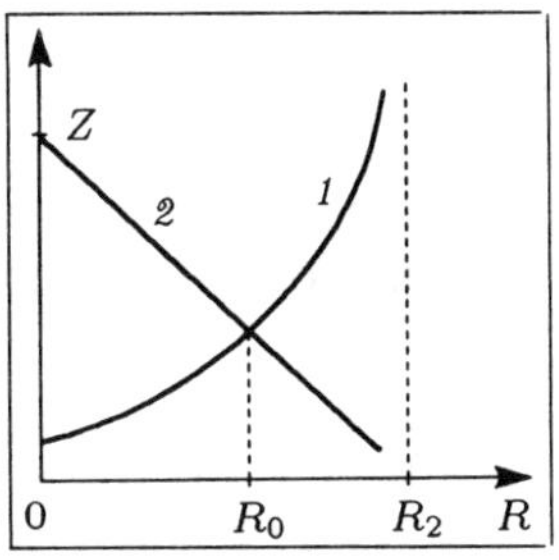

Fig.55.

Combining the diagrams of the left (curve 2) and right (curve 1) hand sides of equation (21) in Fig. 55, we are convinced that for an enough high value of control parameter Z the curves are intersecting in some point R_0, $0 < R_0 < \infty$. The point of crossing is singular by virtue of the monotony of the diagrams. Hence, the model (20) really has the only solution describing the equilibrium situation in economy.

However, the value of the model is not limited to this. It can be used for a comparative analysis of different but close equilibrium states (not answering, naturally, of the question how the system reaches that equilibrium or leaves it). Assume that the equilibrium parameters s_0 and Z_0 have changed by small values δs_0 and δZ_0 during the transition from one equilibrium state to another (parameter τ is considered constant). Then all other characteristics of the system will change as well. They can be found from (20), in view of the fact, that both compared states are equilibrium ones. For example, from the second equation (20), using Taylor expansion, we obtain

$$\frac{s_0}{p_0}\frac{\delta p}{p} = \frac{\delta s}{p_0} - F''(R_0)\,\delta R\,.$$

Performing a similar procedure with other equations of system (20) and unifying the results, we have

$$\frac{\delta p}{p_0} = a_1 (\delta A + \delta \omega) + a_2 (\delta Z - \delta I) + a_3 \delta s - a_4 \delta Y \,. \tag{22}$$

In (22) all characteristics of the investigated system are present (coefficients $a_i < 0$, $i = 1, \ldots, 4$, are determined by equilibrium values of s_0, p_0, Y_0, r_0, by functions R, ω, A, I and their derivative). Therefore, it is possible to analyze the whole complex of changes occurring during the transition from one equilibrium state to another (a so-called *system approach*). Let, for example, for a constant number of engaged workers (i.e. $\delta R = 0$, $\delta Y = 0$), a constant salary ($\delta s = 0$) and consumption level ($\delta \omega = 0$) be required to lower the price ($\delta p < 0$), i.e. to increase the real salary of workers. Then it is necessary to aspire to reduce the investments ($\delta A < 0$), to lower the total amount of money ($\delta Z < 0$) and to increase the speculative demand ($\delta I > 0$). Note that the requirements following from the analysis of relation (22), can generally be inconsistent.

Certainly, this and other systems of measures following the from constructed model, are not realized automatically by the corresponding variation of Z or s (or both). The model (20) only indicates necessary changes in behavior of the economic agents. The question of how to really ensure these changes, by convincing the participants of the marketing process to accept them, is beyond the scope of the considered model. Its decision is connected with the study of even more hardly formalizable objects. In researching similar objects, we have used approaches developed for natural scientific problems, such as the idea of saturation, transitions from micro to macro level, use of "conservation laws", concepts on stationary state and an equilibrium, etc.

4. Macromodel of economic growth. In a growing economy the number of workers $R(t)$ is not constant, and increases in time. In the simplest model it is considered that the rate of gain of the employed workers is proportional to the number of employees already working

$$\frac{dR}{dt} = \alpha R(t) \,.$$

Therefore $R(t) = R_0 e^{\alpha t}$ is a known function of time (α is given, $R_0 = R(0)$ is the number of workers in the initial moment $t = 0$). The workers produce the national income $Y(t)$, which is partially spent on consumption and partially on savings

$$Y(t) = \omega + A \,. \tag{23}$$

The saved part of a product A returns into the economy to compensate the withdrawal of industrial facilities, as well as to create new capacities.

Capacity $M(t)$ means the largest possible production by an economy. The actual production of goods naturally depends on the number of workers and is given by production function

$$Y(t) = M(t) \cdot f(x(t)). \tag{24}$$

In (24) the quantity $x(t) = R(t)/M(t)$ consisted of workers on unit power. Concerning the function $f(x)$ the following assumptions are made: $f(0) = 0$, $f' > 0$ (the production grows as the number of engaged workers increases) and $f'' < 0$ (saturation). The function $f(x)$ is determined for values of x on an interval $0 \leq x \leq x_M$, where $x_M = R_M/M$, while $R_M(t)$ is the number of workplaces in the economy with capacity $M(t)$. If all places are filled, the function $Y(t)$ by definition is equal to $M(t)$, i.e. for $f(x)$ the condition $f(x_M) = 1$ should be satisfied.

One of main problems of *the theories of economic growth* is finding the optimal ways of separating the produced goods, the consumed and saved parts. As a criterion of optimality one can choose, for example, the consumption per person (amount of product consumed by one worker), i.e. $c(t) = \omega(t)/R(t)$.

The product $A(t)$ saved in a unit of time is spent on the creation of a *new capacity*

$$A(t) = a\, I(t),$$

where $a > 0$ is considered a given and constant quantity of fund forming, necessary for the creation of a unit of new capacity, $I(t)$ is the number of units of new capacity.

The rate of withdrawal of the existing capacity is assumed proportional to the capacity, to $\beta M(t)$, coefficient of withdrawing of $\beta > 0$ is set as a constant.

As a result for the variation of function $M(t)$ we obtain the following balance relation

$$\frac{dM}{dt} = I(t) - \beta\, M(t). \tag{25}$$

Equations (23)–(25) contain four unknown variables – $Y(t)$, $\omega(t)$, $M(t)$, $I(t)$. To close the model we assume that the rate of inserting new capacity is proportional to the already existing capacity: $I(t) = \gamma\, M(t)$, where $\gamma > 0$ (a quantity inverse to characteristic time of growth of capacity) is considered given and constant (naturally, $\gamma > \beta$). Then the solution of equation (25) is obtained easily

$$M(t) = M_0\, e^{(\gamma-\beta)t}, \tag{26}$$

and hence, all other unknown variables are determined as well.

We now analyze a simple but indicative example of economic growth, in which the capacity is increased in time at the same rate as the number of workers. Then, obviously, the following equality has to be fulfilled

$$\gamma - \beta = \alpha \,. \tag{27}$$

It also means that the function $Y(t)$ grows with the same rate (in so far as $f(x(t)) = f(x = R_0/M_0) = \text{const}$), the same is valid for functions $\omega(t)$, $I(t)$.

We shall obtain the number of workers and the ratio between consumption and accumulation, when the consumption per capita of workers is maximal. By definition

$$c(t) = \frac{\omega(t)}{R(t)} = \frac{Y(t) - A(t)}{R(t)} \,.$$

Taking into account that $Y(t) = M(t)\, f(x)$, $A(t) = a\gamma M(t)$, and considering (26), (27), we obtain

$$c(t) = c = \frac{f(x) - a\,(\alpha + \beta)}{x} \,, \tag{28}$$

i.e. per capita consumption is not changed in time. Its maximum, as is obvious from (28), is achieved at condition

$$\frac{dc}{dx} = \frac{d}{dx}\left[\frac{f(x) - a\,(\alpha + \beta)}{x}\right] = 0 \,,$$

which gives the equation for the sought x_m

$$x_m f'(x_m) - f(x_m) + a\,(\alpha + \beta) = 0 \,. \tag{29}$$

This equation always has the unique solution $0 < x_m \leq x_M$ (exercise 9). Note that besides all the assumption made, to realize a considered model of economic growth, it is necessary to coordinate the number of workers R_0 with capacity M_0 in the initial moment of time, so that $R_0/M_0 = x_m$.

The norm of accumulation ensuring the maximal value of c_m,

$$n_m = \frac{A_m}{Y_m} \,,$$

is obtained from equality $Y_m = M_m f(x_m)$, $A_m = a\gamma M_m$ and from (27), (29)

$$n_m = 1 - x_m \frac{f'(x_m)}{f(x_m)} \,, \tag{30}$$

and is called *a norm of Solow's golden rule of growth.*

If the condition (27) is not fulfilled, the modes of economic growth become more complicated, and the optimization of their characteristics will be performed within all the considered time interval. Recall that the constructed model and the models similar to it do not take into account changes of economic relations (they are considered constant) and operate mainly with basic technological connections, giving in particular the upper technological restrictions on the rate of economic growth (exercise 11). Analogies with natural scientific objects have also been widely used to derive these.

EXERCISES

1. Let the parameters α_1, N_0, p, s be fixed. Reveal the value of the parameter α_2, at which the function $P_m(\alpha_2)$ reaches its minimum, and prove that in this case current profit $P(t)$ is maximal at $t = 0$, i.e. at the start of the advertising campaign.

2. The inequality $P_m > \alpha_1 s$ (depending on $\alpha_1 > 0$ nonlinearly) is a necessary condition for profitable advertising. Considering as fixed the parameters α_2, N_0, p, s find the areas of values of parameter α_1, when at the growth of α_1 the given inequality is strengthened (is weakened).

3. Assuming that the effect of saturation from advertising occurs at $N(t) \approx N_0$, find from equation (6) the moments of time, when the continuation of campaign will become obviously unprofitable.

4. Prove the validity of inequalities (11).

5. Using the formulae (7), (9), (10), and with the help of equality (13), check that for the solution of (14) the property $x'_{nm} = -x'_{mn}$, condition 2) and the criterion of an optimality $X' = S$ are fulfilled.

6. Show that according to the described procedure the initial matrix of debts had a form

$$x_{nm} = \begin{vmatrix} 0 & +1 & \dots & +1 & \dots & +1 & +1 \\ -1 & 0 & \dots & +1 & \dots & +1 & +1 \\ & \dots & \ddots & \dots & & \dots & \\ -1 & -1 & \dots & 0 & \dots & +1 & +1 \\ & \dots & & \dots & \ddots & \dots & \\ -1 & -1 & \dots & -1 & \dots & 0 & +1 \\ -1 & -1 & \dots & -1 & \dots & -1 & 0 \end{vmatrix}$$

where N is even, and $X = N(N-1)$, has a form

$$x'_{nm} = \left| \begin{array}{cccccccc} 0 & 0 & \dots & 0 & 0 & \dots & 0 & 0 \\ 0 & 0 & \dots & 0 & 0 & \dots & N-3 & 0 \\ & \dots & \ddots & & \dots & & \dots & \\ 0 & 0 & \dots & 0 & +1 & \dots & 0 & 0 \\ 0 & 0 & \dots & -1 & 0 & \dots & 0 & 0 \\ & \dots & & & \dots & \ddots & \dots & \\ 0 & -(N-3) & \dots & 0 & 0 & \dots & 0 & 0 \\ -(N-1) & 0 & \dots & 0 & 0 & \dots & 0 & 0 \end{array} \right|$$

so that $X' = N^2/2 = S < X$.

7. The expression $\Delta Y = \Delta A/(1-c)$, where ΔA is the variation in the level of investments, and ΔY is the corresponding gain (loss) of production, is called *a relation of Keynes multiplier*. Derive it from model (20).

8. Obtain the expressions for coefficients a_i, $i = 1, ..., 4$ in (22) and prove that all of them are positive.

9. Using the properties of function $f(x)$, prove the existence of a single solution of equation (29).

10. By means of the considerations applied at the derivation of (30), show that for the value of the maximal per capita consumption c_m the equality $c_m \doteq f'(x_m)$ is valid (the value $f'(x_m)$) called *limiting labor productivity* is the gain of production at the increase of unit employment).

11. Show that if all production is left to accumulate and the capacities are loaded completely, the economic growth is realized with rate $Y(t) = Y_0te^{\gamma m^t}$, $\gamma = 1/a - \beta$.

3 Some Rivalry Models

We will now construct models of various types of rivalry – two-kind struggle in populations, arms race, military operations. We will show the generality of methodological approaches used to derive and analyze these models.

1. Mutual relations in the system "predator – victim". Strictly speaking, these relations (as well as the relations in a similar system "parasite – host") cannot be called rivalry. The "rivalry" of victims with a predator is expressed in the change of number of victims, which in turn influences the number of predators themselves. Indeed, any organism (especially population) does not live in isolation, but interacts with its environment. The widespread type of interaction is the use by one living organism (animals, birds, fishes, insects) of the others organisms as a food.

The mathematical model of the simplest, two-kind "predator–victim" system is based on the following assumptions:

1) the number of populations of victims N and predators M depend only on time (point model not taking into account the spatial distribution of population in an occupied territory; compare with the model of a community of amoebas in section 1);

2) in the absence of interaction the number of kinds is changed by the Malthus model from subsection 3, section 1, Chapter I; thus the number of victims is increased, while the number of predators falls, as they have nothing for food:

$$\frac{dN}{dt} = \alpha N\,, \qquad \frac{dM}{dt} = -\beta M, \qquad \alpha > 0, \quad \beta > 0\,;$$

3) the natural mortality of victims and the natural birth rate of predators are considered insignificant;

4) the effect of saturation of both populations is not taken into account;

5) the growth rate of the number of victims decreases proportionally to the number of predators, i.e. to quantity cM, $c > 0$, while the rate of growth of predators is increased proportionally to the number of victims, i.e. to dN, $d > 0$.

Unifying the assumptions 1) –5), we come to a system of *Lotki-Volterra equations*

$$\begin{aligned} \frac{dN}{dt} &= (\alpha - cM)\,N\,, \\ \frac{dM}{dt} &= (-\beta + dN)\,M\,, \end{aligned} \tag{1}$$

From here via the initial number $N(0) = N(t = 0)$, $M(0) = M(t = 0)$ we determine the population at any moment $t > 0$.

The nonlinear system (1) is convenient for investigations in a plane of variables N, M; we divide the first equation by the second

$$\frac{dN}{dM} = \frac{(\alpha - cM)\,N}{(-\beta + dN)\,M}\,. \tag{2}$$

Equations (1), (2) have the equilibrium state (or a stationary, non time-dependent solution)

$$M_0 = \frac{\alpha}{c}\,, \qquad N_0 = \frac{\beta}{d}\,. \tag{3}$$

We are interested in the stability of equilibrium state (3). This means the following. If the initial numbers are precisely equal to values (3), then how do they vary in time? If for any reason the numbers depart a little from the values M_0, N_0, will the system return to the equilibrium state? Finally, if the initial values $N(0)$, $M(0)$ essentially differ from the equilibrium ones, how they vary in time with respect the values N_0, M_0?

To understand the temporal dynamics of functions $N(t)$, $M(t)$, we shall transform equation (2) to a form

$$dN\,(-\beta + dN)\,M = dM\,(\alpha - cM)\,N\,,$$

and divide both parts of the resulting equality by NM, transferring all members to the left hand side

$$\beta\,\frac{dN}{N} - d\,dN + \alpha\,\frac{dM}{N} - c\,dM = 0\,. \tag{4}$$

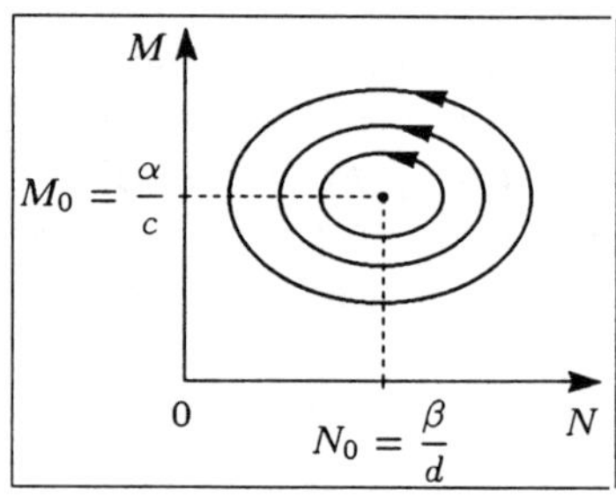

Fig.56.

Equation (4) can be easily integrated and the following relation can be obtained

$$\beta\ln N - dN + \alpha\ln M - cM = \text{const}\,,$$

where the constant on the right hand side is determined by initial values $N(0)$, $M(0)$. In other words, equation (2), and hence, system (1) have an integral

$$\ln N^{\beta} + \ln e^{-dN} + \ln M^{\alpha} + \ln e^{-cN} = C\,.$$

From the latter expression we obtain

$$N^{\beta}e^{-dN} = C_1 M^{-\alpha}e^{cM}\,, \qquad C_1 > 0\,. \tag{5}$$

The existence of integral (5) enables one to answer the posed questions (in Fig. 56 the phase trajectories of system (1) are represented; the arrows show the direction of motion by trajectories in time).

a) If $N(0) = N_0$, $M(0) = M_0$, then at all moments in time the number of population does not vary;

b) For a small variation from the equilibrium, both the numbers of predators, and of victims do not return to their equilibrium values (thus from

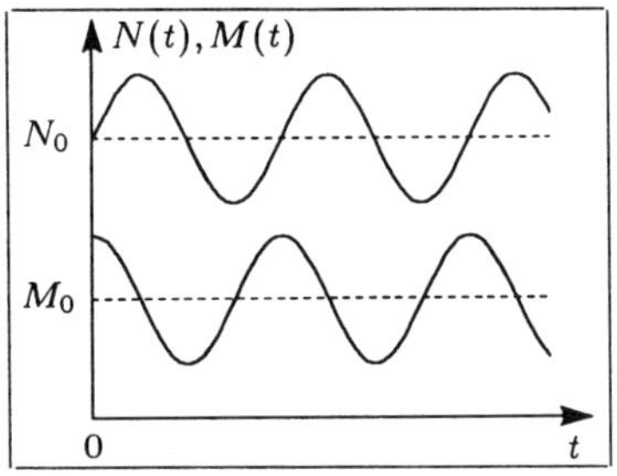

Fig.57.

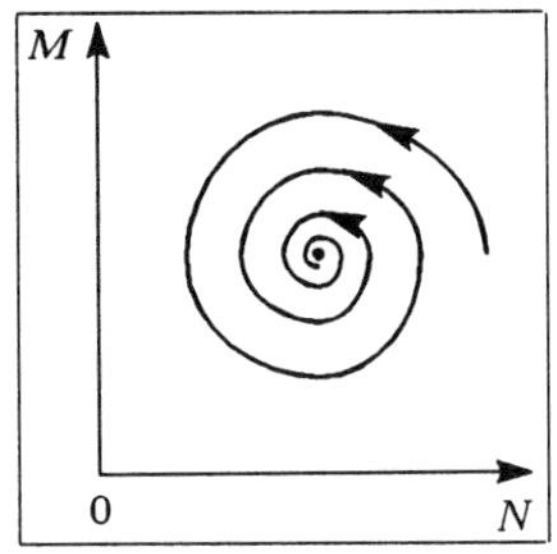

Fig.58.

model (1) the standard equation of oscillations is obtained; see subsection 3, section 5, Chapter I);

c) If the variation from equilibrium is large, the behavior of functions $N(t)$, $M(t)$ is the same as in the case b).

These conclusions mean that the numbers of victims and predators perform periodic oscillations with respect to the equilibrium state. The amplitude of oscillations and their period are determined by initial values; $N(0)$, $M(0)$ (exercise 1, 2), they are not in phase: the minimal value of $M(t)$ corresponds to the average value of $N(t)$, and vice versa (Fig. 57). Oscillations with quite understandable content (and they are really observed in nature) mean the appearance in two-kind population systems of much more complex processes than in single-kind systems (compare with the Malthus and logistic model in Chapter I).

More precise mathematical descriptions of two-kind interactions take into account the non-uniformity of distribution of populations in occupied territories (systems of equations in partial derivatives correspond to them), the temporary delay between the birth of individuals and their maturity and so on. There are much more complex descriptions both in time, and in space. For example, taking into account the saturation of the number of victims in the first equation of (1), we have

$$\frac{dN}{dt} = (\alpha - cM - aN)\,N\,. \tag{6}$$

In this case the phase trajectories look like spirals (Fig. 58), converging in time to an equilibrium state, and the amplitude of oscillations decreases in time (exercise 3).

2. Arms race between two countries. Assume that the total amount of arms of each country changes in time depending on three effects: amount of the opponent's arms, aging of already existing arms and the degree of mistrust between the opponents. The rates of gain and reduction of arms are proportional to the factors mentioned, that is

$$\begin{aligned} \frac{dM_1}{dt} &= \alpha_1(t)\, M_2 - \beta_1(t)\, M_1 + \gamma_1(t)\,, \\ \frac{dM_2}{dt} &= \alpha_2(t)\, M_1 - \beta_2(t)\, M_2 + \gamma_2(t)\,. \end{aligned} \tag{7}$$

In equations (7) $M_1(t) \geq 0$, $M_2(t) \geq 0$ are the amounts of arms, the coefficients $\alpha_1(t) > 0$, $\alpha_2(t) > 0$, $\beta_1(t > 0$, $\beta_2(t) > 0$ characterize the rates of growth and "aging" of arms (analog of process of amortization of industrial capacities in economic models), functions $\gamma_1(t) \geq 0$, $\gamma_2(t) \geq 0$ describes the level of mutual mistrust of the competitors, which is considered independent of the amount of arms, but is determined by other reasons.

The Richardson's model (7) does not take into account many important effects influencing the dynamics of an arms race, but, nevertheless, enables us to analyze a number of essential properties of this process. The analysis is especially simple in a particular case, when the functions α_i, β_i, γ_i, $i = 1, 2$, do not depend on time

$$\begin{aligned} \frac{dM_1}{dt} &= \alpha_1 M_2 - \beta_1 M_1 + \gamma_1\,, \\ \frac{dM_2}{dt} &= \alpha_2 M_1 - \beta_2 M_2 + \gamma_2\,, \end{aligned} \tag{8}$$

We study the system (8) in a plane M_1, M_2 with the purpose of determining the qualitative behavior of functions $M_1(t)$, $M_2(t)$ in time. The equations (8) have an equilibrium state at $dM_1/dt = 0$ and $dM_2/dt = 0$. The equilibrium values M_1^0, M_2^0 are obviously obtained from the conditions

$$\alpha_1\, M_2 - \beta_1\, M_1 + \gamma_1 = 0\,, \qquad \alpha_2\, M_1 - \beta_2\, M_2 + \gamma_2 = 0.$$

and are equal to

$$M_1^0 = \frac{\alpha_1\gamma_2 + \beta_2\gamma_1}{\beta_1\beta_2 - \alpha_1\alpha_2}\,, \qquad M_2^0 = \frac{\alpha_2\gamma_1 + \beta_1\gamma_2}{\beta_1\beta_2 - \alpha_1\alpha_2}\,. \tag{9}$$

From (9) follows the first important conclusion: to have an equilibrium for positive values of M_1^0, M_2^0 (by their content the function $M_1(t)$, $M_2(t)$ are non-negative), the following inequality should be fulfilled

$$\beta_1\, \beta_2 > \alpha_1\, \alpha_2\,. \tag{10}$$

The meaning of condition (10) is revealed from the following considerations. Let, for example, the parameters α_1, β_1 and β_2 be constant, and

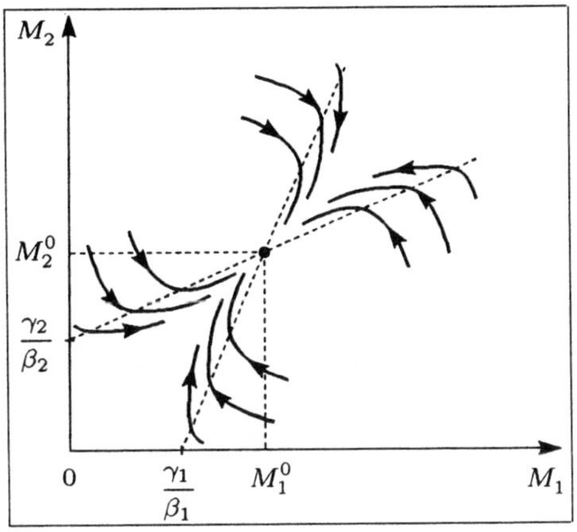

Fig.59.

parameter α_2 be increasing. This means that the first country does not change its strategy concerning arms, while the second increases the arms with constant rate of aging of weapons (parameter β_2). Then for a rather large value of α_2 the balance will obviously become impossible, and the inequality (10) be violated. Note that if both parameters γ_1, γ_2, describing the mutual mistrust are equal to zero, the balance implies the absence of arms for both parties.

We now study the stability of balance (9) at condition (10). In this case the integral curves of equations (8) in plane M_1, M_2 look like those represented in Fig. 59. The dashed lines indicate the isoclines of zero ($M_2 = \alpha_2/\beta_2\, M_1 + \gamma_2/\beta_2$) and infinity ($M_2 = \beta_1/\alpha_1\, M_1 - \gamma_1/\alpha_1$). The isocline of zero has an inclination which is smaller than the isocline of infinity (it follows from inequality (10)). The continuous lines correspond to integral curves. The arrows show the direction of motion on integral curves in time. The functions $M_1(t)$ and $M_2(t)$ with increase of t tend to their equilibrium values. Thus, the equilibrium is stable: any variation from it becomes negligibly small in sufficiently large time scales (exercises 4, 5).

From the constructed model it is easy to determine some characteristics of possible behavior of the competitors during the transition from one state of balance to another. Let, for example, the escalating rate of arms in the first and second countries change by small amount $d\alpha$ ($d\alpha = d\alpha_1 = d\alpha_2$). Thus, the amount of arms changes as well, and both parties wish that the increase dM_1^0 and dM_2^0 were equal and the interests of both parties were not suppressed. For dM_1^0, dM_2^0 from (9) we obtain

$$dM_1^0 = \frac{\alpha_1\alpha_2\gamma_2 + \alpha_2\beta_2\gamma_1 + \alpha_1^2\gamma_2 + \alpha_1\beta_2\gamma_1}{(\beta_1\beta_2 - \alpha_1\alpha_2)^2}\, d\alpha\,,$$

$$dM_2^0 = \frac{\alpha_1\alpha_2\gamma_1 + \alpha_1\beta_1\gamma_2 + \alpha_2^2\gamma_1 + \alpha_2\beta_1\gamma_2}{(\beta_1\beta_2 - \alpha_1\alpha_2)^2}\, d\alpha\,.$$

Assume for simplicity that the mistrust of the partners is equal ($\gamma_1 = \gamma_2$). Then from equality $dM_1^0 = dM_2^0$ we obtain a condition of parity of the parties at small change of balance

$$\alpha_1\,(\alpha_1 + \beta_2 - \beta_1) = \alpha_2\,(\alpha_2 + \beta_1 - \beta_2)\,,$$

which can be put at the basis of corresponding negotiations between the countries, if the quantities α_1, α_2, β_1, β_2 are known. Thus, for example, let $\alpha_2 = \sigma\alpha_1$, $\sigma > 0$. In this case, from the previous equality we have

$$\alpha_1\,(1 - \sigma) = \beta_1 - \beta_2\,. \tag{11}$$

At $\sigma < 1$ (the rate at which the second party gains arms is less, than that of the first) for preservation of parity it is necessary that $\beta_2 < \beta_1$, i.e. for the second party (according to formula (11)) the rate of amortization of arms should be less. At the opposite inequality $\sigma > 1$ we have, naturally, an inverse ratio between the rates of amortization.

3. Military operations of two armies. Both regular armies and partisan groups can take part in a conflict. The main characteristics of the rivals in the considered models are the numbers of the parties $N_1(t) \geq 0$ and $N_2(t) \geq 0$. If in any moment of time one of the quantities is zero, the given party is considered defeated (when in that moment the number of the other party is positive).

In case of actions between regular armies the dynamics of their number is determined by three factors:

1) rate at which people are lost due to reasons not directly associated with military operations (illness, trauma, desertion);

2) rate of losses caused by military operations of the the rival party (which in turn are determined by the quality of its strategy and tactics, level of morale spirit and professionalism of fighters, weapons, etc.);

3) speed at which reinforcements arrive, considered as certain given the function of time.

At these assumptions for $N_1(t)$, $N_2(t)$ we obtain the system of equations

$$\begin{aligned}
\frac{dN_1}{dt} &= -\alpha_1\,(t)\,N_1 - \beta_2\,(t)\,N_2 + \gamma_1\,(t)\,,\\
\frac{dN_2}{dt} &= -\alpha_2\,(t)\,N_2 - \beta_1\,(t)\,N_1 + \gamma_2\,(t)\,.
\end{aligned} \tag{12}$$

From here, for the given functions α_i, β_i, γ_i, $i = 1, 2$, and initial values $N_1(0)$, $N_2(0)$ the solution is determined uniquely at any moment of time

$t > 0$. In (12) the coefficients $\alpha_{1,2} \geq 0$ characterize the rates of losses due to usual (non-military) reasons, $\beta_{1,2} \geq 0$ are the rates of losses due to rival actions, $\gamma_{1,2}$ is the speed with which reinforcements arrive.

The war between regular and partisan formations is described by another model. The main difference is that the partisan formations are less vulnerable than the army, since they are acting hidden, often invisible to the rival, forced to act without much choice on the areas occupied by partisans. Therefore it is considered that the rate of losses of partisans acting in different sites on a known territory is proportional not only to the population of the army $N_1(t)$, but also to that of the partisans $N_2(t)$, i.e. is determined by a term like $\beta_1(t)\, N_1\, N_2$. Hence, the model becomes nonlinear

$$\begin{aligned} \frac{dN_1}{dt} &= -\alpha_1\,(t)\,N_1 - \beta_2\,(t)\,N_2 + \gamma_1\,(t)\,, \\ \frac{dN_2}{dt} &= -\alpha_2\,(t)\,N_2 - \beta_1\,(t)\,N_1 N_2 + \gamma_2\,(t)\,. \end{aligned} \tag{13}$$

All variables in (13) have the same content as in (12). Consider the models (12), (13) (*Lanchester's model*) in a particular case: $\gamma_1 = \gamma_2 = 0$ (the parties do not gain reinforcements); $\alpha_1 = \text{const}$, $\alpha_2 = \text{const}$; $\beta_1 = \text{const}$, $\beta_2 = \text{const}$ (the latter means, in particular, that the rivals always will have enough weapons for fighters to fit the military needs).

The model (12) becomes autonomous and has a form

$$\begin{aligned} \frac{dN_1}{dt} &= -\alpha_1\,N_1 - \beta_2\,(t)\,N_2\,, \\ \frac{dN_2}{dt} &= -\alpha_2\,N_2 - \beta_1\,N_1\,. \end{aligned} \tag{14}$$

From equations (14) it is seen that in this case the number of fighters of the parties can only decrease in time. Which is the temporal character of this process and which party will be defeated? To solve this problem, we introduce one more simplification (quite justified for short-term campaigns): assume that $\alpha_1 = \alpha_2 = 0$. In other words, the losses of parties are determined only by actions of the opponent. Then the system (14) becomes simpler

$$\frac{dN_1}{dt} = -\beta_2\,N_2, \qquad \frac{dN_2}{dt} = -\beta_1\,N_1, \tag{15}$$

and its integral is obtained easily

$$\beta_1\,N_1^2(t) - \beta_2\,N_2^2(t) = \beta_1\,N_1^2(0) - \beta_2\,N_2^2(0) = C\,. \tag{16}$$

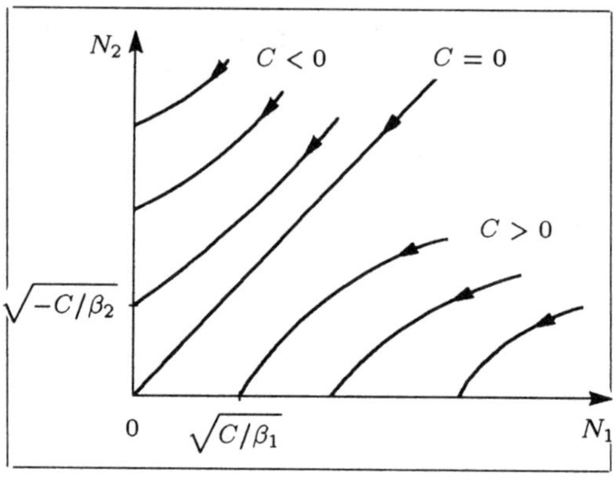

Fig.60.

From (16) the winner (Fig. 60) is uniquely determined. At $C > 0$ the first army wins, at $C < 0$ the second wins in the case $C = 0$ where the parties destroy each other simultaneously, and no winner exists. The sense of these results is quite clear from the form of the constant in (16). To secure victory, not only the number of the parties at the beginning of military actions is important ($N_1(0)$, $N_2(0)$), but also their training, the quality of their arms and so on (i.e. the coefficients β_1, β_2). So, if $C > 0$, from (16) follows

$$\beta_1 N_1^2(0) > \beta_2 N_z^2(0),$$

and for the second party to win it is necessary for them either to increase their initial number, or to improve the quality of military actions, or both them simultaneously. Note that the effect of increasing β_2 is less, than that of equally increasing $N_2(0)$ presenting in the latter inequality in the second order (a so-called *square-law law of battle actions*).

Differentiating the first of the equations (15), taking into account the second, we obtain the equation for $N_1(t)$

$$\frac{d^2 N_1}{dt^2} = \beta_1 \beta_2 N_1 . \tag{17}$$

From (17), in view of the initial conditions $N_1(t = 0) = N_1(0)$ and $dN_1/dt(t = 0) = -\beta_2 N_2(0)$, we find the number of the first army as a function of time

$$N_1(t) = N_1(0)\,\mathrm{ch}\,\sqrt{\beta_1\beta_2}\,t - N_2(0)\sqrt{\beta_2/\beta_1}\;\mathrm{sh}\,\sqrt{\beta_1\beta_2}\,t\,, \tag{18}$$

whence it is also easy to find the function $N_2(t)$ (exercise 7).

Consider now the actions of the regular army against the partisans for the same simplifying assumptions as in the previous case: then, the model

(13) has a form

$$\frac{dN_1}{dt} = -\beta_2 N_2, \qquad \frac{dN_2}{dt} = -\beta_1 N_1 N_2 . \tag{19}$$

The number of parties, as previously, decreases in time, but by another law. Multiply the first equation in (19) by $\beta_1 N_1$, and the second by β_2, and subtract the second equation from the first. In a result we come to equation

$$\frac{d}{dt}\left[\frac{\beta_1}{2} N_1^2(t) - \beta_2 N_2(t)\right] = 0 ,$$

with an integral,

$$\frac{\beta_1}{2} N_1^2(t) - \beta_2 N_2(t) = \frac{\beta_1}{2} N_1^2(0) - \beta_2 N_2(0) = C_1 . \tag{20}$$

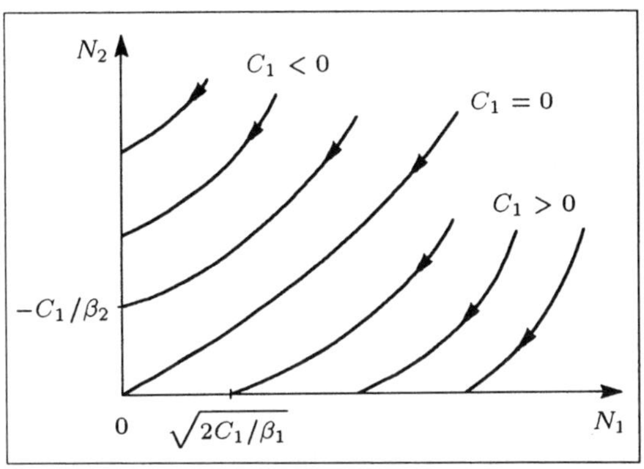

Fig.61.

We now study the phase trajectories of system (19) with the help of integral (20). From Fig. 61. it is seen that at $C_1 > 0$ the army wins, at $C_1 < 0$ the partisans win, and at $C_1 = 0$ there is no winner. Similarly, as with action of regular armies, the victory is reached not only due to the initial number, but also due to military training and quality of weapons. Let, for example, $C_1 > 0$, that is

$$\frac{\beta_1}{2} N_1^2(0) > \beta_2 N_2(0) . \tag{21}$$

Then the partisans have to ensure the increase of coefficient β_2 and to raise the initial number $N_2(0)$ correspondingly, otherwise they will be defeated. This increase with the growth of $N_1(0)$ should not be linear, but proportional to the second order $N_1(0)$ (*parabolic law of battle actions*). One can say that in some sense the regular army is in a more favorable situation, since the inequality (21) for them is fulfilled for a smaller growth of initial number, than the inequality opposite to (21) for the number of partisans.

The behavior of functions $N_1(t)$, $N_2(t)$ in time is obtained from (19) using integral (20). So, for $N_1(t)$ we have

$$\frac{dN_1}{dt} = C_1 - \frac{\beta_1}{2} N_1^2 ,$$

which is equivalent to

$$\frac{dN_1}{C_1 - \beta_1 N_1^2/2} = dt . \tag{22}$$

Integrating (22), it is easy to find $N_1(t)$ and then $N_2(t)$, as implicit functions of time (exercise 7).

In summary we note that the simplest models of rivalry considered here correspond to the systems of ordinary second order differential equations (in general, non-autonomous and nonlinear equations), which are widely used to describe numerous natural scientific objects. It is natural, as the approaches used to construct the models (saturation, proportionality of rates of growth of a quantity to its value, etc.) are similar to the approaches used in mechanics, physics, chemistry.

EXERCISES

1. Show, by linearizing the system (1) in the vicinity of equilibrium (3), that the singular point corresponding to it is a center.
2. Calculate the period of oscillations in the system "predator – victim" depending on its characteristics (α, β, c, d) and the initial condition.
3. Prove that a linearized system of equations consisting of equation (6) and the second equation in (1), that the equilibrium point is a stable node.
4. Show that the equilibrium point (9) is a stable node.
5. Find an evaluation for the time scale of arrival of system (8) at equilibrium within 1% to an accuracy of depending on its characteristics and the initial deviation.
6. Establish the character of curvature of the phase trajectories in Fig. 60.
7. Find the function $N_2(t)$, using either equation (15), or the integral (16), and compare it with function $N_1(t)$ from (18). Prove that the conclusions drawn on the basis of analyzing the integral (16), are correct (i.e. at $C > 0$ the first party wins, at $C > 0$ the second party wins).
8. Find the solution of equation (22) and determine the moment at which the function $N_1(t)$ turns to zero (when, naturally, $C_1 < 0$). Establish the character of curvature of the phase trajectories in Fig. 61.

4 Dynamics of Distribution of Power in Hierarchy

We will now construct and study a macromodel describing some key interactions in the system "state power – civil society". Consider the influence of reaction of a civil society and other characteristics of system on the dynamics of distribution of power within the hierarchy, and also some properties of these distributions.

1. General statement of problem and terminology. The study of power structures is one of the main problems of sciences of society, first and foremost of political science. The concept of "power" is mysterious and multi-faceted, and is hardly formalizable and quantitatively measurable. It is natural therefore, that the mathematical models of political science are mainly descriptive and phenomenological and are applied to monitor a rather narrow circle of problems: statistical processing of expectations and results in elections definition; of ratings of various political forces; forecasts for future parliamentary voting on a current basis, etc.

When constructing of mathematical models of general political science, it is reasonable to start by studying just state hierarchies having power on official, easily formalizable bases. This is the important difference between state authority and the authority of independent mass media information (MMI), intellectual and moral authorities and other kinds of authority.

Hierarchy, or *hierarchy structure*, means the set of institutes ordered by seniority (hierarchy levels, positions, posts, grades, etc.), possessing powers on behalf of the state (i.e. by Constitution, laws, charters, decrees, rules, instructions, etc.).

We mean not only federal ministries, but also interregional, regional and local bodies officially having the appropriate power. The word "hierarchical" emphasizes that an order of subordination is determined inside the structure. Each of the parts (except the highest) has the seniors, giving orders and (except the lowest link) subordinates which are realizing the orders, outgoing both from the given level, and from other senior parts. Certainly, the orders move only from the seniors to the subordinates.

The civil society is part of a society with no direct state power. This includes the citizens (including the government officials beyond the framework of their official duties) and their various associations (political, cultural, professional), family and private enterprises, etc. Obviously, the members of a civil society cannot give orders on behalf of the state either to each other, or to any part of the power structure. For example, a non-state corporation, however large it may be, has no official right to force anybody to a certain behavior. At the same time any level of hierarchy has such a possibility with respect to certain part of a civil society, and some – with respect to the whole society. Note, that the same "physical person" can simultaneously belong both to the power structure, and to a civil society.

The reaction of society is the response (positive, negative or neutral) of the civil society to the actions of certain institutes of power (by means of elections, referendums, plebiscites, through MMI, interrogations of public opinion, meetings, strikes and so on).

The concept "reaction of hierarchy" is introduced into the model as well; the content of this, as well as the term "distribution of power" and other conceptual details, are explained below.

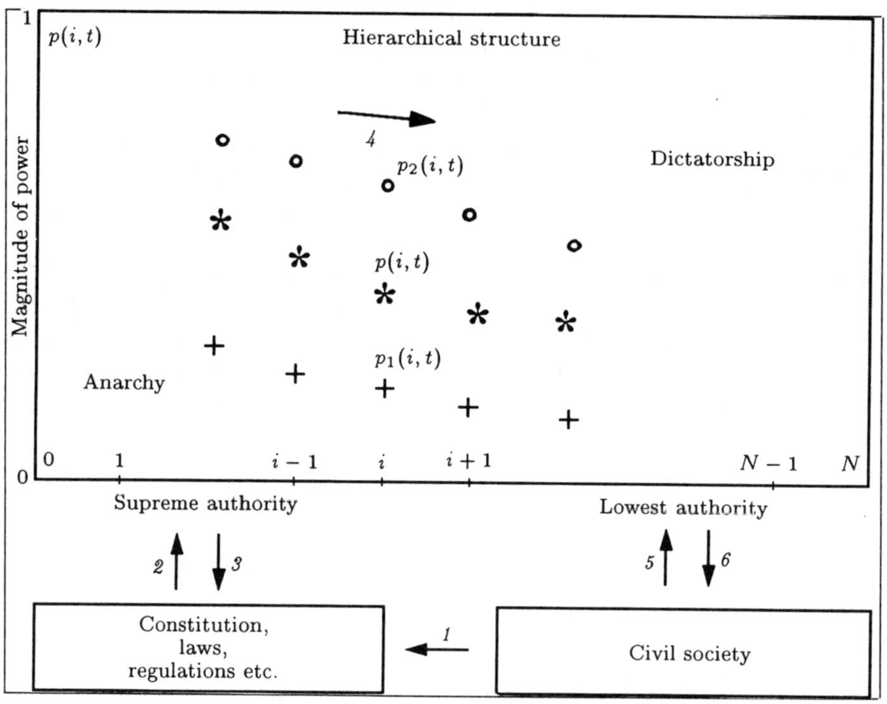

Fig.62.

The hierarchical structure consists of $N+1$ levels (Fig. 62), each with an attributed number i $(0 \leq i \leq N)$. The highest level has a number 0, the lowest has the number N. The arrow *4* denotes the direction of subordination within the structure "from above downwards" (or "from left to right"), i.e. the direction of movement of instructions of power (orders) transmitted via the hierarchical ladder.

Certainly, for any particular and extended hierarchical structure (especially in scales of a country, even of a small one) the determination of the arrangement of all components by subordination is a very difficult and labor-consuming task. It is also due to the fact that power bodies have in reality "a tree" structure, can belong to various administrative departments and can be either personalities or commissions, etc. The concept of "subordination" has therefore a conditional averaged sense.

However to construct a mathematical model and analyze the fundamental properties of imperious structures it is enough, that this problem can be solved basically, however, on the basis of some quantitative criteria. To clarify this question it is necessary to specify the various understandings of the word *power*.

It is frequently used in the sense *bodies of power* (Supreme Court, municipal assembly, local police branch, etc.). Equivalent replacements of the given concept are the terms "level", "hierarchical structure", "hierarchy".

Another important meaning of this word is described by the term *imperious powers*. It is considered, that the imperious powers of any level can be generally determined with the help of some cumulative quantitative criterion including a nominal (formal) situation of the level in structure; the volume of human, financial, material, informative, intellectual, legislative, circumspect and other kinds of resources at its disposal; the scale and location of controlled territory; the prestige in the eyes of public opinion and experts, etc. In this sense the word "power" means a possible level of influence of the given institute on the behavior of other levels and the life of the civil society. Institutes with great powers, naturally have a higher place (smaller number) in the hierarchy in comparison with the those with smaller powers. Thus when considering common questions there is no need to introduce any absolute units of measurements of power; it is enough to consider the imperious power of the highest level as a unit (or for 100%), then the powers of any other level will be expressed in fractions (or percentages) relative to the highest institute.

The further specification of the term "imperious powers" is connected with concepts of *maximal powers* and *minimal powers*. The former describes actions, which the authority can execute in a maximal way in a certain situation according to the legislation. For example, the governor has the right under known circumstances to announce an extreme situation in a territory controlled by him, but is not able to make decisions on any question of war and peace. The minimal powers describe actions always to be undertaken by an authority (the President is obliged to represent the annual budget message to Congress). Both these concepts are well illustrated, for example, by the articles of the criminal Code determining maximal and minimal terms of punishment for the same kind of crimes. In other words, the maximal and minimal powers give upper and lower legal limits of power for each hierarchical level.

Note that the limits of power, as is known from the practice of many countries, can be determined not only by legislation, but also by traditions and systems of precedents. However, for brevity we shall mean below only the legislatively established powers.

In mathematical model the maximal imperious powers are set by some positive function $p_2(i, t)$, $0 \leq i \leq N$ (circles in Fig. 62), monotonously decreasing with the growth of number i, i.e. at any moment of time t is valid $p_2(i+1, t) < p_2(i, t) < p_2(i-1, t)$, $1 \leq i \leq N-1$. The minimal powers are set by positive function $p_1(i, t)$, $0 \leq i \leq N$ (crosses in Fig. 62), also monotonously decreasing with the growth of i, $p_1(i+1, t) < p_1(i, t < p_1(i-1, t)$, $1 \leq i \leq N-1$ (obviously, always $p_1(i, t) < p_2(i, t)$, $0 \leq i \leq N$. Both functions, generally speaking, depend on time t, since the legislation, territorial division and so forth can vary in time.

Purely speculative situations, when in the legislation it is stated that "the authority can do everything" or "the authority is not obliged to do certain actions" are not considered, i.e. the limits of power (even formally) are always defined, hence, the functions p_1, p_2 are determined as well.

Let us emphasize once again: for considered (in generalized manner) hierarchical structure the situation of any level is determined not only by its nominal place, but also by all relevant (and frequently more powerful) factors. Thus, the number of hierarchical levels corresponding to some "average official" of the imperious tree, is not a formal coordinate, but a coordinate "in essence". It is easy to give more detailed and strict mathematical formulation (microdescription) for the mentioned imperious powers and for the arrangement of levels by "subordination"; however it has no decisive role for the study of basic properties of the "power – society" system.

Finally, one more understanding of the word "power" used in the model is connected with *the level of imperious influence* (or *the amount of power*) actually carried out in the given moment by the given level. Indeed, the imperious powers determine only the upper and lower legal limits of a level or the amount of power (in this sense one can to understand the known expression "amount of imperious powers"). These limits, generally speaking, are not always reached and are not in evidence everywhere. Assume, for example, that in accordance with some article of the criminal Code providing a jail sentence of three to five years, the courts during some time scale have given verdicts with an average term of four years. Then, the power realized by them under the given article is 80% from the maximal and 133% from the minimal power.

In a mathematical model, the actually achievable power corresponds to a non-negative function $p(i, t)$, $0 \leq i \leq N$, dependent from "spatial" coordinates i and time t (asterisk in Fig. 62). If for any values i, t is fulfilled $p(i, t) > p_2(i, t)$ (or $p(i, t) < p_1(i, t)$), it is natural to speak about the "exit" of authorities for frameworks of powers, or about *excess (or diminution) of authority.*

The basic difference between functions $p_1(i, t)$, $p_2(i, t)$ and function

$p(i, t)$ is, that, as distinct from known, given (in general form) imperious powers $p_1(i, t)$, $p_2(i, t)$, function $p(i, t)$ is the unknown, sought quantity describing *the current distribution of power in hierarchical structure*. The construction of the corresponding mathematical model and the study of spatial-temporal dynamics for function $p(i, t)$ (distribution of power) depending on all factors present in the investigated system, is represented below.

Note that the concept of *a real distribution of power* means we have to introduce the following assumption 1:

all partners in the system "power – society" are following the rules: the laws are preserved, the taxes are paid, the orders are carried out (otherwise function $p(i, t)$ – the magnitude of power – becomes rather uncertain or loses its sense at all).

Such an approach "from the simple to the complex", i.e. constructing a model for a strongly idealized situation and its further improvement, is typical for the mathematical modeling of complex objects (and frequently the only possible approach).

We will represent the general description of interactions in the system "power – society". It consists in the following

a) the civil society accepts the Constitution (the arrow *1* in Fig. 62) directly or through the representatives. It thus acts as a source of power for the hierarchical structure, interacting with it based on the existing legislation;

b) the hierarchical structure exists not by itself, but as an "opened" system interacting with the Constitution and the civil society. The Constitution (in wide context including the laws, charters, etc.) serves as a kind of reservoir for the hierarchy, from which its parts either can extract additional "portions" of power (arrow *2*), or return certain extra "fractions" (arrow *3*). Thus, between hierarchy and Constitution an "exchange of power" is realized and, in an implicit form as though exchange of power ("of freedom") exists between hierarchy and civil society – creator of the Constitution;

c) inside the hierarchical structure itself there is a redistribution of current power between the components of its levels (arrow *4*) according to the mechanisms of transfer of the imperious orders accepted in the hierarchy;

d) in relation to a civil society the imperious structures act as forbidding (arrow *5*) or permitting (arrow *6*) institutes introducing and cancelling certain regulations (typical example – an annual appeal to the army and demobilization, when part of the population starts to perform an additional duty, while the other part is released from it).

One of the key questions for the description of interaction in system "hierarchy – society" is the definition of the "amount" of exchange of power between the hierarchical structure and the Constitution (and in fact between the hierarchy and society). The following assumption 2 is introduced:

the sign and amount of exchange of power between the hierarchical structure and the Constitution is determined by the reaction of the system.

Reaction of the system denotes the total reaction of both partners (hierarchy and society) to the current distribution of power $p(i,\ t)$. For example, concerning the *reactions of society*, this implies that if at a given moment of time it is denying certain actions of the given link of hierarchy (resistance), it forces the hierarchical level to reduce its power, as if storing its some "fraction" in a constitutional reservoir (to reduce or even avoid certain negative, from the point of view of society, consequences). A suitable example is the requirement to reduce certain types of taxes. A positive reaction of society (i.e. its support) induces the institution to increase the level of its power by taking necessary "resources" from the legislation (for example, the requirement to strengthen the actions against crime). The qualitative character of the reaction of a society is related to the dominant type of consciousness (legal, anarchic, authoritarian, mixed).

In the considered model the reaction of a society is described by a given function $F_S(i,\ t,\ p,\ p_1,\ p_2)$, generally depending on all the quantities introduced earlier: the number of level i, time t, magnitude of power performed by the institute $p(i,\ t)$, imperious powers $p_1(i,\ t)$, $p_2(i, t)$. It enables a rather complete and precise reflection of the the structural and varying in time attitude of society to hierarchy. For example, if for all conceivable meanings of the arguments we have $F_S < 0 (F_S > 0)$, then it is obvious that anarchic (totalitarian) consciousness is dominant in society. The reaction of society is represented as a quite observable and measurable quantity, and serves as the basis of its behavioral characteristics in some average sense. Thus it is meant, that a reaction expressed in time is correctly interpreted by institution and is taken into account in its activity.

Similarly in the model of behavioral characteristics of an imperious structure, *the reaction of hierarchy* is introduced. This concept characterizes the aim of hierarchy levels to raise or lower the level of power they possess at a given moment of time. It is described by a function F_H, depending on the same arguments as F_S, and with the same sense, but concerning the hierarchy (it is possible to call it *degree of "powering" of institutions*).

For simplicity, we shall further consider $F_H \equiv 0$, i.e. the hierarchy is indifferent to the level of its power, and the reaction of system $F = F_S + F_H$ is determined only by the "civil" component. The given simplification does not change the purely mathematical properties of model, as the functions F_S, F_H, F depend on the same arguments. However when interpreting results, it is certainly necessary to take into account the role of each component F_S, F_H in the total reaction F.

In the given general description the system "power – society" appears as a closed self-consistent and self-organizing object with various direct connections and feedbacks. The level of power which is possessed by any level of the hierarchy at any moment in time is by no means arbitrary, but is a result of self-consistent interaction between all the components of the object: the imperious orders moving through the hierarchical ladder, the reaction of the society, current legislation, initial conditions of the system and so on. To derive a mathematical model on the basis of this general scheme, we shall consider in more detail the imperious structure itself.

2. Mechanisms of redistributing power inside the hierarchical structure. Inside an hierarchy any level accepts the fact that certain imperious orders are given by the senior powers, and in turn transfers certain orders to the lower members. Thus there is some redistribution of power between the levels of the hierarchy (recall that the point is not the imperious powers, but the actual current magnitude of power – $p(i, t)$). Hence, the following behavioral postulate:

in a hierarchy the power can be transferred only from levels with greater current power to levels with smaller current power (the rate of transfer is higher, the bigger the difference between the current power of those levels).

Two main mechanisms responsible for redistribution of power within the hierarchy are considered.

a) *Short-ranging.* Conditionally, this mechanism can be called a *transfer of power along a ladder*, when a chief is passing the order to the nearest subordinate who also interacts only with their direct subordinates. The given mechanism corresponds to a well known bureaucratic procedure (the word "bureaucratic" and all other concepts are used as working terms without any emotional context).

Let the i-th level has transferred a certain order to the $(i + 1)$th (for example, has charged to prepare a project of an annual report about financial activity of one of the subordinate establishments). What has taken place in such an elementary act of interaction between the neighbors within the hierarchy? Together with the order, the subordinated has received some small and temporal portion of power, in addition to the power which he really possessed (for example, having the order he studies financial documents in more detail than previously). On the other hand, the i-th level has lost some part of the current power, by shifting the control on the given task to $(i + 1)$th level.

The set of transmitted orders forms a kind of flux of power from the ith to the $(i + 1)$th level. We define *the power flux* $W(i, t)$ as the amount of power received in a unit of time by $(i + 1)$th level from the i-th level.

According to the postulate $W(i, t)$ is positive at $p(i, t) > p(i + 1, t)$, is negative at $p(i, t) < p(i + 1, t)$, and is equal to zero at $p(i, t) = p(i + 1, t)$. Consider a rather general expression for $W(i, t)$:

$$W(i, t) = -\kappa[p(i, t) - p(i + 1, t),\, p(i, t),\, p(i + 1, t),\, i,\, i + 1,\, t] \times \\ \times [p(i + 1, t) - p(i, t)]\,, \tag{1}$$

where the function κ is positive at all values of arguments.

Formula (1) is a concrete representation of the postulate concerning the considered mechanism of short-ranging. The positiveness of function κ provides the necessary sign for $W(i, t)$. In particular, if in any moment of time t it appears that $p(i+1, t) > p(i, t)$, the lower level has greater power than the higher one, a flux of power automatically "acts" in the direction to reduce the difference $p(i, t) - p(i+1, t)$, supporting to overcome the anomalous situation (indicating troubles at the given levels of hierarchy). The flow of power grows with the difference $p(i, t) - p(i+1, t)$, which corresponds to the second part of the postulate (it is required to *impose* an appropriate restriction on κ as function on $p(i, t) - p(i+1, t)$). This means, in particular, that the chief (with rest equal conditions) gives more orders to subordinates, the less is his currently work load.

The generality of expression (1) is in the fact that for a flux of power it is determining, besides the dependence on $p(i, t) - p(i+1, t)$, also dependence on all variables introduced above: values of sought function $p(i+1, t)$, $p(i, t)$; coordinates i, $i+1$; time t. The dependence on t and i, $i+1$ also means an implicit dependence on the given imperious powers p_1, p_2 and the reaction of society F (however, this dependence is easy to introduce in explicit form). The generality also indicates the absence of essential restrictions posed by these dependencies, except the requirement that at $p(i+1, t) = p(i, t)$ the condition $W(i, t) = 0$ should be fulfilled. Note, however, that in view of (1) the flux of power is determined only by quantities directly concerning the interaction with institutions (in this case with the neighbors).

The function κ has a certain bureaucratic sense, describing properties of the power "medium", some aspects of mutual relation inside the hierarchy. Let, for example, flux of power $W(i, t)$ and $p(i, t)$ be fixed, and $\kappa = \kappa_0 =$ const so that function κ does not depend on time and other quantities. Then, in view of (1) we see that with increase of κ_0 the difference $p(i+1, t) - p(i, t)$ both in absolute and relative values is decreasing. One can say that with growth of κ_0 "*the responsibility*" of the senior level decreases, and the latter becomes more positive concerning the equipartition of its power with those of lower levels. At the reduction of κ_0, the situation, naturally, in the opposite.

b) *Long-ranging.* The figurative description of this mechanism can be given by known expression it "an order over a head". This means, that the ith level passes the orders to levels with numbers exceeding the nearest subordinates. To such actions one can attribute, for example, the order for an army to transfer to the summer clothes; its realization does not actually need a procedure of transferring an order along a ladder, it is enough to publish the appropriate order for the attention (and execution) of the whole army simultaneously.

The flux of power, appeared due to the mechanism of long-ranging (the sense is the same, as in case (a)), received by $i+1$th level from jth link, can be expressed by a rather general formula

$$V(i+1,\,j,\,t) = \chi\,(p\,(i+1,\,t),\,p(j,\,t),\,i+1,j,\,t)\cdot[p(j,\,t)-p(i+1,\,t)]\,, \quad (2)$$

where function $\chi \geq 0$ is in accordance with the postulate. By its content the behavioral characteristic χ is close to function κ, but, first, it describes the interaction not of neighboring but of distant institutions, i.e. $j \neq i+2$, i; second, the quantity χ can turn to zero, i.e. the possibility of absence of orders over a head between some levels of hierarchy is taken into account.

As with the function κ, it depends only on quantities directly concerning the interacting parts. These dependencies a priori are not exposed to any restrictions (the only requirement is that $V(i+1,\,j,\,t) = 0$ with $p(i+1,\,t) = p(j,\,t)$) and can be complemented by explicit dependence on other quantities introduced earlier.

We stress again that both considered mechanisms represent an averaging of the detailed microdescription of interactions inside an hierarchical structure with a complex configuration, namely, with topology of plenty of trees with crossed crowns.

One more important explanation: in mechanisms a) and b) the orders transmitted from ith (jth) to $(i+1)$th level in a normal situation (i.e. at $p(i,\,t) > p(i+1,\,t)$ or $p(j,\,t) > p(i+1,\,t)$, $j < i$) carry some positive fraction of power for $(i+1)$th level, according to a postulate and its concrete definitions (1), (2). In the already mentioned anomalous case, when $p(i,\,t) < p(i+1,\,t)$ or $p(j,\,t) < p(i+1,\,t)$, $j < i$, then in accordance with the postulate some portion of power passes from the lower to the senior level. This assumption naturally indicates the essence of hierarchical structure (if, certainly, the particular chief is not mistaken in his real power in comparison with that of his subordinates).

At the same time the postulate describing a situation whereby the lower cannot receive a fraction of power from the senior level, if the latter in the given moment has smaller or equal real power, does not describe a possible situation in which the senior has more power than the subordinate, but captures a portion the subordinate's power (this would contradict the postulate).

Consider this latter situation in more detail to prove that in precise interpretation is not actually possible (at least, in some "average" sense).

Indeed, by capturing part of the power of some his subordinates (for example, by withdrawing an order), the chief will sooner or later "transmit" the same order but in a different way. Since in any hierarchy a certain level is not capable directly of realizing an appreciable fraction of orders to be necessarily carried out by his subordinates (we mean a rather extended hierarchical structure). On the other hand, the task has to be performed anyway, and the order, with a part of the power, will be transferred to the subordinates. Thus, the introduced postulate is natural and reasonable at least "in average".

Another point is the more particular mutual relation between parts of hierarchies externally similar to a discussed above situation. They can be described by certain earlier functions. Let, for example, in a conditional pair "chief – deputies of the chief", the former for any reason be dissatisfied with the work of one of his deputies. The senior can, for example, undertake the following actions

1) displacing "the physical person", by replacing him with a more suitable figure;

2) directing a flux of orders through other deputies.

In these two cases for the hierarchy as a whole and in its model, nothing actually is changed, except the persons. In particular, the level of "deputies" as a whole receive the same flux of power.

The following methods can also be used:

3) strengthening the role of orders over the head of the given deputy;

4) changing his imperious powers;

5) abolishing the given post.

The case (3) corresponds to a reduction of κ and increase of χ, in case (4) the functions p_1 and p_2 are changed correspondingly, in case (5) the number of levels N in the hierarchy is decreased. Certainly, the various combinations of these actions (including with respect to the whole level of "deputies") can be realized as well. If the chief, on the contrary, is quite satisfied with a deputy, his actions will be, naturally, the opposite. All these changes do not mention an essence of the postulate and are described with the help of changing the above introduced functions p_1, p_2, κ, χ, F. The corresponding formalization should certainly be performed carefully and accurately.

3. Balance of power in a level, conditions on boundaries of hierarchy and transition to a continuous model. Now, when the basic model concepts are introduced, and the interrelations and assumptions are described, it is possible to make the final step in constructing the model and answering a question: how to find out the distribution of power in hierarchy, i.e. the function $p(i, t)$? For any ith link of an imperious structure at any moment of time t it satisfies an equation, deduced using a kind of local power conservation law:

the rate of change (reduction or increase) of power of level is determined by flux of power and reaction of the system.

We now estimate the quantity of power received by ith level during a time interval Δt between the moments t and $t + \Delta t$. This quantity is formed as:

a) flux of power received from $(i-1)$th level via the mechanism (1), that is

$$\Delta p_- = W(i-1, t)\,\Delta t\,;$$

b) flux passed to $(i+1)$th level via a similar mechanism,

$$\Delta p_{+} = -W(i,\, t)\,\Delta t\,;$$

c) sum of the fluxes received and given via the mechanism (2) from farther levels $(j \neq i+1,\, i-1)$,

$$\Delta p_{\Sigma} = \sum_{j=0}^{N} V(i,\, j,\, t)\,\Delta t\,;$$

d) by the rate of an exchange of power between the institutions and legislation, which is determined in accordance with assumption 2, by the reaction of the civil society,

$$\Delta p_F = F(i,\, t,\, p(i,\, t),\, p_1(i,\, t),\, p_2(i,\, t))\,\Delta t\,.$$

Summing the quantities Δp_{-}, Δp_{+}, Δp_{Σ}, Δp_F, we obtain the total variation

$$\Delta p = p(i,\, t+\Delta t) - p(i,\, t) =$$

$$= \left[W(i-1,\, t) - W(i,\, t) + \sum_{j=0}^{N} V(i,\, j,\, t) + F(i,\, t,\, \ldots)\right] \Delta t\,. \tag{3}$$

When deducing the equation of balance (3), the intervals of time Δ are assumed, as usual, to be rather small, so that it was possible to consider W, V, F to be constant. We also used (and this is essential) the assumption 1 about following regulations, meaning in particular that in any part of the hierarchy the order is always realized and the portion of power associated with it is realized as well (and does not disappear without trace). Recall that the above introduced hierarchical postulate has found its mathematical formulation in (1), (2).

By dividing both parts of (3) by Δt, we obtain the equation for the rate of change of power of ith level in time

$$\frac{\Delta p(i,\, t)}{\Delta t} = -[W(i,\, t) - W(i-1,\, t)] + \sum_{j=0}^{N} V(i,\, j,\, t) + F(i,\, t,\, \ldots)\,. \tag{4}$$

The content of (4) is clear from considerations preceding the derivation of balance (3).

The equations (3), (4) are written for an arbitrary number $0 < i < N$. To achieve a balance in a point $i = 0$ it is necessary to determine $W(-1,\, t)$ of power flux received by the most senior level (with number 0) from a nonexistent higher level. Obviously, this flux is always equal to zero (nobody can order to the highest level). Similarly, $W(N, t)$ – the flux of power transmitted by the lowest level of hierarchy to a nonexistent lower one, is also equal to zero (there is nobody to order). So, the flux of power on boundaries of hierarchy is always equal to zero

$$W(-1, t) = W(N, t) \equiv 0 . \tag{5}$$

Considering only a part of the hierarchy, not including the highest and (or) the lowest levels, one has to set the power flux as functions of time on the boundaries (from various reasoning).

To close the equation (4) it is necessary, besides conditions (5), also to set the distribution of power in some initial moment of time $t = t_0$, i.e. the function

$$p(i, t_0) = p_0(i) \geq 0 , \qquad 0 \leq i \leq N . \tag{6}$$

Equation (4) with conditions (5), (6) and given functions κ, χ, F, p_1, p_2 represents a closed discrete model of the distribution of power in the system "hierarchical structure – civil society", mathematically realizing the general scheme given in subsection 1. It enables us for any moments of time $t > t_0$ and any point $0 \leq i \leq N$ to find the solution (power distribution in hierarchy), i.e. the function $p(i, t)$.

We now carry out a transition to a continuous model, considering imperious hierarchy as a "continuous medium", i.e. assuming the number of levels to be rather large ($N \gg 1$), and all functions included in the model to be rather smooth. A continuous coordinate x is put into correspondence with the discrete coordinate i (numbers i correspond to points x_i, chosen, for example, by a rule $x_i = i$), and integer piece $[0, N]$ – piece $[0, l]$ (l is an analog "of length" of the hierarchical structure, and the coordinate x characterizes the place of the level in hierarchy: the more is x, the lower is the institution). The differences of a kind $p(i+1, t) - p(i, t)$ in (1) and $p(i, t+\Delta t) - p(i, t)$, $W(i, t) - W(i-1, t)$ in (4) are replaced by the appropriate first derivative, the sum in the right hand side of (4) is replaced by an integral, while the dependence of all functions in (4) – (6) on arguments i, t is replaced with dependence on x, t.

In the result for function $p(x, t)$ (the detailed calculations are not represented, as they are rather obvious; compare with the derivation of the model of heat transfer in section 2, Chapter II) the following equation is obtained

$$
\begin{aligned}
\frac{\partial p}{\partial t} &= \frac{\partial}{\partial x}\left[\kappa\left(p,\, \frac{\partial p}{\partial x},\, x,\, t\right)\right] + \\
&+ \int_0^l \chi[p(x'),\, p(x),\, x',\, x] \cdot [p(x',\, t) - p(x,\, t)]\, dx' + F(p,\, p_1,\, p_2,\, x,\, t) \\
& 0 < x < l\,, \qquad t > t_0\,;
\end{aligned} \tag{7}
$$

with boundary conditions of the second kind

$$
W(0,\, t) = -\kappa\, \frac{\partial p}{\partial x}\bigg|_{x=0} = 0\,, \quad W(l,\, t) = -\kappa\, \frac{\partial p}{\partial x}\bigg|_{x=l} = 0\,, \quad t \geq t_0\,, \tag{8}
$$

and initial conditions

$$
p(x,\, t_0) = p_0(x) \geq 0\,, \qquad 0 \leq x \leq l\,, \tag{9}
$$

so that $\kappa > 0$, $\chi \geq 0$, and the positive functions p_1, p_2 monotonously decrease by x.

The equation (7) represents a "parabolic" integro-differential equation (in a sense that in the absence of the integral member it is similar to equation of heat transfer; see also exercise 1). From a mathematical point of view the model (7)–(9), "approximating" the initial model (4)–(6), is closed and is correct, i.e. uniquely determines the solution – a smooth non-negative function $p(x,\, t)$ – for all $0 \leq x \leq l$ and $t \geq t_0$ (at some insignificant for generality restrictions on input data). Note that in the numerical modeling of problem (7)–(9), the inverse transition to (4)–(6) is made (see exercise 2).

Recall the content of quantities included in model (7)–(9). The function $p(x,\, t)$ (decision) describes the spatial-temporal dynamics of power distribution in a hierarchical structure, i.e. the dependence of the magnitude (level) of real power of a link from its location (coordinate x) and time t. The rate of change of function $p(x,\, t)$ (left hand side of equation (7)) is determined by the following characteristics:

1) difference of flux of power either received via the mechanism (a) from the nearest neighbors in hierarchy, or returned to them (the first, differential term in the right hand side of (7));

2) sum of fluxes of power received by a level via the mechanism (b) from remote levels of hierarchy or returned to them (the second, integral member in the right hand side of (7));

3) reaction of a civil society, i.e. the function $F(x,\, t,\, p(x,\, t),\, p_1(x,\, t),\, p_2(x,\, t))$ (the third member in the right hand side of (7));

4) minimal and maximal imperious powers of hierarchy – functions $p_1(x,\, t) > 0$, $p_2(x,\, t) > 0$, monotonously decreasing by x;

5) internal behavioral properties of hierarchical structure – functions

$$\kappa\left(p(x,\,t),\,\frac{\partial p}{\partial x},\,x,\,t\right) > 0\,, \qquad \chi(p(x',\,t),\,p(x,\,t),\,x',\,x,\,t) \geq 0\,;$$

6) initial power distribution in structure – function $p_0(x)) \geq 0$.

Thus, the model (7)–(9) is the mathematical realization of a general scheme of a system "power – society", described in subsection 1 and corresponding to a self-organizing object with various direct connections and feedbacks. The power distribution described by it in hierarchy is established not arbitrarily, but as a result of interactions of all elements of the system.

Let us outline the scope of applicability of the obtained model. In its construction we have essentially used a postulate, assumption 1 on the following of regulations, assumption 2 on the determination of the influence of reaction of a civil society on the rate of exchange of power between hierarchy and legislation, and also was considered, that the character of current interaction between two institutions of hierarchy depends only on the current condition of those institutions.

All these assumptions were formulated in possible general form, three of them seem to be quite natural and reasonable (concerning the assumption about following the regulations, it was a necessary idealization at the initial stage of research, and can be modified). We call (7)–(9) *the general mathematical model of dynamics of power distribution* in state hierarchical structures interacting with a civil society.

The above constructed model enables one to analyze a number of rather general problems connected to the function and evolution of the system "power – society". Some of them are as follows:

1) derivation of necessary and sufficient conditions of existence (or non-existence) of stationary (i.e. constant in time) power distributions and conditions of the stability of this distributions;

2) analysis and forecast of various crises of power: excess of authority; anomalous, non-monotonous power distributions in hierarchy, when certain lower levels undertake higher real power than senior ones; anarchic or totalitarian evolution of power distribution, etc;

3) study of results of changes occurring in the reaction of a civil society, in the legislation, in mutual relations within the hierarchy and so on, on spatial-temporal dynamics of power distribution.

It is also possible also to study the various combinations of problems (1–3). Obviously, the analysis of these and other problems should be accompanied by their careful formulation and accurate interpretation of results in the terms of the used model.

The model (7)–(9) enables one to give a rather general answer to a number of global questions arising in the study of problems (1–3). For example, using the theorems of comparison of the solutions of parabolic equations (see section 2, Chapter V), it is easy to establish that for an increase of positive reaction of the civil society with the actions of all levels, the power in any level of hierarchy is also increasing, while for an increase of negative reaction the opposite is true and the large initial power distributions in hierarchy result (at equal rest conditions) in greater level of power in all levels in all subsequent moments of time, and vice versa.

However, of paramount interest are more specific questions, the answers to which can be obtained at certain concrete definitions of system "power–society". Some of them are studied below.

4. The legal system "power–society". Stationary distributions and exit of power from its legal scope. One of the most naturally formalizable systems "power–society" is a legal system. Therefore the simplest concrete case of general model (7)–(9) corresponds just to this case.

The system "power–society" is called *legal*, if the reaction of society to actions of any level of hierarchy is always directed on deduction of power distribution within the framework of the powers. It is considered that the maximal and minimal powers are legal in a standard sense of this concept (a similar type of reaction corresponds to a legal public consciousness).

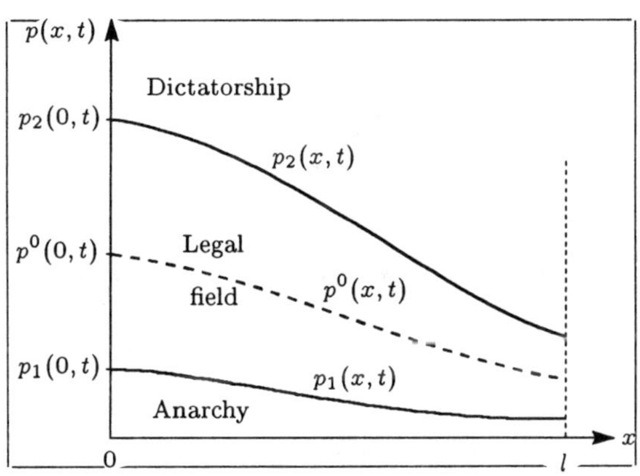

Fig.63.

In a general model the above formulated definition is realized by the given appropriate function $F(p, p_1, p_2, x, t)$. In Fig. 63 the qualitative form of functions $p_2(x, t)$, $p_1(x, t)$ – the maximal and minimal imperious powers of levels of hierarchy in some moment of time t – is shown depending on their place (coordinate x). By their sense $p_1(x, t)$, $p_2(x, t)$ are monotonously decreasing on x positive functions, and $p_2(x, t) > p_1(x, t)$, $0 \le x \le l$, $t \ge t_0$ (subsection 1). The area between the curves p_1, p_2 is the legal field, area above values p_2 is conditionally called dictatorship, the area lower than p_1 is called anarchy. The sought solution – the power distribution $p(x, t)$ – a priori in the different moments of time can either be completely within the legal area, or partially or completely outside it.

The qualitative form of dependence of function $F(p, p_1, p_2, x, t)$ from power $p(x, t)$ (in moment t for some x) is shown in Fig. 64. If the power exceeds the maximal powers ($p > p_2$), the reaction of the society is negative and increases with the growth of this excess. If $p < p_1$, i.e. the power is less than the minimal powers, then, on contrary, the reaction is positive and grows with the increase of the deviation. Within an interval $p_1 \le p \le p_2$ the reaction of the society with growth of p monotonously changes from positive to negative, turning to zero in $p^0(x, t)$, $p_1 < p^0 < p_2$, $0 \le x \le l$, $t \ge t_0$.

One can call $p^0(x, t)$ *an ideal* (average) power distribution in a sense that with realization of such distribution the reaction of society to actions of hierarchy would universally be zero (the function $p^0(x, t)$ is considered monotonous by x, see Fig. 63)

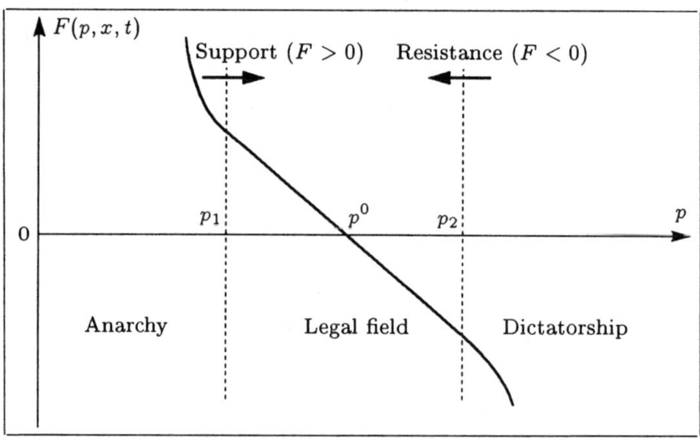

Fig.64.

Thus, in a legitimate system the society negatively reacts to positive deviation of power from its average, strengthening the reaction with the increase of this deviation (for negative deviations – vice versa) By its quantitative amplitude the reaction depends not only on a difference $p - p^0$, but, generally speaking, also on time t; it can be different for different parts of the hierarchy (coordinate x) and can include more subtle dependence on $p(x, t)$ (for example, in Fig. 64 the intensities of reaction for areas of dictatorship $p > p_2$ and anarchy $p < p_1$ are essentially higher than for legal area $p_1 < p < p_2$).

Now we move on to study stationary (not time-dependent) power distri-

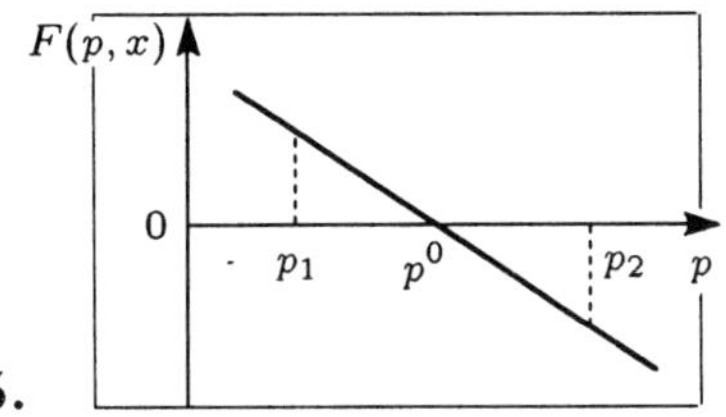

Fig.65.

butions in a legal system. We start from the analysis of most simple model, but at the same time containing all essential components of a legal system. We shall consider, that in general model (7)–(9):

1) the function κ corresponding to the mechanism of transfer of power by an order along a ladder, is constant, $\kappa = \kappa_0 = \text{const} > 0$;

2) in a hierarchy there is no mechanism of over the heads' orders, $\chi \equiv 0$,

3) the reaction of society is a linear function of deviation from an ideal level of power and also does not depend on t, that is

$$F(p,\, x) = k_1\, (p^0(x) - p)\,,$$

where $k_1 > 0$ characterizes the amplitude of reaction (Fig. 65);

4) ideal power distributions p^0 and imperious powers p_1, p_2 do not vary in time and linearly decrease with growth of coordinate x (Fig. 66), that is

$$p^0 = H - kx, \qquad H > 0, \qquad p_1 = (1-\alpha)\, p^0, \qquad p_2 = (1+\alpha)\, p^0\,.$$

Here $k > 0$ is the degree of decrease of function p^0 on x, $0 < \alpha < 1$ is the relative difference between p^0 and p_1, p_2 (and $p^0 = (p_1 + p_2)/2$ is the average arithmetic from the minimal and maximal powers), $p^0(0) = H$, $p^0(l) = H - kl > 0$. Quantity $(H - kl)/H = p^0(l)/p^0(0) = p_1(l)/p_1(0) = p_2(l)/p_2(0) < 1$ denotes the relation of powers of the lowest and supreme institutions of hierarchy, or overfall powers.

As a result of simplifying 1)–4) we obtain *a base model* of a legal system

$$\frac{\partial p}{\partial t} = \frac{\partial}{\partial x}\left(\kappa_0\, \frac{\partial p}{\partial x}\right) + k_1 (p^0(x) - p), \qquad 0 < x < l, \qquad t > t_0,$$

$$\kappa_0\, \frac{\partial p}{\partial x}\bigg|_{x=0} = \kappa_0\, \frac{\partial p}{\partial x}\bigg|_{x=l} = 0, \qquad t \geq t_0, \tag{10}$$

$$p\,(x,\, t_0) = p_0\,(x) \geq 0, \qquad 0 \leq x \leq l\,,$$

where $k_1 > 0$, $p^0(x) = H - kx$, $H > 0$, $k > 0$. Note that model (10) is linear, and hence, its general solution can be obtained rather easily. The stationary solutions ($p(x,\, t) = p(x)$) of (10) are obtained from a problem

$$\kappa_0 p'' = k_1\, [p - (H - kx)]\,, \quad p'(0) = p'(l) = 0\,, \quad 0 < x < l\,, \quad p' = \frac{dp}{dx}\,,$$

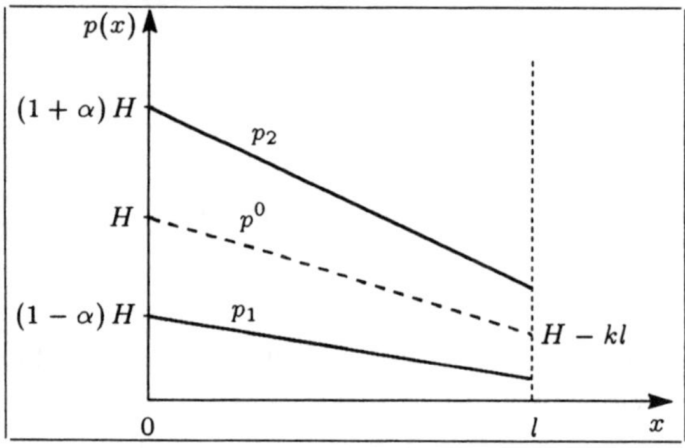

Fig.66.

containing four parameters: κ_0, k_1, k, l, and being reduced via scaling (see section 1, Chapter V) $\bar{p} = p/H$, $\bar{x} = x/l$ to a problem with two parameters

$$ap'' = p - (1 - bx)\,, \qquad 0 < x < l\,, \qquad p'(0) = p'(1) = 0\,, \tag{11}$$

where $a = \kappa_0/(k_1 \cdot l^2)$, $b = kl/H < 1$ (the bars above p and x in (11) are omitted).

The problem (11) has a well known unique (non-negative) solution, given via monotonously decreasing by x function,

$$p(x) = \frac{a^{1/2}b}{e^{a^{-1/2}} - e^{-a^{-1/2}}} \left[\left(1 - e^{-a^{-1/2}}\right) e^{xa^{-1/2}} + \right. \\ \left. + \left(1 - e^{a^{-1/2}}\right) e^{-xa^{-1/2}}\right] + (1 - bx)\,. \tag{12}$$

The qualitative behavior of solution (12) is determined by parameter a (parameter $b \approx 1$, so $1-b \ll 1$, as the rather extended hierarchical structures with essential overfall of powers) are considered. The form of solution for $b = 0.9$ and three values $\alpha = 10^{-3}$, 10^{-2}, 10^{-1} is given in Fig. 67. With growth of a the solution (curves *1–3*) becomes more flat, and at some $a > a_{cr}$ the solution is partially located within area $p > p_2$ (curve *3*). Analyzing (12) and Fig. 67, one can draw the following preliminary conclusions:

1) the stationary power distribution in a legal system does exist;

2) there is an area of parameters of a system, where the power distribution escapes from the scope of power (in this case for part of the lower levels the power is exceeded). It is easy to write down a necessary and sufficient condition of the location of power within a legal area

$$a^{1/2}b\left(e^{a^{-1/2}} - 1\right) / \left(e^{a^{-1/2}} + 1\right) \le (1 - b)(1 + \alpha)\,. \tag{13}$$

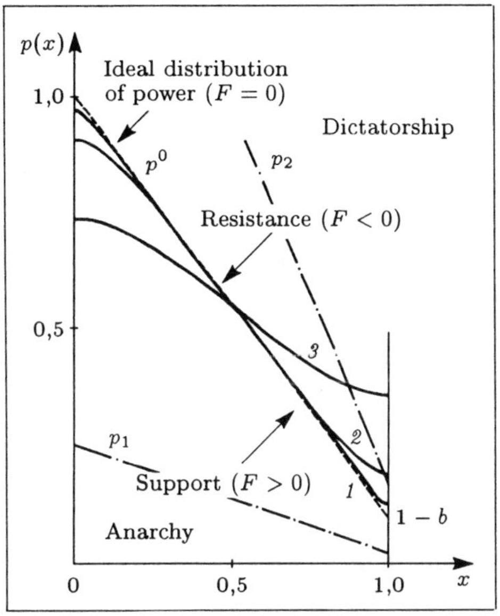

Fig.67.

With growth of a the criterion (13) is violated more strongly, so that if $a \to \infty$, then $p(x) \to 1 - b/2$. The maximal excess (achieved in a point $x = 1$) is equal to $p(1) - p_2(1) = 1 - b/2 - (1 + \alpha)(1 - b)$, which at $b \approx 1$ yields $p(1) - p_2(1) \approx 0.5$, i.e. the excess is 50% from the maximal powers;

3) The ideal power distribution (dashed line in Fig. 67) is not realized, the solution always has "an excess" to $p^0(x) = 1 - bx$ - to the second term in the right hand side of (12).

In view of the importance of conclusions 1) - 3), it is necessary to find the degree of their generality, i.e. to understand whether are they properties only of the simplified model (10).

5. Role of basic characteristics of system in a phenomenon of power excess (diminution). Let us introduce some generalizations into the base model (10).

1) *The mechanism of over the head orders.* On the right hand side of equation (10) there is an additional term of a form

$$\int_0^l \chi[p(x') - p(x)]\, dx'$$

(here is adopted $\chi = \chi_0 = \text{const} > 0$). The problem for the stationary solution (analog of the problem (11)) after the corresponding scaling has the form

$$ap'' = [p - (1 - bx)] - \sigma \int_0^1 [p(x') - p(x)]\,dx', \qquad (14)$$
$$p'(0) = p'(1) = 0, \qquad 0 < x < 1,$$

where $\sigma = \chi_0 l/k_1 > 0$. By simple replacement (14) is reduced to (11) with non-zero boundary conditions (exercise 3), with a solution of type (12). The condition (13) turns to

$$\frac{b}{1+\sigma}\left[z^{-1/2}\frac{e^{z^{1/2}} - 1}{e^{z^{1/2}} + 1} + \frac{\sigma}{2}\right] \leq (1-b)(1+\alpha), \qquad z = \frac{1+\sigma}{a},$$

which, as (13), is always violated for rather large values of a and (or) σ (the maximal excess is the same as in the case $\chi_0 = 0$, since $p(x) \to 1 - b/2$ at large a, σ).

2) *The specified reaction of a civil society.* In a society there is a part which basically disagrees with any actions of hierarchy (in the base model as though it was assumed that no such component is present). A similar reaction can be described by adding a term $k_2 p$, $k_2 < 0$ into the right hand side of equation (10) (i.e. this part of reaction is always negative, and more, the larger is the power). Then for function $p(x)$ we have a problem similar to (11). However for rather large values of k_2 and a the solution is leaving the area $p > p_2$, entering the area $p < p_1$, i.e. the power distribution leaves the scopes of the minimal powers (in this case it happens with highest levels).

3) *Various mechanisms of transfer of power along a ladder.*

a) In the first term on the right hand side of equation (10) we now consider not the function $\kappa = \kappa_0 = \text{const}$, but the function $\kappa = \kappa_0|\partial p/\partial x|^\alpha$, $\alpha > -1$. This means that the degree of responsibility of hierarchy depends on a gradient of $p(x, t)$ (the inequality $\alpha > -1$ provides parabolicity of the corresponding equation). For a stationary problem there is a unique monotonous solution. For rather large values of a parameter similar to parameters α and σ from (11) and (14), the power distribution leaves the scope of the powers (since at its growth the solution tends to a constant).

b) In the first term on the right hand side of equation (10) instead of function $\kappa = \kappa_0 = \text{const}$, we take the function $\kappa = \kappa_0 p^\beta$, $\beta > -1$, i.e. the responsibility of parts of the hierarchy depends on the value of their power. In comparison with the previous cases the results are not changed essentially.

In cases a), b) these conclusions follow from the analysis of fields of integral curves corresponding to nonlinear equations of the first order (see exercise 4).

6. Interpretation of results and conclusions. Thus, from 1) -3) from subsection 5, it follows that conclusions 1), 2) from subsection 4 (about the existence of stationary power distributions in legal system and the presence of phenomenon of excess of power even in such system) have much in common. If the conclusion 1), subsection 4 is unconditionally positive, the conclusion 2), subsection 4, has to be discussed in more detail, as it indicates the presence of an obvious power crisis.

In all cases considered in subsections 4, 5, the exit of power from the scope of power occurs for rather large values of parameters a and (or) σ (or their analogs). To understand the essence of the problem, we turn to the base model in subsection 4. In (11) the parameter a is determined by the formula

$$a = \frac{\kappa_0}{k_1 l^2}, \tag{15}$$

being the *system parameter* (containing some characteristics of the object), and also the *parameter of similarity* (for systems with various values of κ_0, k_1, l, but with identical values of a, the structure of power has the same qualitative form).

From (15) it is seen that a is increased with the growth of κ_0, characterizing the intensity of the mechanism of power transfer by the order along a ladder, and with a reduction of parameters k_1, l, i.e. of the reaction of society and of the length of hierarchical structure respectively.

Let k_1, l be fixed, and let κ_0 grows (a also grows with it). When κ_0 grows the responsibility of all parts of hierarchy decreases in the sense that the senior levels more easily "transfer" the fluxes of power to lower neighbors (see subsection 2). The limiting irresponsibility ($\kappa_0 = \infty$) means that any level becomes a kind of teletype, transmitting the orders and received portions of power "downwards" without any change. As a result an excess of power is being accumulated at the lower institutions of hierarchy (curve *3* in Fig. 67), which cannot be reduced by means of the reaction of society (it is fixed). Therefore the lowest levels are forced to leave their scopes of maximal imperious powers. A similar picture is observed at the change of parameter $\sigma = \chi_0 l/k_1$ (order over a head).

The bureaucratic content of condition (13) and its analogs is that if the characteristics of hierarchy and of civil society correlated in a certain way, the power distribution remains within a legal area (and vice versa).

The reasoning accompanying the analysis of relation (15) expresses the general contradiction inherent to hierarchical structure. It is as follows. On one hand, in a hierarchy there should by necessity be a sufficient overfall of imperious powers corresponding to its highest and lowest parts, otherwise the hierarchy is meaningless. On the other hand, the mechanisms of power transfer within a hierarchy always act to smooth the power difference between levels. If the given mechanisms become even stronger, the reaction (remaining without changes) of civil society even in a legal system will not be capable of competing with these processes and compensating the smoothing. The distribution of power becomes more and more flat and is forced to partially leave the scope of powers. Note that the given property occurs not only for the legal system "power–society".

Thus, the following two conclusions are true:

1) there is always a unique stationary power distribution in a hierarchical structure for a legal system, and the magnitude of power monotonously falls with transition from the higher to lower levels;

2) even in a legal system there is always an area of parameters, in which the realization the power distribution oversteps the boundaries of the legal field.

Equally common is the conclusion 3) of subsection 4 about the impossibility of realizing ideal power distribution as well, i.e. such a distribution, when the reaction of society would universally be zero. This conclusion is established in a simple way, and its content is rather clear: if there was zero reaction the hierarchy would cease to interact with the partner – civil society. At the same time, according to the model, the hierarchical chain is a subject of opposing influences (support and resistance of society, "gain" and "loss" of power in levels). In the conditions of a legal system their dynamic competition forms in a result of stationary structure of power distribution just as the static competition between forces of gravity and elasticity determines the form of a cable, hanging between pillars. In the absence of a reaction the only possible solution in this case would be the power distribution which was constant in space and in time (exercise 4). From this conclusion, in particular, it follows that when the difference between the maximal and minimal imperious powers is reduced (in a limiting case it means the equality $p_1 \equiv p_2 \equiv p^0$) power distribution always leaves the legal area. In other words, for the actions of a hierarchy it is not necessary to allocate too narrow a legal field, otherwise it will be "forced" to leave the legal areas of power.

When constructing the considered model of a system "power–society" and its generalizations, analogies with natural scientific objects were widely used: the transition from discrete to continuous model (a hierarchy is treated as continuous environment); analog of Fourier's law for description of mechanisms of redistribution of power within hierarchy; "the law of conservation of power" at the derivation of the basic equation; the concept of "power flux", etc. These analogies allow reasonable interpretation, their application enables deeper understanding of the basic properties of hardly formalizable object.

EXERCISES

1. The equation of a type $u_t = L(u_{xx}, u_x, u, x, t)$ is called *parabolic*, if $\partial L/\partial u_{xx} > 0$. Obtain the conditions for function $\kappa(p, p_x, x, t)$ in equation (7),

written down without the integral term, when it is parabolic.

2. By putting $\chi = 0$, using the expansion of functions via Taylor series and keeping the main terms, perform the transition from model (4)–(6) to (7)–(9). Replacing the derivative by differences, execute the inverse transition from model (7)–(9) to (4)–(6).

3. Via replacement $v(x) = p(x)(1+\sigma) - (1-bx) - \sigma B$, where $B = \int_0^l p(x')\,dx'$, obtain from (14) the boundary problem for function $v(x)$.

4. Representing the stationary equation (10) as a system of two equations, find for cases $\kappa = \kappa_0$, $\kappa = \kappa_0|\partial p/\partial x|^\alpha$, $\alpha > -1$, and $\kappa = \kappa_0 p^\beta$, $\beta > -1$ the replacements reducing it to an equation of first order.

5. Check that at $F \equiv 0$, $\chi \equiv 0$ the solution of equation (7) at conditions (8) in a stationary case has a form $p(x) = \text{const}$.

Bibliography for Chapter IV: [1, 5, 9, 10, 37, 38, 40, 48–50, 56, 58, 85, 86].

Chapter V

STUDY OF MATHEMATICAL MODELS

1 Application of Similarity Methods

We will now give the characteristics of ways of simplifying of mathematical models based on the properties of their symmetry. We will represent the description of self-similar (automodel) phenomena and study some automodel processes for nonlinear parabolic and hyperbolic equations.

1. Dimensional analysis and group analysis of models. One of the fundamental properties of natural, technological, many economic and social objects – symmetry (similarity, repeatability, reproducibility) – is reflected in their mathematical models. The presence of any kind of symmetry for the investigated phenomenon means essential simplicity of the object in comparison with its less symmetrical analog. This is the basis of widely used methods of simplifying mathematical models and, hence, of methods of simplifying of their analysis. They include the reduction of the order of the system of equations describing the model, of the number of variables on which the sought quantities depend, or of the number of constant parameters determining the process, etc. (thus the symmetry of the three-variable function with respect to those variables enabled us to find the largest possible speed of a three-stage rocket; see subsection 4, section 1, Chapter 1).

A typical approach to the use of properties of symmetry is a dimensional analysis of the quantities included in the model. The part of characteristics of objects is measured in certain units with direct (mechanical, physical, economic, etc.) content. For example, the mass in grams, temperature in Kelvin degrees, national gross product in roubles. Such quantities are called *dimensional*, their numerical value depends on a choice of units of measure. Among them are distinguished the quantities with independent (basic) dimension, or *dimension independent* ones. For example, if for the description of a mechanical phenomenon the CGS system of units (centimeter, gram, second) is used, then the dimensions of length x, mass m and time t are independent and are not expressed through each other. On the contrary, the dimension of kinetic energy $E = mv^2/2$ is determined through the dimensions of the basic quantities via the formula $[E] = [m][x]^2[t]^{-2} = \mathrm{g} \cdot c^2 \cdot \mathrm{s}^{-2}$, called *the dimension formula* (here $v = dx/dt$, the symbol $[f]$ denotes the dimension of quantity f). Such quantities are called *dimension dependent.* Recall that the phenomena and processes can be described also by dimensionless variables, say, by the ratio of the length of water carrier layer to its width, power index in a formula defining the dependence of the coefficient of thermal conductivity from temperature, annual bank percentage, etc.

The systems of units of measure can be chosen differently, so that the connections between quantities describing the object (derived from the laws of nature or from other considerations), should not change along with the units of measure. For example, Newton's second law $F = ma$ (F is the force, a is the acceleration) in the SI system is written precisely in the same way as in the CGS system. The invariance of phenomena and processes with respect to the change of units of measurements is reflected via so-called Π-theorem.

Consider a functional connection

$$a = F(a_1, a_2, \ldots, a_k, a_{k+1}, \ldots, a_n) \tag{1}$$

between $n+1$ dimensional quantities $a, a_1, \ldots, a_n$, where $a_1, \ldots, a_k$ have an independent dimension, and let this connection not depend on a choice of system of units of measure (a is the sought quantity, the rest are given).

Then, the connection (1) can be rewritten as

$$\Pi = F(\underbrace{1, \ldots, 1}_{k}, \underbrace{\Pi_1, \ldots, \Pi_{n-k}}_{n-k}), \tag{2}$$

i.e. as a relation between $n+1-k$ quantities $\Pi, \Pi_1 \ldots, \Pi_{n-k}$, representing a dimensionless combination from $n+1$ dimensional quantities $a, a_1, \ldots, a_n$.

Thus the quantities $\Pi, \Pi_1, \ldots, \Pi_{n-k}$ are connected with $a, a_1, \ldots, a_n$, via a simple ratio

$$\begin{aligned} a &= \Pi a_1^{m_1} a_2^{m_2} \ldots a_k^{m_k}, \\ a_{k+1} &= \Pi_1 a_1^{l_1} a_2^{l_2} \ldots a_k^{l_k}, \\ &\ldots\ldots\ldots\ldots\ldots\ldots, \\ a_n &= \Pi_{n-k} a_1^{p_1} a_2^{p_2} \ldots a_k^{p_k}. \end{aligned} \tag{3}$$

Here the power indices $m_1, \ldots, m_k; l_1, \ldots, l_k; \ldots, p_1, \ldots, p_k$ are the same as in the corresponding formulae of dimensions for dimension dependent quantities $a, a_{k+1}, \ldots, a_n$, for example in the formula $[a] = [a_1]^{m_1} [a_2]^{m_2} \ldots [a_k]^{m_k}$.

The proof Π-theorem is based on the invariance of connection (1) concerning the units of measure. First of all we rewrite (1) in *dimensionless form*, in view of the fact, that any dimension independent quantity a_i, $i = 1, \ldots, k$, can be represented as $a_i = \bar{a}_i \alpha_i$. Here, the dimensionless coefficient $\bar{a}_i$ is the numerical value of a_i in the used system of units, and the multiplier α_i has a dimension a_i, and characterizes the scale of measurement (tens or hundreds of pounds, hundreds or thousands of degrees, millions or billions of roubles, etc.). The numerical values of dimensionless multipliers for dimensional dependent quantities $a, a_{k+1}, \ldots, a_n$ are calculated using *scale multipliers* α_i, $i = 1, \ldots, k$, by a rule

$$\bar{a} = \frac{a}{\alpha_1^{m_1} \alpha_2^{m_2} \ldots \alpha_k^{m_k}}, \quad \bar{a}_{k+1} = \frac{a_{k+1}}{\alpha_1^{l_1} \alpha_2^{l_2} \ldots \alpha_k^{l_k}}, \quad \bar{a}_n = \frac{a}{\alpha_1^{p_1} \alpha_2^{p_2} \ldots \alpha_k^{p_k}},$$

directly following from the formulae of dimensions for each of them. The ratio (1) can be treated also as a connection between numerical values of $a, a_1, \ldots, a_n$, (i.e. a connection between dimensionless quantities $\bar{a}, \bar{a}_1, \ldots, \bar{a}_n$), not depending, by assumption, on units of measure. Thus, for any sample of scale multipliers α_i, one has

$$\bar{a} = F(\bar{a}_1, \bar{a}_2, \ldots, \bar{a}_k, \bar{a}_{k+1}, \ldots, \bar{a}_n),$$

or

$$\frac{a}{\alpha_1^{m_1} \alpha_2^{m_2} \ldots \alpha_r^{m_k}} =$$

$$= F\left(\frac{a_1}{\alpha_1}, \frac{a_2}{\alpha_2}, \ldots, \frac{a_k}{\alpha_k}, \frac{a_{k+1}}{\alpha_1^{l_1} \alpha_2^{l_2} \ldots \alpha_k^{l_k}}, \ldots, \frac{a_n}{\alpha_1^{p_1} \alpha_2^{p_2} \ldots \alpha_k^{p_k}}\right).$$

Let us put now $\alpha_1 = a_1, \alpha_2 = a_2, \ldots, \alpha_k = a_k$. In other words, we choose the scale multipliers so that in the obtained system of units of measure of quantities $\bar{a}_1, \ldots, \bar{a}_k$ are identically equal to unity. Then (2) and (3) immediately follow from the latter ratio.

The application of the Π-theorem reduces the number of quantities appearing in the description of an object, and gives an obvious way of representing the sought variable a (and of $a_{k+1}, \ldots, a_n$) through Π, Π_1, Π_{n-k} and $a_1, \ldots, a_k$. It is "indifferent" to a particular kind of functional dependence (1), only the sufficient smoothness of function F is required.

In particular, if $n = k$, then as it readily follows from (2), $\Pi = \text{const}$, and

$$a = \text{const} \cdot a_1^{m_1} a_2^{m_2} \dots a_k^{m_k} ,$$

i.e. the solution is a simple expression of given parameters (to obtain the exact value of a, one has to determine the constant). Let, for example, it be known, that the period of small oscillation of a pendulum T (see subsection 3, section 3, Chapter 1) does not depend on its initial deviation and velocity, and is determined only by the length l, mass m and gravity acceleration g. The functional dependence $T = T(l, m, g)$ contains four dimensional variables, three of them have independent dimensionalities. We select those as T, l and m, then for the dimensionality g we have $[g] = [l][T]^{-2}$, or $[T] = [l]^{1/2}[g]^{-1/2}$, whence

$$T = \text{const} \cdot \sqrt{l/g} .$$

Within the accuracy of a dimensionless coefficient this formula coincides with that obtained from the solution of the equation of oscillations of a pendulum (incidentally, it is clear that their period does not depend on m).

Note that the dimensionless parameters describing the object at a variation of units of measurement do not vary, and consequently they do not appear in the Π-theorem.

The procedure exclusion of dimensions (*process of scaling*) is always useful in the study of mathematical models, since it can provide important preliminary information about the object. For example, by scaling the problem in subsection 4, section 4, Chapter IV, it became clear that its solution is determined actually not by four, but only by one parameter.

The dimensionless quantities $\Pi, \dots, \Pi_{n-k}$ (obtained using the Π-theorem) can be called *similarity parameters (criteria)* in the sense that phenomena and processes which are different in scale but identical in essence behave qualitatively uniformly for a given sample of parameters $\Pi_1, \dots, \Pi_{n-k}$ (and vary uniformly in their modification).

The invariance of models relative to a system of units of measurement is a particular case of more common properties of their symmetry. The most detailed and widely used approach using the similarity of models is based on the so-called *invariant-group method* of analyzing of differential equations. Indeed, the majority of differential equations representing the constituent of mathematical models of various phenomena remain unchanged *(invariant)* for some transformations of independent variables and sought functions.

For example, the equation of heat transfer (5) from section 2, Chapter II

$$c \frac{\partial T}{\partial t} = \operatorname{div}(\kappa \cdot \operatorname{grad} T) \tag{4}$$

is invariant relative to a "shift" of time $t' = t + t_0$ and, if the functions c and κ do not depend on $\vec{r}$, also to a "shift" of coordinates $\vec{r}' = \vec{r} + \vec{r}_0$. In the particular case $c = c_0$, $\kappa = \kappa_0 T^\sigma$, i.e. power dependence of thermal conductivity on temperature, (4) is not changed at transformations of "expansion–compression": $t' = \alpha t$, $\vec{r}' = \beta\vec{r}$, $T' = \gamma T$ (the numbers α, β, γ should satisfy some conditions; see exercise 1).

It is easy to see that one-dimensional equations of gas dynamics for an ideal polytropic gas have similar properties

$$\frac{\partial}{\partial t}\left(\frac{1}{\rho}\right) = \frac{\partial v}{\partial m}, \qquad \frac{\partial v}{\partial t} = -\frac{\partial p}{\partial m}, \qquad \frac{\partial}{\partial t}\left(p\rho^{-\gamma}\right) = 0, \tag{5}$$

rewritten in mass coordinates (see subsections 5 and 7, section 4, Chapter II). It is easy to prove that in a certain sense invariance is inherent to the majority of models constructed in Chapters I–IV.

The above considered transformations of variables and their various combinations and generalizations concern a class of so-called point transformations, or Lie transformations (the derivatives of quantities are not transformed.) Let them also satisfy some additional conditions (represent themselves a group). Then the differential equation adopting a Lie group can be simplified: either its order is reduced, or the number of independent variables determining the sought functions is reduced. One of the aims of the group analysis is the definition of all groups of transformations allowed by the given equation or system of equations, and also to define the corresponding *invariant solutions* of equations. From the point of view of the study of mathematical models, the most important aspect is not the complicated and tedious procedure of group analysis, but rather the final results concerning the concrete equation (for majority of basic mathematical models these results are obtained).

We represent as an important particular example, the outcome of the group analysis of equation (4) at $c = 1$, $\kappa = \kappa(T)$ in the one-dimensional case (Table 3). The numbers in the left column indicate various group transformations, and in the Table their corresponding invariant solutions for various forms (specifications) of function $\kappa(T)$ are represented. The sign $\ll$ indicates repetition of expression standing in a line to the left; the dash indicates the absence of invariant solutions. The invariant, i.e. the combination not varying for a given transformation, is denoted by ξ.

At substitution of the invariant solution $T(x, t)$ into the one-dimensional equation (4), an ordinary differential equation for the function $f(\xi)$ (also an invariant of transformation) is obtained. Its study is much easier than that of the initial equation in partial derivatives. Hence, it is not difficult to write down the function $T(x, t)$ through $f(\xi)$ and to study its properties, if the properties of $f(\xi)$ are known.

The group transformations 1–4 are the basic ones, since they are allowed at arbitrary specifications of function $\kappa(T)$. The case 1 is the stationary solution (invariance of (4) relative to a shift of t); 2 is the constant in space solution (invariance to a shift of x); 3 is a solution of a travelling wave type (invariance to a simultaneous shifts of t and x); 4 is a solution with constant value of function $T(x, t)$ in a point $x = 0$ (invariance relative transformations of type $t' = \alpha^2 t$, $x' = \alpha x$, $T'(x', t') = T(x, t)$).

Table 3

Group classifications of the equation

$$\frac{\partial T}{\partial t} = \frac{\partial}{\partial x}\left(\kappa(T)\frac{\partial T}{\partial x}\right)$$

N	$\kappa(T)$ arbitrary	$\kappa(T) = e^T)$	$\kappa(T) = T^\sigma$ $(\sigma \neq -4/3)$
1	$T = f(\xi),\, \xi = x$	$\ll$	$\ll$
2	$T = f(\xi),\, \xi = t$	$\ll$	$\ll$
3	$T = f(\xi),\, \xi = x - t$	$\ll$	$\ll$
4	$T = f(\xi),\, \xi = x^2/t$	$\ll$	$\ll$
4	–	$T = \frac{1}{\alpha}\ln x + f(\xi)$, $\xi = tx^{\frac{1-2\alpha}{\alpha}}$; $T = \ln t + f(\xi)$, $\xi = x(\alpha = 0)$	$T = t^{1/(2\alpha)} f(\xi)$, $\xi = xt^\beta, \beta = -(\alpha + \sigma/2)/(2\alpha)$; $T = t^{-1/\sigma} f(\xi)$, $\xi = x(\alpha = -\sigma/2)$
5	–	$T = x + f(\xi)$, $\xi = te^x$	$T = e^t f(\xi)$, $\xi = xe^{-\sigma t/2}$
6	–	$T = 2t + f(\xi)$, $\xi = xe^{-t}$	$T = e^t f(\xi)$, $\xi = te^{\sigma x}$

The extensions of the basic group (called *subgroups*) occur at concrete definitions $\kappa(T) = e^T$, $\kappa(T) = T^\sigma$ of the function $\kappa(T)$. Thus, for solutions of the fifth subgroup the quantity ξ is invariant with respect to a transformation of expansion–compression, given in this case by an input equation (we have to explain: the solution 4, in fact, corresponds to a particular case of transformations of expansion–compression for any function $\kappa(T)$, while for $\kappa(T) = T^\sigma$ they correspond to solutions with separating variables). The solutions for subgroups 6 and 7 are generated by a combination from transformations of shift and expansion–compression.

Note: the form of almost all invariant solutions from Table 3 was obtained even before the group analysis from other considerations (from the theory of dimensionalities, passage to the limit, by guessing solutions and direct substitution into the equation (4)). However it does not affect the advantages of the given method, for at least for two reasons. It gives the strict and exhaustive answer on the basic group of transformations and those specifications of equations (models), when this group is extended (beforehand it is known that other specifications of $\kappa(T)$ in equation (4), say, of the form $\kappa(T) = \ln(1 + T)$, do not allow solutions additional to the basic ones). Besides, the symmetry properties of the equations can be rather diverse and unexpected. For example, for case $\kappa = T^\sigma$ $\sigma = -4/3$ not described in Table 3, the equation (4) has a large number of nontrivial transformations, defined only with the help of group analysis, representing a regular way of simplifying mathematical models.

We have to outline also that this method studies properties of the differential equation as such, i.e. the properties of only of the part of a model of an initial object (without other input data, for example without boundary conditions). Therefore the applicability of an obtained invariant solution to describe a concrete phenomenon should be investigated additionally. As distinct from the group analysis, the theory of dimensionalities (it can be considered as an application of particular case of a group symmetry implying the expansion–compression of units of measurement) and Π-theorem following from it, are dealing with it in all its completeness.

2. Automodel (self-similar) processes. Among invariant solutions of differential equations an important class of *self-similar*, or *automodel* solutions is distinguished (the reason for this name will become clear below). They include such widely used solutions as that of the travelling wave type (case 4 in Table 3 for $\kappa = T^\sigma$, see also exercise 4, section 2, Chapter II), power law automodel solutions (case 5; see also exercise 3, section 2, Chapter II) and exponential automodel solutions (cases 6, 7).

We shall prove the usefulness of construction and analysis of automodel solutions for the study of mathematical models. As a first example we consider solutions of the travelling wave type for a set of equations (5). They are searched as

$$\rho\,(m,\,t) = \rho\,(\xi) = \rho\,(m - Dt)\,,$$

$$v\,(m,\,t) = v\,(\xi) = v\,(m - Dt)\,,$$

$$p\,(m,\,t) = p\,(\xi) = p\,(m - Dt)\,,$$

where $D > 0$ is a constant. Substituting these expressions into (5) and replacing (for uniformity) the third equation by the divergence equation of energy (21), section 4, Chapter II, instead of partial equations, we obtain a system of three ordinary differential equations

$$D\,\frac{d}{d\xi}\,\frac{1}{\rho} + \frac{dv}{d\xi} = 0, \qquad -D\,\frac{dv}{d\xi} + \frac{dp}{d\xi} = 0, \qquad -D\,\frac{d}{d\xi}\left(\varepsilon + \frac{v^2}{2}\right) + \frac{d}{d\xi}\,(pv) = 0,$$

where $\varepsilon(\xi) = \varepsilon\,(m - Dt)$ is the gas internal energy ($\varepsilon = \varepsilon(p,\,\rho)$). Integrating them within arbitrary limits ξ_0, ξ_1, we prove that the travelling wave corresponds to flows, when three integrals are conserved

$$D\frac{1}{\rho} + v = C_\rho\,, \qquad -Dv + p = C_p\,, \qquad D\varepsilon + D\frac{v^2}{2} - pv = C_\varepsilon\,. \qquad (6)$$

In case of continuous flows the integrals (6) give a unique solution – constants values of ρ, v, p, εat all ξ.

A nontrivial result is obtained if one takes into account that the gas dynamics equations due to existence of a "gradient catastrophe" can have *discontinuous solutions* (see subsection 7, section 4, Chapter II). Let the jump of functions $\rho(\xi)$, $v(\xi)$, p, (ξ) be located at $\xi = 0$. In areas $\xi > 0$, $\xi < 0$, i.e. to its right and left, the flow is obviously constant and is characterized by a set of quantities ρ_0, v_0, p_0 and ρ_1, v_1, p_1 (see Fig. 68, where the solution is represented as function of a coordinate m at t_1, t_2, t_3; $t_1 < t_2 < t_3$). The jump of gas dynamical parameters is not arbitrary. We show it, analyzing the content of the above written integrals (valid also for discontinuous flows).

The shock described by an automodel solution is moving through the gas with constant velocity D. The invariance of mass velocity ensures the equality of the matter flux $I_\rho = D$, "inflowing" from the one side of the shock and "outflowing" from its other side. The violation of this natural physical requirement would mean the appearance or disappearance of the matter at transition through it. The mass "inflow" is equal by definition to $I_{\rho_1} = D = \rho_0\,(u - v_0)$, while the "outflow" is $I_{\rho_1}(u - v_1)$, where u is the Eulerian velocity of a moving shock, and $u - v_0 > 0$. The first integral (6), as follows from equality $I_\rho = I_{\rho_0} = I_{\rho_1}$, represents an identity $u = u$.

Turn to the second integral (6), rewriting it for quantities with subscripts 0 and 1

$$C_{p_0} = p_0 + \rho_0\,(u - v_0)^2 - I_\rho u \equiv I_{p_0} - I_\rho u\,,$$

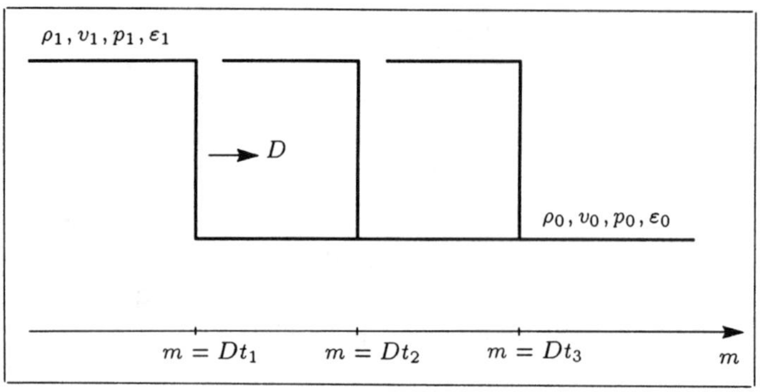

Fig.68.

$$C_{p_1} = p_1 + \rho_1 (u - v_1)^2 - I_\rho u \equiv I_{p_1} - I_\rho u \, .$$

The quantities appearing in these equalities, I_{p_0}, I_{p_1} are the fluxes of momentum I_p from the right and left of a shock. Since $C_p = C_{p_0} = C_{p_1}$, they also are equal $I_p = I_{p_0} = I_p$. Otherwise at transition through an infinitely thin surface of the shock the particle of substance would gain an increment of momentum, which is possible only under the influence of infinitely large forces.

Finally, the third integral (6) after simple calculations can be rewritten for both sides of the shock in the form

$$C_{\varepsilon_0} = I_{\varepsilon_0} - u \left(I_{p_0} + I_{\rho_0} \frac{u}{2} \right) , \qquad C_{\varepsilon_1} = I_{\varepsilon_1} - u \left(I_{p_1} + I_{\rho_1} \frac{u}{2} \right) ,$$

where $I_\varepsilon = \rho(u-v)\varepsilon + \rho(u-v)(u-v^2)/2 + p(u-v)$ is the energy flux. In so far as in a travelling wave C_ε is fixed and is constant, and as was already established above, constants are also I_ρ, I_p, the energy flux I_ε is constant as well, i.e. $I_{\varepsilon_0} = I_{\varepsilon_1}$. The energy "inflowing" into the shock is equal to the "outflowing" energy (otherwise this would mean the existence of sources of infinite intensity within the medium).

In view of the results following from the analysis of integrals (6), we come to a conclusion about the continuity of fluxes of mass, momentum and energy at transition through the surface of shock of gas dynamical parameters

$$I_{\rho_0} = I_{\rho_1} \, , \qquad I_{p_0} = I_{p_1} \, , \qquad I_{\varepsilon_0} = I_{\varepsilon_1} \, .$$

Note, that this conclusion is true also at $D = 0$ (*tangent shock*).

For the gas moving through a shock (*shock wave*) using the established equalities it is easy to find using the known velocity D *the Hugoniot conditions* – connection between the quantities before and after the jump (see exercises 2, 3). These conditions can also be obtained with the help of direct evaluation of fluxes on both sides of the shock wave. However the automodel solutions of the travelling wave type by virtue of their properties allow us to automatically obtain Hugoniot conditions for a large number of models of continuous media, in particular for situations, when the jump happens not in a infinitely thin layer, but is spatially extended due to the dissipative processes always present in the medium. Thus the equations for a travelling wave also describe the structure of a transition layer.

Note that the Hugoniot conditions permit jumps both of compression, when the pressure, density and internal energy of gas behind the shock wave are increasing, but also, formally, of rarefaction jumps. The latter, however, can be realized only for appropriate physical processes, for example, for special types of chemical reactions.

Self-similarity or automodelity, of the constructed solution is well seen in Fig. 68: it is reproduced without modifications in various instants at various points of in a substance. The solution is understood in a general sense, as it satisfies the differential equations (5) only in the areas of continuous flow. Considered in area $-\infty < 0 < \infty$, it represents one of *the generalized solutions* of Cauchy problems for the equations of gas dynamics. If in any point with a fixed mass coordinate, say, in $m = 0$, the appropriate boundary conditions are given, it can be treated as a solution in area $m > 0$ of the problem about the piston, moving with constant velocity in a medium with constant parameters.

We will now study a power law automodel solution for a particular case of the equation (4)

$$\frac{\partial T}{\partial t} = \frac{\partial}{\partial x}\left(k_0 T^\sigma \frac{\partial T}{\partial x}\right), \tag{7}$$

describing the heat propagation within a infinite medium ($-\infty < x < \infty$) from an instantaneous point source. In a moment $t = 0$ in a point $x = 0$ an amount of heat Q_0 is released. The given statement completely coincides with the statement of a similar problem from subsection 5, section 2, Chapter II, with the difference, that the coefficient of thermal conductivity (more precisely, the *temperature conductivity*) $k(T) = k_0 T^\sigma$, $\sigma > 0$ is not constant, and is a growing power law function of temperature.

To determine an automodel solution we use the Π-theorem. Obviously, it depends on four parameters x, t, k_0, Q_o, i.e. $T = Tx, (t, k_0, Q_0)$. The given equality links the five dimensional quantities, three of which have independent dimensions. Therefore, in accordance with the Π-theorem it is reduced to a functional dependence $\Pi = F(\Pi_1)$ between two dimensionless quantities Π, Π_1.

Consider as dimensional independent quantities Q_0, x, t. Then from equation (7) for k_0 the formula of dimensionality $[k_0] = [Q_0]^{-\sigma}[x]^{\sigma-2/[t]}$

follows. The formula of dimensionality $[T] = [Q_0]/[x]$ for T is obtained from a condition

$$Q_0 = \int_{-\infty}^{\infty} T(x,\, t)\, dx\,, \qquad t \geq 0\,,$$

implying the conservation of an originally released energy in an infinite medium without sources and losses of heat. From Π-theorems and formulae of dimensionality we have

$$\Pi = T(x,\, t)\, x\, Q_0^{-1}\,, \qquad \Pi_1 = k_0\, t x^{-(2+\sigma)}\, Q_0^{\sigma}$$

and in view of the connection $\Pi = F(\Pi_1)$,

$$T(x,\, t)\, x\, Q_0^{-1} = F(k_0\, t x^{-(2+\sigma)}\, Q_0^{\sigma})\,.$$

We rewrite the latter equality in the form of Table 3, introducing the notation $\xi = \Pi_1^{-1/(\sigma+2)}$:

$$T(x,\, t) = Q_0 x^{-1} F(\Pi_1) = Q_0 x^{-1} \Phi(\xi) = Q_0^{\frac{2}{2+\sigma}} k_0^{-\frac{1}{2+\sigma}} t^{-\frac{1}{2+\sigma}}\, \xi \Phi(\xi)\,,$$

or denoting $\xi\Phi(\xi)$ through $f(\xi)$,

$$T(x,\, t) = Q_0^{\frac{2}{2+\sigma}} k_0^{-\frac{1}{2+\sigma}} t^{-\frac{1}{2+\sigma}}\, f(\xi)\,, \qquad \xi = Q_0^{-\frac{\sigma}{2+\sigma}} k_0^{-\frac{1}{2+\sigma}} t^{-\frac{1}{2+\sigma}}\, x\,. \tag{8}$$

The dimensionless invariant ξ is called *the automodel variable*; the function $f(\xi)$ is called *dimensionless function* (representative) of temperature; the dimensional multiplier in front of $f(\xi)$ in (8) is the scale multiplier. *A fixed automodel state* denotes the state corresponding to a fixed automodel coordinate $\xi = \xi_0$.

The analysis of dimensionalities enables us to obtain valuable preliminary information about the process, without deriving the complete solution of the problem. So, the rate of change of solution in time in a point $\xi = \xi_0$, in particular, in a point $\xi = 0$, is determined only by the scale factor. Therefore, in the beginning of coordinates $x = 0$ $(\xi = 0)$ the temperature decreases by the known law

$$T\,(0,\, t) \sim t^{-\frac{1}{2+\sigma}}\, f(0)\,.$$

The growth rate in time of coordinate x for state $\xi = \xi_0$ is also known

$$x\,(\xi_0) \sim t^{\frac{1}{2+\sigma}}\, \xi_0\,.$$

If $T(0,\, t)$ is the characteristic temperature of the heated area, and $x(\xi_0)$ is its characteristic scale, then their product $T(0,\, t) \cdot x(\xi_0)$ does not depend on time, in agreement with the condition of constance of thermal energy Q_0.

Substituting (8) into (7) and differentiating, we obtain a nonlinear ordinary differential second order equation with respect to $f(\xi)$

$$\frac{f}{1+\sigma} + \xi \frac{f'}{2+\sigma} = -(f^\sigma f')' .$$

Transforming it to a form

$$\frac{1}{(2+\sigma)} (\xi f)' + (f^\sigma f')' = 0$$

and integrating once,

$$\frac{1}{(2+\sigma)} \xi f + f^\sigma f' = C_1 .$$

Here $C_1 = 0$, since at $\xi = 0$ the solution is considered infinite ($f(0) < \infty$), it is smooth and is symmetrical by construction, i.e. $f'(0) = 0$. At $C_1 = 0$ the variables in the latter equation are separated and it is easily integrated

$$f(\xi) = \left[(\xi_\Phi^2 - \xi^2) \frac{\sigma}{2(2+\sigma)} \right]^{1/\sigma} , |\xi| \le \xi_\Phi .$$

In so far as the obtained formula at $|\xi| > \xi_\Phi$ gives negative values, in this area $f(\xi)$ has to be equal to zero (in points $|\xi| = \xi_\Phi$ the solution is "matched" with a trivial solution). The quantity $\xi_\Phi = \xi_\Phi(\sigma) > 0$ is determined from equality $\int_{-\infty}^{\infty} f(\xi)\, d\xi = \int_{-\xi_\Phi}^{\xi_\Phi} f(\xi)\, d\xi = 1$ – the dimensionless analog of condition of constancy of energy within the medium

$$\xi_\Phi = \left[\frac{(2+\sigma)^{\sigma+1} \, 2^{1-\sigma}}{\sigma \pi^{\sigma/2}} \frac{\Gamma^\sigma(1/2 + 1/3)}{\Gamma^\sigma(1/\sigma)} \right]^{\frac{1}{\sigma+2}} ,$$

where Γ is a gamma-function.

Using (8), we come to a final form of the solution of the problem on an instantaneous point heat source in a nonlinear medium

$$T(x, t) = \\ = \begin{cases} Q_0^{\frac{2}{2+\sigma}} k_0^{-\frac{1}{2+\sigma}} t^{-\frac{1}{2+\sigma}} \left\{ \left[1 - \frac{x^2}{x_\Phi^2} \right] \frac{\xi_\Phi \sigma}{2(2+\sigma)} \right\}^{1/\sigma} , & |x| \le x_\Phi(t) , \\ 0 , & |x| > x_\Phi(t) . \end{cases} \tag{9}$$

Here $x_\Phi(t) = \xi_\Phi (Q_0^\sigma k_0)^{1/(2+\sigma)} \, t^{1/(2+\sigma)}$.

The solution (9) is a generalized one, in so far as for rather large values of σ its derivatives by x and t in points $|x| = x_\Phi(t)$ do not exist and therefore it does not satisfy equation (7) in classical sense (see exercise 4). However, a natural physical requirement of continuity of a heat flux $W(x, t) = -k_0 T^\sigma \, \partial T/\partial x$ in points $|x| = x_\Phi(t)$, as well as for solution (21), section 2, Chapter II, is fulfilled (otherwise it would mean the presence in these points of sources or losses of energy of infinite intensity).

The heat propagates by a wave, covering in time newer and newer sites of substance (Fig. 69). It moves through the medium with a *finite velocity* (compare with solution (20), section 2, Chapter II, for case $\sigma = 0$). *The wavefront* is the point separating the heated part of space from the cold one. It is propagating by the law $x_\Phi(t) \sim t^{\frac{1}{2+\sigma}}$, corresponding simultaneously both to a fixed automodel state $\xi = \xi_\Phi$, and a fixed physical state $T(x_\Phi(t), t) = 0$.

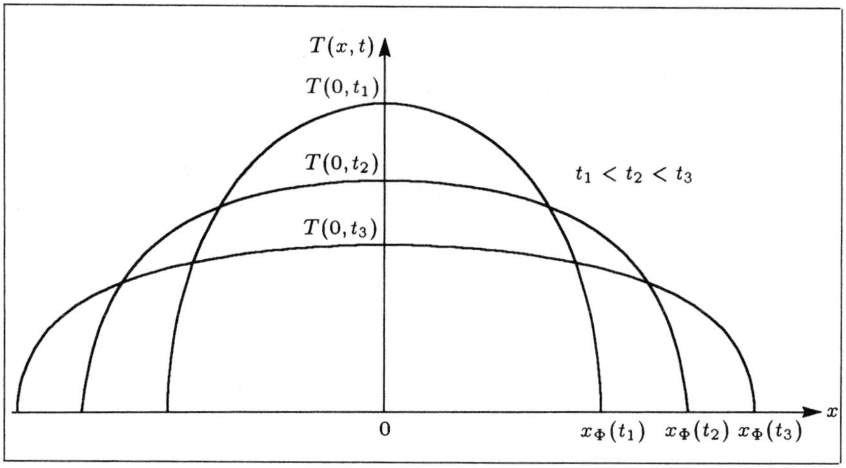

Fig.69.

The self-similarity of solution (9) has a slightly different geometric expression, than in the previous example. The curves in Fig. 69 can be combined in one (in function $f(\xi)$) after their "automodel handling" – corresponding transformations of expansion-compression both of arguments x, t, and function $T(x, t)$.

Automodel solutions play a special role in nonlinear models as important particular solutions revealing certain properties of objects. Their study stimulates the development of a kind of elementary language of nonlinear phenomena (shock waves, thermal waves, etc.). However their value is by no means limited to this. In certain conditions they serve as *intermediate asymptotics* for a large class of formally non-automodel processes. Obviously, the release of thermal energy in any part of a substance cannot be either instantaneous, or point like. Nevertheless its propagation in a sufficiently cold medium (at $k(T) = k_0 T^\sigma$, $\sigma > 0$), in a moment $t = 0$, after some large time scale and for large enough values of $x_\Phi(t)$ with good accuracy is described by solution (9). The details of the initial stage are "forgotten" and the process obeys the self-similar behavior (intermediate asymptotics).

3. Various cases of propagation of perturbations in nonlinear media. If the nonlinear model possesses rich group properties, there is a possibility of obtaining not separate solutions, but families of intermediate asymptotics with various behavior.

Consider various cases of propagation of heat described by equation (7). The process is considered in a semilimited medium $0 < x < \infty$, cold in a moment $t = t_0$

$$T\,(x,\, t_0) = T_0(x) = 0\,, \qquad 0 < x < \infty\,, \tag{10}$$

and heated from the left boundary in such a manner that the temperature in a point $x = 0$ grows in time by the law

$$T\,(0,\, t) = A_0\,(t_f - t)^n\,, \qquad n < 0\,, \quad t_0 \le t \le t_f < \infty\,. \tag{11}$$

The problem (7), (10), (11) is meaningful only at $t < t_f$, as in a final moment $t = t_f$ the boundary temperature turns to infinity. Similar types of solutions are called *blowing-up solutions* and are mathematical idealizations of many really observable phenomena determined by the geometry of objects, their nonlinearity, etc. They include, for example, the convergence of spherical shock waves to the center, closure of air bubbles in fluids and other forms of cumulation (geometric factor). Another example: the growth of population of the Earth $N(t)$ has been well described ones during the last few centuries by the law $N(t) = N_0(t_f - t)$, $t_f = 2026$ (years), following from the population model $dN/dt = \alpha_0 N^2$ (see subsection 2, section 6, Chapter I; strong nonlinearity).

The solution of problem (7), (10), (11) is obviously determined by quantities x, t, k_0, A_0, t_0, t_f. The functional dependence $T = T(x,\, t,\, k_0,\, A_0,\, t_0,\, t_f)$ includes seven parameters, three of which are dimension independent. We exclude the parameters t_0, t_f disturbing the "similarity". Consider (without a loss of generality) *the moment of blow-up* $t_f = 0$ and the process at $-\infty \le t < 0$, i.e. assume $t_0 = -\infty$ (the time is again increasing, directed from the past to future). The formulated problem in accordance with Π-theorem, is self-similar. Conducting calculations similar to those used in the derivation of (8), we obtain the sought solution

$$T\,(x,\, t) = A_0(-t)^n\, f(\xi)\,, \qquad \xi = k_0^{-1/2}\, A_0^{-\sigma/2}(-t)^{-\frac{1+n\sigma}{2}}\, x\,, \quad \xi \ge 0\,. \tag{12}$$

The substitution of (12) in (7), for determination $f(\xi)$ gives an ordinary differential second order equation (with boundary conditions following from (10), (11))

$$\frac{d}{d\xi}\left(f^{\sigma}\frac{df}{d\xi}\right)-\frac{1+n\sigma}{2}\,\xi\,\frac{df}{d\xi}+nf=0\,,\qquad f(\infty)=0\,,\quad f(0)=1\,. \tag{13}$$

The equation (13) in turn permits a similarity transformation $\xi'=a\xi$, $f'=\beta f$, $\beta=\alpha^{2/\sigma}$ (expansion-compression) and consequently is reduced to a first order equation (exercise 5), studied by standard methods. The analysis shows that the solution of problem (13) exists at all $n<0$, $\sigma>0$, and is unique and monotonous.

Its properties are determined by the relation between parameters n (growth rate of boundary temperature) and σ (nonlinearity of medium). The value $n=-1/\sigma$ is critical, when in equation (13) the second term is cancelled and it is integrated explicitly (exercise 6). Rewriting it with the help of (12) in initial variables, we have

$$T_S\,(x,\,t)=\begin{cases} A_0\,(-t)^{-1/\sigma}\left(1-\frac{x}{x_\Phi}\right)^{2/\sigma}, & x\le x_\Phi\,,\\ 0\,, & x>x_\Phi\,,\end{cases} \tag{14}$$

where

$$x_\Phi=x_S\equiv\left(2k_0A_0^{\sigma}\,\frac{\sigma+2}{\sigma}\right)^{1/2} \tag{15}$$

The first of boundary conditions (13) is fulfilled "in advance" (i.e. at $x=x_\Phi(k_0,\,A_0,\,\sigma)<\infty$), the heat flux on the front is equal to zero (the same as at solution (9)). The solution (14) is a generalized one, as its derivatives in a point $x=x_\Phi$ for a sufficiently high value of σ do not exist.

At $n<-1/\sigma$ the conditions on the boundary between the heated and cold substance are also fulfilled in a finite point $\xi_\Phi=\xi_\Phi(n,\,\sigma)<\infty$ (the value of the coordinate of front for concrete n, σ is estimated numerically). In its neighborhood the behavior of the solution is described by an asymptotic expression; its first term – the solution of a simplified equation following from (13) in assumption $f(\xi)\to 0$, $\xi\to\xi_\Phi$

$$f(\xi)=\begin{cases}\left(-\frac{1+n\sigma}{2}\,\sigma\xi_\Phi\right)^{1/\sigma}(\xi_\Phi-\xi)^{1/\sigma}+\ldots\,, & \xi\le\xi_\Phi\,,\\ 0\,, & \xi>\xi_\Phi\,.\end{cases} \tag{16}$$

Here and further dots denote terms of higher order of smallness. From (16) it is seen that the solution at $n\le 1/\sigma$ is a generalized one with the same order of smoothness as (9) (the smoothness of (14) is higher than that of (9) and (16)). However, heat flux in a point of front, as well as for (9) and (14), is continuous.

As distinct from the case $n \le -1/\sigma$, at $n > -1/\sigma$ the conditions at the front of waves are fulfilled only at $\xi \to \infty$. The solution is a smooth function at all $\xi \ge 0$, its asymptotic expansion near the front is given by the formula

$$f(\xi) = C\xi^{\frac{2n}{1+n\sigma}} + C_1\xi^{\frac{2n-2}{1+n\sigma}} + \dots\,, \qquad 1 + n\sigma > 0\,, \tag{17}$$

where $C = C(n, \sigma) > 0$(is obtained numerically) and $C_1 = -C^{\sigma+1}+[2n(2n+n\sigma-1)]/(1+n\sigma)^2 < 0$.

For all n, σ at $\xi \to \infty$ the solution has an asymptotics

$$f(\xi) = 1 + f'(0)\,\xi + \dots\,, \qquad f'(0) = \frac{df}{d\xi}(n, \sigma)\bigg|_{\xi=0} < 0\,,$$

meaning that the heat flux on the boundary of the medium is positive and the energy enters into the substance.

We will now analyze the physical properties of the constructed solutions, distinguishing cases $n = -1/\sigma$, $n < -1/\sigma$ and $n > -1/\sigma$.

1) At $n = -1/\sigma$ the front of a thermal wave is motionless. *The effective size* of the heated area, for example, *half-width*, i.e. a point $x_{\rm ef}(t)$ is such that $T(x_{\rm ef}(t), t)/T(0, t) = 1/2$ also does not vary in time. Self-similarity of the given solution, as well as of all solutions in separated variables (of the form $u(x, t) = U(t)\,V(x)$), expressed in a uniformity of its spatial profiles in various instants. The profiles differ in this case only by growing in time amplitude $U(t)$. The regular Bussinesque solution has the same property, when the amplitude is a decreasing function of time (exercise 2 of section 1, Chapter II).

It is possible to call solution (14) *a standing thermal wave* (see Fig. 70, where the profiles of a solution for medium with $\sigma = 2$ are shown, the crosses denote the half-width). At this regime at $t \to 0$, an infinite amount of energy is supplied into the substance; the temperature and thermal conductivity at all $0 \le x < x_S$ tend to infinity.

Nevertheless the heat does not penetrate further the point with coordinate $x = x_s$, defined by the intensity of a boundary condition A_0 and properties of medium k_0, σ. The solution (14) shows that the heat propagation can be *localized* within the area of finite scales – *localization area* (x_S is the *depth of localization*). Therefore there is a principal possibility of concentrating any amount of energy within limited areas of substance without its propagation outside the boundaries of the zone of localization.

2) In the case where $n < -1/\sigma$ the coordinate of wavefront of heat increases infinitely at $t \to 0$

$$x_\Phi(t) = k_0^{1/2} A_0^{\sigma/2} (-t)^{\frac{1+n\sigma}{2}} \xi_\Phi \to \infty\,.$$

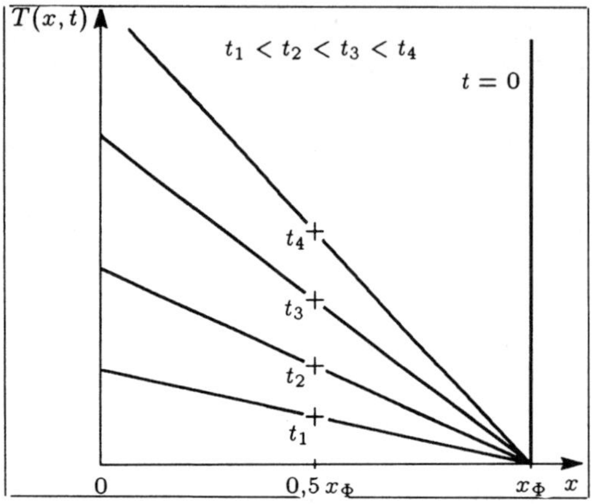

Fig.70.

The same is true for the wave half-width $x_{ef}(t)$ and coordinates of all points $x(t,\, \xi_0)$ with a fixed self-similar state $\xi = \xi_0$.

We now trace the variation of solution in a fixed point of the substance $x = x_0 < \infty$. The similar coordinate corresponding to it, $\xi(t,\, x_0)$, as it is seen from the second formula (12), tends to zero at $t \to 0$. Expanding the first formula (12), we obtain at $t \to 0$

$$T\,(x_0,\, t) = A_0(-t)^n\, f(\xi(x_0,\, t)) \to A_0(-t)^n f(0) = A_0(t)^n \;\to \infty\,.$$

Thus, while approaching the moment of sharpening, the wave covers the whole space, the temperature in any point of the medium increases infinitely, no localization is present. In this respect the solution is similar to the travelling wave (21) of section 2, Chapter II, however infinite parameters are reached not at $t \to \infty$, but in a finite time scale. A superfast heating of the medium is occurring.

3) The heat propagation is realized in an absolutely different manner at $n > -1/\sigma$. From (12) it is seen, that the half-width is reduced in time by the law

$$x_{ef}(t) = k_0^{1/2}\, A_0^{\sigma/2}(-t)^{\frac{1+n\sigma}{2}}\xi_{ef} \to 0\,, \qquad t \to 0\,.$$

The solution represents a thermal wave with decreasing effective depth of heating. The energy dissipating into the substance by time is being localized in the contracting area near the boundary (Fig. 71). The "front" of the thermal wave, as follows from (12) and (17), is located in a point $x = \infty$ (at $x_{ef} < \infty$ the size of the heated area would have decreased by time, which is not possible in a medium without absorption of energy).

For a fixed point of substance $x = x_0$ $(0 < x_0 < \infty)$ we have $\xi(x_0,\, t) \to \infty$, $t \to 0$, i.e. the coordinate $\xi(x_0,\, t)$ tends to the front's coordinate as distinct from the previous case. Using the asymptotics (17) and formula

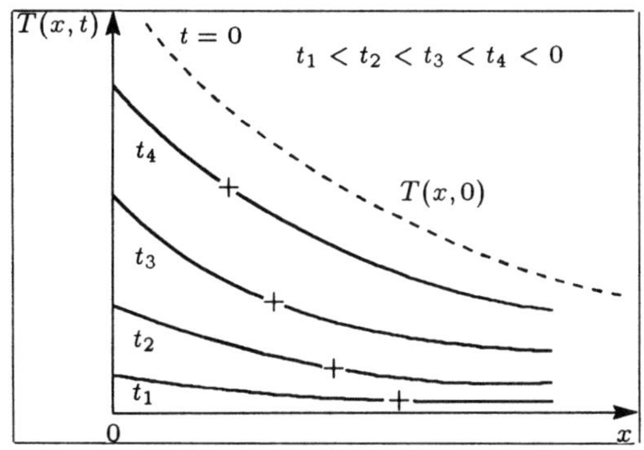

Fig.71.

(12), we obtain at $t \to 0$

$$T(x_0,\, t) \to C(k_0^{-n}\, A_0)^{\frac{1}{1+n\sigma}} x_0^{\frac{2n}{1+n\sigma}} + \\ + C_1\, (k_0^{1-n} A_0^{1-\sigma}) x_0^{\frac{2n+2}{1+n\sigma}} (-t) + \dots \tag{18}$$

Despite the infinite growth of temperature at a point $x = 0$, in all the remaining medium at all $t \le t_f$ it is limited from above via a limiting curve (first term in (18); the dashed line in Fig. 71). In the case of (3) heat localization occurs again, but in different regime than at $n = -1/\sigma$.

The analysis of similar solutions of the problem (7), (10), (11) has, with the help of simple mathematical tools enabled us to reveal the existence in a thermal conducting medium of three basically different situations of heat propagation and to study a series of their important properties. They are determined mainly by the velocity of energy supply to the medium at $t \to 0$ and the degree of its nonlinearity. For "slow" heating ($n = -1/\sigma$, *S-regime*, $n > -1/\sigma$, *LS-regime*) localization of heat occurs, at for "fast" heating ($n < -1/\sigma$, *HS-regime*) this effect is absent.

The sets of various self-similar solutions can be obtained not only for mathematical models reduced to the quasilinear parabolic equations, but also for equations of gas dynamics. Let us study the blowing-up regimes of propagation of perturbations for an elementary quasilinear hyperbolic equation – the Hopf equation (32) of section 4, Chapter II

$$\frac{\partial p}{\partial t} + k_0 p^\sigma \frac{\partial p}{\partial m} = 0\,, \tag{19}$$

rewritten with respect to the pressure $p = p(m,\, t)$, $\sigma = (\gamma+1)/(2\gamma) < 1$.

Consider for it the problem of a piston contracting the gas occupying a half-space $m > 0$, with zero initial pressure

$$p(m,\, t_0) = p_0(m) = 0\,, \qquad 0 < m < \infty\,. \tag{20}$$

The piston is located at a point $m = 0$, the pressure on it is blowing-up

$$p(0, t) = A_0(t_f - t)^n, \qquad n < 0, \quad t_0 \le t < t_f < \infty. \tag{21}$$

The problem (19)–(21) is similar, if similarly to the previous case we put $t_0 = -\infty$ (as before, without any loss of generality, we adopt $t_f = 0$). Its solution in accordance with the Π-theorem has the form

$$p(m, t) = A_0(-t)^n f(\xi), \qquad \xi = k_0^{-1} A_0^{-\sigma}(-t)^{-(1+n\sigma)} m, \quad \xi \ge 0. \tag{22}$$

From (19) and (22) we obtained for the determination of $f(\xi)$ the first order equation

$$\frac{df}{d\xi}(f^\sigma + (1 + n\sigma)\xi) - nf = 0, \tag{23}$$

which, rewritten in the form

$$\frac{d\xi}{df} = \frac{f^\sigma + (1 + n\sigma)\xi}{nf},$$

is linear with respect to ξ. It has a general solution

$$\xi = f^{\sigma + 1/n} - f^\sigma, \tag{24}$$

satisfying the condition $f_0 = 1$ following from (21) (in the analysis of (24) it is also necessary to fulfill condition $f(\infty) = 0$, following from (20)). The properties of the solutions depend on the relations between n and σ.

1) At $n = -1/\sigma$ from (22) and (24) we obtain an explicit solution in separating variables (*the standing compression wave*, or gas dynamical S-regime)

$$p_S(m, t) = \begin{cases} A_0(-t)^{-1/\sigma}\left(1 - \dfrac{m}{m_\Phi}\right)^{-1/\sigma}, & m \le m_\Phi, \\ & m > m_\Phi, \end{cases} \tag{25}$$

where $m_\Phi = m_S \equiv k_o A_0^\sigma$ is the coordinate of a wavefront of compression, separating the moving gas from the unperturbed substance. The front of wave and its half-width are fixed in time, the pressure and all the other gas dynamical quantities – the velocity, density, etc. – increase infinitely within area $0 \le m < m_S$ at $t \to 0$. However at $m > m_S$ the gas remains motionless and cold at all $t < 0$ (in Eulerian coordinates this means that the piston contracts the nearby finite mass of gas m_S into an infinitely thin layer, in no way affecting the rest of the substance). The solution (25) demonstrates the effect of *localizations of compression* on finite depth m_S, determined by parameters of the problem, i.e. an effect quite similar to the localization of heat in the S-regime.

2) The same almost complete analogy is valid also in the case where $n > -1/\sigma$ (gas dynamical LS-regime). The solution, as is seen from (24), monotonously decreases with an increase of ξ and turns to zero at $\xi = \infty$ (the condition $f(\infty) = 0$ is fulfilled). The wavefront of compression is located at a point $m = \infty$, its half-width is

$$m_{\text{ef}}(t) = k_0 A_0^{\sigma} (-t)^{1+n\sigma} \xi_{\text{ef}}$$

tending to zero at $t \to 0$ (ξ_{ef} is found easily from (24) in view of equality $f(\xi_{\text{ef}}) = 1/2$). The energy passed to the gas by the piston is concentrated within an area decreasing in time near the boundary.

For arbitrary $0 < m_0 < \infty$ we have $\xi(m_0, t) \to \infty$, $t \to 0$. Therefore, solving (24) relative $f(\xi)$ at $\xi \to \infty$

$$f(\xi) = \xi^{\frac{n}{1+n\sigma}} + n(1+n\sigma)^{-1} \xi^{\frac{n-1}{1+n\sigma}} + \dots ,$$

and passing to initial variables with the help of (22), we obtain at $t \to 0$

$$p(m_0, t) \to \left(\frac{A_0}{k_0^n}\right) m_0^{\frac{n}{1+n\sigma}} + \frac{n}{1+n\sigma} \left(\frac{A_0^{\sigma+1}}{k_0^{n-1}}\right)^{\frac{1}{1+n\sigma}} m_0^{\frac{n-1}{1+n\sigma}} (-t) + \dots .$$

The solution tends from below to a limiting curve (first term in the obtained formula), turning at $t = 0$ to infinity only in a point $m = 0$.

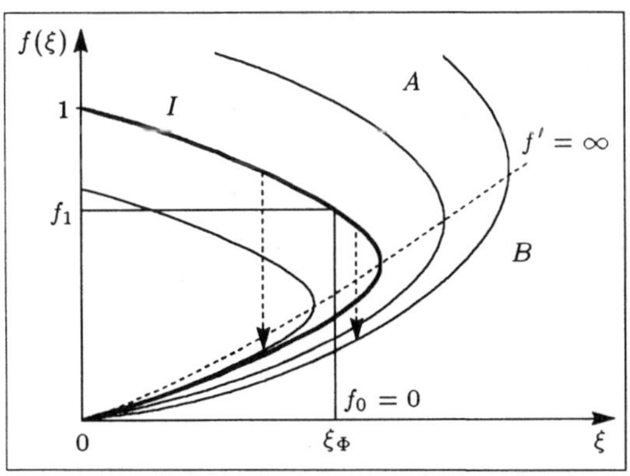

Fig.72.

3) For the analysis of the solution at $n < -1/\sigma$ we consider the field of the integrated curves of equation (23), shown in Fig. 72 (the bold curve I, for which $f(0) = 1$, the dashed line denotes the isocline of infinity dividing the plane into areas A and B). From this figure it is seen that there is no continuous solution fulfilling the requirement $f(\infty) = 0$, in so far as the

curve I cannot be continued into the area B for a line $f' = \infty$. For the same reason it is impossible to construct discontinuous solutions obtained with the help of jumps denoted by arrows. Thus, the only possibility is a transition from the curve I by jumping to the abscissa axis and continuing at $\xi > \xi_\Phi$; ξ_Φ is the coordinate of the jump, of null solution (the value of function $f(\xi) = 0$ satisfies equation (23)).

The condition of a jump of the solution to equation (19) can be obtained by analogy with equations (5), by presenting (19) in a divergence form

$$\frac{\partial p}{\partial t} + \frac{\partial \varphi(p)}{\partial m} = 0\,,$$

where $\varphi(p) = k_0 p^{\sigma+1}/(\sigma+1)$, and by constructing its discontinuous solution of the travelling wave type $p(m,\,t) = p(m - Dt)$. The values of function p before and after the shock are connected by a relation

$$D = \frac{\varphi(p_0) - \varphi(p_1)}{p_0 - p_1}\,.$$

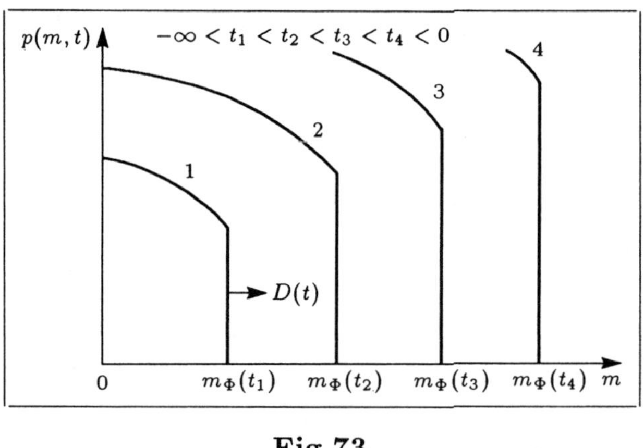

Fig.73.

From here for the similar solution (22) we have

$$\bar{D} = -(1 + n\sigma)(\sigma + 1)\,\xi_\Phi = \frac{f_0^{\sigma+1} - f_1^{\sigma+1}}{f_0 - f_1}\,,$$

where ξ_Φ is the similar coordinate of the jump, D is the dimensionless velocity connected to the instantaneous velocity $D(t)$ by relation $D =$

$(\sigma+1)^{-1}k_0A_0^\sigma(-t)^{n\sigma}\,\bar{D}$. For a case where $f_0 = 0$ we obtain the connection between f_1 and ξ_Φ

$$f_1 = [-(1+n\sigma)(\sigma+1)\xi_\Phi]^{1/\sigma},$$

using the obvious fact that the point $(f_1,\, \xi_\Phi)$ lies above the line $f' = \infty$, i.e. the sought discontinuous solution really does exist.

The solution as a function of m is represented in various instants in Fig. 73. Its front $m_\Phi(t)$ and half-width m grow infinitely in time by the law

$$m_\Phi(t) \sim m_{\text{ef}}(t) \sim (-t)^{1+n\sigma} \to \infty\,, \qquad t \to 0\,.$$

Finally the perturbation covers the whole space, the solution in any point $m_0 < \infty$ tends to infinity at $t \to 0$ (exercise 7).

Thus, as for the processes of thermal conduction, during "slow" blowing-up regimes the perturbations are localized, while for the "fast" case, the localization is absent.

Note that equation (19) is deduced from equations (5) under the assumption that the flow is continuous and isentropic (simple wave). In so far as the constructed solutions in case $n \geq -1/\sigma$ are continuous, they satisfy (5) and permit direct gas dynamical interpretation (as distinct a discontinuous solution at $n < -1/\sigma$).

EXERCISES

1. Check by a direct substitution that the equation (4) at $c = c_0$, $\kappa = \kappa_0T^\sigma$ is an invariant relative transformation of expansion-compression under the condition of $\alpha\beta^{-2}\gamma^\sigma = 1$.
2. Using (6), prove that on a shock not moving along the mass of gas (tangent shock, $D = 0$), the velocity and pressure are continuous.
3. The shock wave for which $p_1 \gg p_0$, is called *strong*. Obtain from (6) the Hugoniot condition for the density of such a shock (in case of ideal gas $\varepsilon = p/((\gamma-1)\rho)$), assuming for simplicity that $v_0 = 0$. Show that on a strong shock wave the density jump does not depend on D and is given by the formula $\rho_1/\rho_0 = (\gamma+1)/(\gamma-1)$.
4. Establish that at $\sigma \geq 1$ the first derivatives of solution (9) by x and t are discontinuous in the point $x = x_\Phi(t)$, while at $\sigma \geq 1/2$ the second derivatives are discontinuous.
5. Check that both the change of variables $\eta = \ln\xi$, $\eta = \xi^{2/\sigma}\varphi(\eta)$, $\psi = d\varphi/d\eta$ or the replacement $\varphi = -\xi f^{-1}f^\sigma\, df/d\xi$, $\psi = -\xi f^{1-\sigma}/(df/d\xi)$ reduces (13) to a first order equation of a form $d\psi/d\varphi = \psi F(\psi,\, \varphi)/(\varphi\Phi(\psi,\, \varphi))$ (in the second case the equation is easier to investigate, since the functions F, Φ contain simple nonlinearity of a form $\psi\varphi$).
6. The equation (13) at $n = 1/\sigma$ can be rewritten as $(f^{\sigma+1})'' = -(\sigma+1)nf$. Using replacements $u = f^{\sigma+1}$, $u' = v$, reduce its order and integrate the obtained equation.

7. Be convinced using (22), that at $n < -1/\sigma$ for any point $m_0 < \infty$ is valid $\xi(m_0, t) \to 0$, $t \to 0$ and $p(m_0, t) \to p(0, t) \to \infty$, $t \to 0$.

2 The Maximum Principle and Comparison Theorems

Now we shall consider the continuous dependence of the process from the input data. With the help of comparison theorems and a set of similar solutions we shall construct the closed classification of blowing-up regimes in nonlinear media. Consider the generalizations of the similar method.

1. The formulation and some consequences. A set of intermediate asymptotics, however diverse, cannot describe an object in the general case because of the necessity to make strong simplifying assumptions (zero initial background temperature within a heating medium, constance of flow before and after the shock wave, solution on the boundary – power law function of time, etc.). It in impossible to construct the complete picture without using *the stability* of mathematical models or *continuous dependence of solutions from input data.* This means a property of solutions to vary slightly at slight modification of boundary conditions, coefficients in the equations and other characteristics of the models. The stability, being expressed differently for various situations, is one of the necessary conditions of *correctness* of the models; otherwise one cannot speak about their adequacy for the investigated object (it is assumed that in the defined sense the object is stable as well). For stable models the sets of partial solutions are a kind of orientations or boundaries among the set of all possible solutions. It is especially important if the problem is nonlinear and one cannot construct its general solution from the partial ones.

Concerning the equations of parabolic type this property is reflected in *the maximum principle and comparison theorems.* Consider the Cauchy problem for the equation of nonlinear thermal conductivity

$$\frac{\partial T}{\partial t} = \frac{\partial}{\partial x}\left(k(T)\frac{\partial T}{\partial x}\right), \qquad k(T) > 0, \quad T > 0;$$
$$-\infty < x < \infty, t > 0, \tag{1}$$
$$T(x, 0) = T_0(x) \geq 0, \qquad -\infty < x < \infty,$$

The maximum principle. *The maximum of a solution $T(x, t)$ (temperature at any moment of time does not exceed the maximum of initial data $T_0(x)$) (the initial temperature):*

$$\max_{t>0,\, -\infty<x<\infty} T(x, t) \leq \max_{-\infty<x<\infty} T_0(x). \tag{2}$$

The inequality (2) has an obvious physical sense. The maximum of initial temperature distribution can in no way be increased by time, since by Fourier's law the heat flux is transferring energy from the heated sites of medium to the cold ones.

In the case of the first boundary problem for the equation (1) in a half-space, the maximum principle also has a clear content and means that

$$\max_{t>0,\,0<x<\infty} T(x,\,t) \le \max\left\{\max_{0<x<\infty} T_0(x)\,\max_{t\ge 0} T_0(0,\,t)\right\}, \tag{3}$$

where $T_0(x)$ is the initial temperature of the substance, $T(0,\,t)$ is the temperature on the boundary.

From a maximum principle there follows the comparison theorems. The formulation for the Cauchy problem (1) is as follows.

Let $T^{(1)}(x,\,t)$, $T(x,\,t)$, $T^{(2)}(x,\,t)$ be the solutions of Cauchy problem corresponding to initial data $T_0^{(1)}(x)$, $T_0(x)$, $T_0^{(2)}(x)$. Then if

$$T_0^{(1)}(x) \le T_0(x) \le T_0^{(2)}(x) \qquad \textit{for all } -\infty < x < \infty\,,$$

then

$$T_0^{(1)}(x,\,t) \le T(x,\,t) \le T^{(2)}(x,\,t) \quad \textit{for all } -\infty < x < \infty\,, \quad t > 0\,. \tag{4}$$

In other words, if one has two identical samples of thermal conducting materials such that the initial temperature of one of them is not less than that at a similar point in the other, than in any subsequent instant this property will be fulfilled. For the first boundary problem in a half-space the comparison theorem of solutions means the following.

The fulfilling of inequalities

$$T_0^{(1)}(x) \le T_0(x) \le T_0^{(2)}(x)\,, \qquad 0 < x < \infty\,,$$

$$T^{(1)}(0,\,t) \le T(0,\,t) \le T^{(2)}(0,\,t)\,, \qquad t > 0\,,$$

leads to the inequalities

$$T^{(1)}(x,\,t) \le T(x,\,t) \le T^{(2)}(x,\,t) \quad \textit{for all } 0 \le x < \infty\,, \quad t > 0\,, \tag{5}$$

where $T^{(1)}(x,\,t)$, $T(x,\,t)$, $T^{(2)}(x,\,t)$ are the solutions of the problem corresponding to the *boundary conditions* $T_0^{(1)}(x)$, $T_0(x)$, $T_0^{(2)}(x)$) and $T^{(1)}(0,\,t)$, $T(0)$, $T^{(2)}(0,\,t)$.

As the maximum principle, the comparison theorems have a direct physical content: the larger thermal influence on a fixed object leads to the appearance of larger temperature field in it.

The analogs of the formulated statements are also valid for other problems of the theory of thermal conductivity. Using the partial solutions of equations of parabolic and elliptic types, it is possible to evaluate (to limit from above and below) the solutions of the more general problems and, without knowing them in detail, to draw general conclusions. The continuous dependence of solutions on input data in any form is established also for a broad classes of the hyperbolic equations (for example, for the equation (19) from section 1 it is expressed by comparison theorems similar to the theorems for equation (1)).

From inequalities (4) and properties of solutions (9) the finite velocity of propagation of perturbations in the Cauchy problem (1) for the equation (1) at $k(T) = k_0 T^\sigma$, $\sigma > 0$, follows under condition that initial distribution of temperature $T_0(x)$ is a finite function, i.e. $T_0(x)) \equiv 0$, $|x| \geq R_0 < \infty$. Let T_m be the maximum value of function $T_0(x)$. Then, introducing a shift in time on t_0 into formula (9) of section 1, and choosing adequately large constants Q_0 and t_0, it is easy to satisfy the inequality $T_m \leq \bar{T}(R_0, 0)$, from which the inequality $T_0(x) \leq \bar{T})(x, 0) = \bar{T}_0(x)$, $-\infty < x < \infty$ follows, where $\bar{T}(x, t)$ denotes the solution of the problem of an instantaneous point heat source. As it majorizes the solution of the general Cauchy problem (1) on initial data, then from the comparison theorem (4) follows $T(x, t) \leq \bar{T}(x, t)$, $-\infty < x < \infty$, $t > 0$. Therefore at any moment of time $t > 0$ there exists a quantity $R(t)$ such, that $T(x, t) \equiv 0$ for all $|x| \geq R(t)$. It implies a finite velocity of motion of the front of a thermal wave.

By means of similar constructions from the properties of a solution of a travelling wave type (21), section 2, Chapter II and inequalities (5) one can easily estimate (at finite function $T_0(x)$) the finite velocity of propagation of heat in the case of the first boundary problem in a half-space. Therefore, this effect has a general, not a particular character. It is connected with the singularities of the equation of nonlinear thermal conductivity (1). For many important processes the temperature in certain parts of the substance can be considered practically equal to zero (for example, in the initial stage of a strong explosion in atmosphere the temperature outside its zone is negligibly small as compared with the temperature in the area covered by the explosion). For a rather strong increase in thermal conductivity by temperature the quantity $k(T)$ in these zones also equals zero. Opening the right hand side in equation (1)

$$\frac{\partial T}{\partial t} = k\left(T\right) \frac{\partial^2 T}{\partial x^2} + k'_T \left(\frac{\partial T}{\partial x}\right)^2 ,$$

we prove that in points x, t, in which $T(x, t) = 0$ and, thus, $k(T(x, t)) = 0$, it is degenerated into a first order equation (in the remaining area (1) it is a parabolic second-order equation). This is the actual mathematical situation of the finite velocity of heat propagation in a medium with zero temperature background. In the points of degeneration the solution has to be understood as a generalized one, as was already shown in the partial examples, while in the remaining area it satisfies the equation (1) in the usual (classical) sense.

Consider now in more detail the proof of the existence of heat effect localization in the Cauchy problem. The statement is as follows: the solution $T(x,\,t)$ of the Cauchy problem for the equation

$$\frac{\partial T}{\partial t} = \frac{\partial}{\partial x}\left(k_0 T^\sigma \frac{\partial T}{\partial x}\right), \qquad \sigma > 0, \quad t > t_0\,, \quad -\infty < x < \infty\,, \tag{6}$$

with an initial function

$$T\,(x,\,t_0) = \begin{cases} T_M \left(1 - \dfrac{x}{x_0}\right)^{2/\sigma}, & |x| \le x_0\,, \\ 0\,, & |x| > x_0\,, \end{cases} \tag{7}$$

is localized in the initial area $|x| \le x_0$ during the *localization time* t_{loc} not smaller than

$$t_{\mathrm{loc}} = \frac{x_0^2 \sigma}{2k_0(\sigma+2)\,T_M^\sigma}\,, \tag{8}$$

i.e. $T(x,\,t) \equiv 0,\ |x| > x_0,\ t_0 \le t \le t_0 + t_{\mathrm{loc}}$.

By virtue of (2) the maximum $T_m(t)$ of the solution of the problem (6), (7) does not exceed its initial value, at all $t > t_0$.

$$T_m(t) = T(0,\,t) < T_M\,,$$

so that the equality $T_m(t) = T(0,\,t)$ follows from the symmetry of the problem. Consider a solution $T(x,\,t)$ in the area $x > 0$, and denote it as $T_+(x,\,t)$. Obviously, one can treat the function $T_+(x,\,t)$ at $t > t_0$ as a solution of the first boundary problem for the equation (6) in the area $x > 0$ with the initial condition $T_+(x,\,t_0)$ from (7) and the condition on the boundary $T_+(0,\,t)$, satisfying the inequality,

$$T_+(0,\,t) = T_m(t) = T(0,\,t) \le T_M\,.$$

By its construction, the function $T_+(x,\,t_0)$ is nothing other than solution (14), (15) from section 1 (similar S-regime of $T_S(x,\,t)$, in which $A_0 = [x_0^\sigma/(2k_0(\sigma+2))]^{1/\sigma}$, $x_S = x_0$), taken in a moment $t_0 = -t_{\mathrm{loc}} = -x_0^\sigma/(2k_0(\sigma+2)T_M^\sigma)$.

Compare the solutions $T_+(x,\,t)$ and $T_S(x,\,t)$ at $t_0 \le t \le 0$ (the mutual disposition of function $T(x,\,t_0)$ and functions $T_+(x,\,t)$, $T_S(x,\,t)$ in a moment $t > t_0$ is shown in Fig. 74). The initial data in both cases coincide, and the

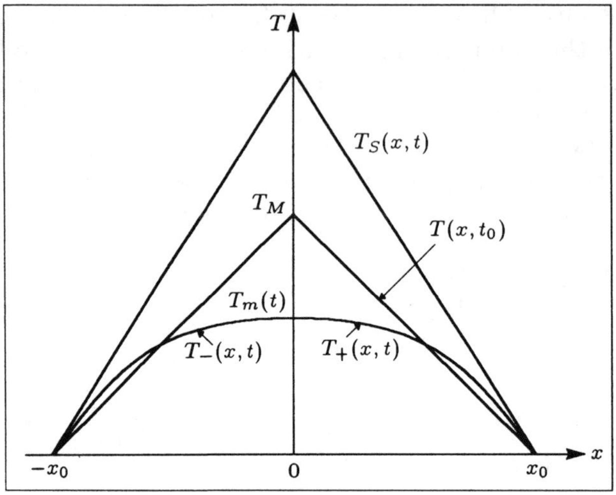

Fig.74.

boundary condition for $T_+(x, t)$ is majored by a boundary condition for $T_S(x, t)$

$$T_+(0, t) \le T_M \le A_0(-t)^{-1/\sigma}, \qquad t_0 \le t < 0.$$

From the comparison theorem (5) and the properties of the S-regime we have

$$T_+(x, t) \equiv 0, \qquad x > x_0, \qquad t_0 \le t < 0.$$

By virtue of a symmetry $T_-(x, t) = T_+(-x, t)$, where $T_-(x, t)$ is the solution $T(x, t)$ in area $x < 0$ (see Fig. 74). In view of this, we finally derive the statement formulated above

$$T(x, t) \equiv 0, \qquad |x| > x_0, \qquad t_0 \le t \le t_0 + t_{\text{loc}}.$$

From the comparison theorem (4) it follows that any initial distribution of temperatures majored by function (7) (and with fronts coinciding with it) is also localized in an area $|x| \le x_0$, and the localization time is evaluated from below by the formula (8). Thus, the following common conclusion is valid: for a medium, where the heat transfer is described by equation (6), it is always possible to specify the profiles of temperatures possessing "inertiality". The fronts of such profiles taken as initial data, do not begin to move in the substance immediately, but after some time (see also exercise 1).

Note that in a medium with a rather strong heat absorption (exercise 2) one can have situations in which the size of the heated area is limited by a constant at all $t_0 \le t \le \infty$ (the physical sense of this effect is obvious).

2. Classification of blow-up regimes. Inertiality is only peculiar to "flat" temperature profiles; it is seen in the comparison of a localized distribution (7) with, say solutions of the problem of an instantaneous point heat source (formula (9), section 1). One of the ways of shaping similar profiles can be the action on a heat-conducting medium of appropriate boundary conditions. For example, in the case of solution (14), (15) of section 1, the inflow of energy from the boundary is matched to properties of a medium in such a way that at any moment of time in substance the inertial profiles of the form (7) are realized (the time of their localization, obviously, decreases with the growth of temperature).

Using the similar solutions investigated in subsection 3, section 1, and the comparison theorems, we shall give a classification of the boundary conditions with blow-up by the results of their action on nonlinear mediums.

Consider for (6) the first boundary problem in a half-space $x > 0$ with a boundary condition

$$T(0,\, t) \to \infty, \qquad t \to 0\,, \qquad t_0 \le t < 0\,, \tag{9}$$

and, for simplicity, with zero initial data

$$T(x,\, t_0) = T_0(x) = 0\,, \qquad x \ge 0\,. \tag{10}$$

The localization in this case, as distinct from the Cauchy problem, means the existence of a constant $l^* < \infty$, such that for a solution $T(x,\, t)$ of the problem (6), (9), (10), it is valid $T(x,\, t) \equiv 0$ at $l \ge l^*$, and $t_0 \le t < 0$. In other words, the thermal perturbations during the whole process of heating do not penetrate further than some final depth l^* (otherwise the localization is absent).

From the comparison theorem (5) and the properties of the similar S-regime, it follows that if

$$T(0,\, t) \le A_0\,(-t)^{-1/\sigma}\,, \qquad t_0 \le t < 0\,,$$

that in the problem (6), (9), (10) one has a heat localization in depth $l^* = x_S$, and its solution is majored by function $T_S(x,\, t)$. If (9) satisfies the inequality

$$T(0,\, t) \le A_0\,(-t)^n\,, \qquad -1/\sigma < n < 0\,, \qquad t_0 \le t \le 0\,, \tag{11}$$

then the solution is localized in depth (exercise 3)

$$l^* \le \left(2k_0 A_0\,(-t_0)^{1/\sigma+n}\frac{\sigma+2}{\sigma}\right)^{1/2}, \tag{12}$$

and at $0 \le x \le l^*$ the following evaluation is valid (exercise 4)

$$T(x,\, t) \le C(n,\, \sigma)(k_0^{-n} A_0)^{\frac{1}{1+n\sigma}} x^{\frac{2n}{1+n\sigma}}\,, \quad x \le l^*\,, t_0 \le t < 0\,, \tag{13}$$

following from the existence of a limiting curve (18), section 1, for a similar LS-regime.

The inequality preceding (11) determines a class of boundary blowing-up conditions leading to a heat localization; the inequality (11) updates this result: at its realization the LS-regime is performed, the temperature at $t \to 0$ grows infinitely only at a point $x = 0$. At functions $T_0(x)$ in (10) which are non-zero but finite, both the above made statements about localization remain valid. It is only necessary to select the rather large quantity A_0 in majoring the solution similar S- and LS-regimes; the concrete evaluations (12), (13) naturally undergo modifications. Note that for a general LS-condition, the wavefront, as distinct from the similar solution, is not at $x = \infty$, but at a finite point.

The effect of localization in boundary problems is not connected to the rate of heating of a substance. To prove this, we analyze the behavior of solution (6), (9), (10) in the case, where

$$T(0,\, t) \geq A_0(-t)^n\,, \quad n < -1/\sigma\,, \quad t_0 \leq t < 0\,. \tag{14}$$

First of all we show that at some moment t_* ($t_0 < t_* < 0$) the solution $T(x,\, t)$ is different from zero in a neighborhood of boundary $x = 0$. For this, we compare it with the solution $\bar{T}(x,\, t)$ of a travelling wave type

$$\bar{T}(x,\, t) = \begin{cases} \dfrac{D\sigma}{k_0}[D(t - t_0) - x]^{1/\sigma}\,, & x \leq D(t - t_0)\,, \\ 0\,, & x > D(t - t_0)\,, \end{cases} \tag{15}$$

for the equation (6), where $D = [A_0^\sigma(-t_0)^{n\sigma - 1}k_0/\sigma]^{1/2}$. By construction $\bar{T}(x,\, t_0) \equiv 0$, and the constant D is selected so that $\bar{T}(0,\, t) \leq T(0,\, t) \leq A_0(-t)^n$, $t_0 \leq t \leq 0$. Therefore, $T(x,\, t) \geq \bar{T}(x,\, t)$, $0 \leq x \leq \infty$, $t_0 \leq t < 0$ by virtue of the comparison theorem (5).

We now turn to solution (12) of section 1 at $n < -1/\sigma$ (similar HS-regime), denoting it through $T_a(x,\, t)$. Select in it A_{0a}, such that at $t = t^*$ it is majored by function (15), taken in a moment $t = t^*$, i.e. $T_a(x,\, t^*) \leq \bar{T}(x,\, t^*)$, $x \geq 0$ and, thus, $T_a(x,\, t^*) \leq T(x,\, t^*)$, $x \geq 0$. From (14) we have $T_a(0,\, t) \leq T(0,\, t)$ for all $t > t^*$. Then from the comparison theorem (5) we obtain the inequality $T_a(x,\, t) \leq T(x,\, t)$ for all $x \geq 0$ and $t \geq t^*$(majorization at $t \geq t^*$ of similar HS-regime by the investigated solution). In so far as $T_a(x,\, t) \to \infty$, $t \to 0$. $x \geq 0$, then $T(x,\, t) \to \infty$ at $t \to 0$ in any point $x \geq 0$

When the inequality (14) is satisfied, the heat localization is absent, the wave of heating at $t \to 0$ covers the whole substance, temperature grows infinitely at any point (this statement is especially true if $T(x,\, t_0) \not\equiv 0$).

This conclusion completes the classification of blowing-up boundary conditions in nonlinear thermal conductive media. When the substance is acted upon by "slow" S- and LS-regimes ($T(0,\,t)\ \leq\ A_0(-t)^{-1/\sigma}$ or $T(0,\,t)\leq A_0(-t)^n$, $-1/\sigma<n<0$), the energy is localized in an area of finite sizes, in "fast" HS-regimes (14) no localization occurs.

To complete the picture we have to explain, that the effect of localization is realized also by omitting the requirement $T(x,\,t_0)\equiv 0$, $x\geq 0$ (or from the requirement of finiteness of the function (10)). In this case the localization has to be understood in a more general *effective* sense as the existence of a constant $L^*<\infty$, such that the solution of the problem (6), (9) at arbitrary limited function $T_0(x)$ is bounded above at $x\geq L^*$, $t\geq t_0$, despite the infinite growth of solution in a point $x=0$. The classification of boundary conditions with blow-up does not depend on the characteristic initial distribution of temperature, and remains the same. In particular, it is not important whether a finite front of a thermal wave exists in the considered process or not.

As an example illustrating these statements, we consider for the equation (6) the problem with boundary condition

$$T(0,\,t)=A_S(-t)^{-1/\sigma}\,,\qquad t_0\leq t<0\,,\tag{16}$$

corresponding to the solution $T_S(x,\,t)$ – similar S-regime, but with constant initial background temperature

$$T(x,\,t_0)=T_0=A_S(-t_0)^{-1/\sigma}\,.\tag{17}$$

Its solution $T(x,\,t)$, obviously, majors $T_S(x,\,t)$:

$$T_S(x,\,t)\leq T(x,\,t)\,,\qquad x\geq 0\,,\qquad t_0\leq t<0\,.$$

In so far as the temperature in $x=0$ is the same for both solutions, from the latter inequality we have the following inequality for derivatives in this point

$$-\left.\frac{\partial T}{\partial x}\right|_{x=0}\leq-\left.\frac{\partial T_S}{\partial x}\right|_{x=0},\qquad t_0\leq t\leq 0\,,$$

and from it – the inequality for heat fluxes at the boundary

$$W(0,\,t)=-k(T)\left.\frac{\partial T}{\partial x}\right|_{x=0}\leq W_S(0,\,t)=-k(T_S)\left.\frac{\partial T_S}{\partial x}\right|_{x=0},\qquad t_0\leq t<0\,.\tag{18}$$

The physical sense of (18) is that with other equal conditions, a medium which has initially been more strongly heated "accepts" the energy supplied from the boundary less readily than a medium which has been strongly heated. Integrating (6) in view of (18) by x from 0 to ∞ and on t from t_0 to $t < 0$ and (as $W_S(\infty, t) = W(\infty, t)$, $t_0 \leq t < 0$) we have

$$\int_0^\infty [T(x, t) - T_0]\, dx \leq \int_0^\infty [T_S(x, t) - T_S(x, t_0)]\, dx\,,$$

or, by dividing the area of integration into a part from $x = 0$ up to $x = x_S$ and from $x = x_S$ up to $x = \infty$ and in view of $T_S(x, t) \equiv 0$, $x \geq x_S$, $t \geq t_0$, we come to an inequality

$$\int_0^{x_S} [T(x, t) - T_0]\, dx + \int_{x_S}^\infty [T(x, t) - T_0]\, dx \leq \int_0^{x_S} [T_S(x, t) - T_S(x, t_0)]\, dx\,,$$

As $T(x, t) \geq T_0$, $x \geq 0$, $t \geq t_0$ (exercise 5), it is possible to rewrite the given inequality as

$$0 \leq \int_{x_S}^\infty [T(x, t) - T_0]\, dx \leq \int_0^{x_S} [T_S(x, t) - T(x, t)]\, dx + \int_0^{x_S} [T_0 - T_S(x, t_0)]\, dx\,,$$

from which follows, by virtue of inequalities $T(x, t) \geq T_S(x, t)$, $x \geq 0$, $t \geq t_0$, and $T_0) \geq T_S(x, t_0)$, $x \geq 0$, the boundedness of function .$T(x, t)$ at all $x \geq x_S$, $t_0 \leq t \leq 0$. In other words, in the problem (6), (16), (17) *the effective localization* of heat occurs at a depth $L^* = x_S$, which is exactly equal to the depth of localization in the absence of temperature background.

The classification of conditions with blow-up for processes described by a hyperbolic equation (19) from section 1 is performed by analogous methods and leads to quite similar results: during the action of "slow" boundary conditions on a medium localization does occur, while in the "fast" case this effect is absent.

3. The extension of "a self-similar method". The approach demonstrated in subsections 1, 2, based on the use of a broad class of self-similar or other partial solutions and of continuous dependence of the process on input data, permits various generalizations. We now show that the effect of heat localization can occur not only in one-dimensional, but also in many- dimensional geometry, by means of constructing a corresponding explicit solution of the equation

$$\frac{\partial T}{\partial t} = \frac{\partial}{\partial x}\left(k_0 T^\sigma \frac{\partial T}{\partial x}\right) + \frac{\partial}{\partial y}\left(k_0 T^\sigma \frac{\partial T}{\partial y}\right) + \frac{\partial}{\partial z}\left(k_0 T^\sigma \frac{\partial T}{\partial z}\right), \quad t_0 \leq t < 0, \tag{19}$$

considered in a quadrant $x \geq 0$, $y \geq 0$, $z \geq 0$ (here, as before, $T = T(x, y, z, t)$ is the temperature).

By analogy with a one-dimensional S-regime (14), (15) of section 1, we shall seek a partial solution (19) in separating variables, i.e. $T(x, y, z, t) = U(t)\, f(x, y, z)$. The substitution of this expression into (19) gives for $U(t)$ the same formula as for one-dimensional geometry: $U(t) = A_0(-t)^{-1/\sigma}$.

For function $f(x, y, z)$ we obtain a complicated elliptical equation. Therefore we shall confine ourselves to the simplest case, in which the spatial part of the solution actually depends on one argument: $f(x, y, z) = f(\xi)$, $\xi = x + y + z$. Then $f(\xi)$ satisfies the equation (13), section 1, at $n = -1/\sigma$, with a solution which is already known. To sum up, we come to a many-dimensional S-regime

$$T_S(x, y, z, t) = \\ = \begin{cases} A_0(-t)^{-1/\sigma}\left(1 - \dfrac{x+y+z}{r_\Phi}\right)^{2/\sigma}, & x + y + z \leq r_\Phi\,, \\ 0, & x + y + x > r_\Phi\,, \end{cases} \tag{20}$$

where $r_\Phi = r_S \equiv (2k_0 A_0^\sigma(\sigma + 2)/\sigma)^{1/2}$ is calculated by the formula for the one-dimensional solution. The solution (20) describes the blowing-up heating of a three-dimensional heat-conducting medium, in so far as the boundary temperature $T(0, y, z, t)$, $T(x, 0, z, t)$, $T(x, y, 0, t)$ turns to infinity (in area $\xi < r_S$) at $t \to 0$. The same thing happens in this area with the solution. However the temperature is equal to zero in all the remaining space $\xi > r_S$ of a quadrant $x \geq 0$, $y \geq 0$, $z \geq 0$ up to a moment $t = 0$. The size of area of localization r_s (distance from the beginning of coordinates to the plane of the front of the thermal wave), as before, depends on the properties of the medium k_0, σ and the intensity of the boundary condition A_0.

In so far as for solutions of equation (19), the comparison theorems are valid, it is easy determine from (20) the class of boundary conditions leading to heat localization in a considered many-dimensional area. The further classification of this is quite similar to the results obtained in subsection 2.

We now turn to a more broad treatment of similar solutions, by considering the concept of *approximate similar solutions*. It is easiest to consider it on an example of the problem of heating a medium with constant thermal physical properties in condition with blow-up

$$\begin{aligned} &\frac{\partial T}{\partial t} = k_0\,\frac{\partial^2 T}{\partial x^2}\,, \qquad 0 < x < \infty\,, \quad t_0 \leq t < 0\,, \\ &T(0, t) \to \infty\,, \qquad t \to 0\,. \end{aligned} \tag{21}$$

Assume without loss of generality $T(x,\, t_0) = 0$, $x \geq 0$. The general solution of the linear problem (21) is well known

$$T(x,\, t) = \frac{x}{2\sqrt{\pi k_0}} \int\limits_{t_0}^{t} \exp\left(-\frac{x^2}{4k_0(t-r)}\right)(t-\tau)^{-3/2}\, T(0,\, \tau)\, d\tau\,. \qquad (22)$$

With the help of (22) it is easy to classify solutions (21) in dependence on the form of $T(0,\, t)$. It is seen that in so far as at $\tau \to t$, $t \to 0$ the first multiplier in the integral expression tends to zero, and the second and the third tend to infinity, then any of the regimes of heat propagation, considered in subsection 2, is possible.

Thus, for $T(0,\, t) = A_0 e^{-\sigma_0/t}$, $a_0 > 0$ is the parameter describing the boundary condition, temperature grows infinitely in all points $0 \leq x \leq x_S = 2\sqrt{k_0 a_0}$ at $t \to 0$, while in area $x \geq x_s$ is limited for all $t \leq 0$. The thermal energy contained to the right of point $x = x_S$ is also limited

$$\lim_{t\to 0} \int\limits_{x}^{\infty} T(x',\, t)\, dx' < \infty\,, \qquad x > x_S\,, \quad t \leq 0\,,$$

i.e. an effective localization of heat occurs in S-regime with the depth $L^* = x_S = 2\sqrt{k_0 a_0}$ computed from (22).

By considering a boundary condition of a more general form

$$T(0,\, t) = A_0 \exp(a_0(-t)^n)\,, \qquad n < 0\,, \qquad (23)$$

we obtain, that at $-1 < n < 0$ the LS-regime, and for $n < -1$ the HS-regime are realized.

The analysis of integral (22) cannot give certain important detailed properties of the process. For example, to obtain the law of change in time of half-width $x_{\text{ef}}(t)$, it is necessary to solve the integral equation (22). The problem (21), (23) on the other hand does not permit the construction of similar solutions, as is easy to show with the help of Π-theorem.

We modify (21), (23) as follows: instead of (23) we shall take a boundary condition $T(0,\, t) = A_0[\exp(a_0(-t)^n) - 1]$, $n < 0$ (a constant is added for convenience and at $t \to 0$ has no role) and we shall conduct a replacement $V(x,\, t) = A_0 \ln(T(x,\, t)/A_0 + 1)$. Then for $V(x,\, t)$ we come to the problem

$$\begin{aligned} &\frac{\partial V}{\partial t} = k_0 \frac{\partial^2 V}{\partial x^2} + \frac{k_0}{A_0}\left(\frac{\partial V}{\partial x}\right)^2, \qquad 0 < x < \infty\,, \quad t_0 \leq t < 0\,, \\ &V(0,\, t) = A_0 a_0 (-t)^n\,, \qquad n < 0\,, \quad t_0 \leq t < 0\,, \qquad (24) \\ &V(x,\, t_0) = 0\,, \qquad 0 \leq x < \infty. \end{aligned}$$

The boundary condition (24) is a power law function of time. If by analogy with subsection 3, section 1, we seek the solution of the problem (24) close to a power law similar solution, then for such a solution at $t \to 0$ the first term on the right hand side of the equation becomes negligibly small in comparison with the second, and it can be neglected (exercise 6).

A more strict analysis shows that the solution of the problem (24) at $t \to 0$ is close to solutions of a simpler problem

$$\begin{aligned}
&\frac{\partial u}{\partial t} = \frac{k_0}{A_0}\left(\frac{\partial u}{\partial x}\right)^2, \qquad 0 < x < \infty, \quad -\infty < t < 0, \\
&u(0,\,t) = A_0 a_0 (-t)^n, \qquad n < 0, \quad -\infty \le t < 0, \\
&u(x,\,-\infty) = 0,
\end{aligned} \tag{25}$$

As distinct from (24) the solution $u(x,\,t)$ of the problem (25) is a power law similar solution, the analysis of which does not represent a difficulty. The function $u(x,\,t)$ is an approximate similar solution of the problem (24) and, hence, of the initial problem (21), (23). The relative precision of description is improved by time

$$\left|1 - \frac{A_0}{T(x,\,t)}\, e^{u(x,\,t)/A_0}\right| \to 0, \qquad t \to 0, \quad x \ge 0.$$

From the properties of $u(x,\,t)$ one has the law for a half-width of a wave of heating $x_{\text{ef}}(t) = (-t)^{(1-n)/2} \ln 2[k_0/\ /(a_0(-n))]^{1/2}$, which decreases at all three cases (compare with a nonlinear medium). Note that the construction and analysis of approximate similar solutions is possible and is used in the study of a broad class of parabolic-type equations.

One more important extension of the self-similar method is based on *the generalization of the concept of a comparison of solutions.* The essence of this approach is in the comparison of problems corresponding not only to different boundary conditions, as in subsections 1, 2, but also to different equations (in the case of equation (1), to different functions $k(T)$). In a certain sense this means the stability of heat transfer process with respect to the *perturbations of thermal physical properties* of the medium. Then, for one of the compared solutions one can select a solution of a well-investigated equation (for example, (6) or (21)) and obtain useful results for solutions of more complicated equations.

We demonstrate an elementary variant of such an approach in the case of the equation (21). Consider solutions $T^{(1)}(x,\,t)$ and $T^{(2)}(x,\,t)$ of two boundary problems in a half-space $x \ge 0$:

$$
\begin{aligned}
&\frac{\partial T^{(1)}}{\partial t} = k_0^{(1)} \frac{\partial^2 T^{(1)}}{\partial x^2}, && x > 0, \quad t > 0, \\
&T^{(1)}(0, t) = T_1^{(1)}(t), && t > 0, \\
&T^{(1)}(x, 0) = T_0^{(1)}(x), && x \geq 0;
\end{aligned}
\tag{26}
$$

$$
\begin{aligned}
&\frac{\partial T^{(2)}}{\partial t} = k_0^{(2)} \frac{\partial^2 T^{(2)}}{\partial x^2}, && x > 0, \quad t > 0, \\
&T^{(2)}(0, t) = T_1^{(2)}(t), && t > 0, \\
&T^{(2)}(x, 0) = T_0^{(2)}(x), && x \geq 0;
\end{aligned}
\tag{27}
$$

For a difference $V(x, t) = T^{(2)}(x, t) - T^{(1)}(x, t)$ from (26), (27) we obtain the following boundary problem

$$
\begin{aligned}
&\frac{\partial V}{\partial t} = k_0^{(1)} \frac{\partial^2 V}{\partial x^2} + (k_0^{(2)} - k_0^{(1)}) \frac{\partial T^{(2)}}{\partial x}, && x > 0, t > 0 \\
&V(0, t) = T_1^{(2)}(t) - T_1^{(1)}(t), && t > 0, \\
&V(x, 0) = V_0(x) = T_0^{(2)}(x) - T_0^{(1)}(x), && x \geq 0.
\end{aligned}
\tag{28}
$$

Let the following requirements of majorization of the solution $T^{(1)}(x, t)$ by a solution $T^{(2}(x, t)$ be satisfied by boundary conditions

$$
\begin{aligned}
&T_1^{(1)}(t) \leq T_1^{(2)}(t), && t > 0, \\
&T_0^{(1)}(t) \leq T_0^{(2)}(t), && x \geq 0,
\end{aligned}
\tag{29}
$$

and by thermal conductivity

$$
k_0^{(1)} \leq k_0^{(2)}. \tag{30}
$$

Let also for a solution $T^{(2)}(x, t)$ be valid

$$
\frac{\partial^2 T^{(2)}}{\partial x^2} \geq 0, \quad \frac{\partial T^{(2)}}{\partial t} \geq 0, \quad \text{by} \quad x \geq 0, \quad t \geq 0. \tag{31}
$$

The non-decreasing property (31) of function $T^{(2)}(x, t)$ in time at any point $x \geq 0$ is ensured (at $T_0^{(2)}(x)) \equiv 0$) with a non-decrease by t of boundary condition $T_1^{(2)}(t)$. We check it, differentiating the equation (27) by t and obtaining for the function $Z^{(2)}(x, t) = \partial T^{(2)}/\partial t = -W^{(2)}(x, t)/k_0$ the problem

$$\begin{aligned} &\frac{\partial Z^{(2)}}{\partial t} = k_0^{(2)} \frac{\partial^2 Z^{(2)}}{\partial x^2}, \qquad x > 0, \quad t > 0, \\ &Z^{(2)}(0, t) \geq 0, \qquad t > 0, \\ &Z^{(2)}(x, 0) \equiv 0. \end{aligned} \tag{32}$$

Its solution, as is easy to show (exercise 7), is non-negative at all $x \geq 0$, $t \geq 0$ (in case $T_0^{(2)} \not\equiv 0$ for the realization of inequality (31) it is enough to also impose a condition $\partial^2 T_0(x)/\partial x^2 \geq 0$, $x \geq 0$).

For the realization of inequalities (29)–(31), the problem (28) for $V(x, t)$, becomes the problem

$$\begin{aligned} &\frac{\partial V}{\partial t} = k_0^{(1)} \frac{\partial^2 V}{\partial x^2} + f(x, t), \qquad x > 0, \quad t > 0, \\ &V(0, t) \geq 0, \qquad t > 0, \\ &V_0(x) \geq 0, \qquad x \geq 0, \end{aligned} \tag{33}$$

with non-negative boundary conditions and non-negative function $f(x, t)$ (heat sources) on the right hand side of the equation (33). Its solution is non-negative at all $x \geq 0$ and $t \geq 0$. The majorization of the solution of the problem (26) with coefficient $k_0^{(1)}$ by a solution of the problem (27) with coefficient $k_0^{(2)}$ in all considered area, then, follows

$$T^{(1)}(x, t) \leq T^{(2)}(x, t), \qquad t \geq 0, \quad t \geq 0. \tag{34}$$

In the case of the equation (1) the inequalities (34) are valid at additional to (29)–(31) condition for the coefficients $k^{(1)}(T)$, $k^{(2)}(T)$, namely, of a form $[k^{(2)}(T)/k^{(1)}(T)]'_T$, $T \geq 0$.

By means of analogous comparison theorems (also valid for parabolic equations of general form) one can establish, for example, the crucially important result: for a medium with arbitrary thermal physical properties it is always possible to specify a class of boundary conditions leading to heat localization, and a class of conditions, when no localization exists in the medium. Thus, this effect has a general character.

EXERCISES

1. Using solution (9) of section 1 and the comparison theorem (4), show that the localization of perturbations in the Cauchy problem is possible only during a finite time scale, and that at any initial function $T_0(x)) \equiv 0$ the coordinate of the front of a thermal wave $x_\Phi(t) \to \infty$, $t \to \infty$.

2. In the equation

$$\frac{\partial T}{\partial t} = \frac{\partial}{\partial x}\left(k_0 T^\sigma \frac{\partial T}{\partial x}\right) - q_0 T$$

($q_0 > 0$) the term $q_0 T$ characterizes the intensity (depending on temperature) of heat absorption in a medium. By replacements $T = e^{-q_0 t} V$, $\tau = (1 - e^{-\sigma q_0 t})/(\sigma q_0)$ reduce it to the equation

$$\frac{\partial V}{\partial \tau} = \frac{\partial}{\partial x}\left(k_0 T^\sigma \frac{\partial V}{\partial x}\right).$$

As a solution $V(x, \tau)$ of the equation without absorption, use the one given by formula (9) of section 1, and perform an inverse transformation from $V(x, \tau)$ to $T(x, t)$. Prove that for the constructed solution $T(x, t)$ of the initial equation $x_\Phi(t) \to \xi_\Phi (Q_0 k_0)^{1/(2+\sigma)} (\sigma q_0)^{-1/(2+\sigma)}$ at $t \to \infty$.

3. Comparing the boundary condition (11) with a solution $T_S(x, t)$ (formulae (14), (15) of section 1) in a point $x = 0$, prove the validity of evaluation (12) for initial data (10).

4. Prove that the solution of the problem (6), (9), (10) is majored in the case of an inequality (11) by the solution considered in subsection 3, section 1, for similar LS-regime.

5. Check the inequality $T(x, t) \geq T_0$, $x \geq 0$, $t \geq t_0$, for the solution of the problem (6), (16), (17), using (5) and taking into account that the constant satisfies equation (6).

6. Check that if we seek the solution of equation (25) at $n = -1$ in a form $V(x, t) \approx A_0 a_0(t)^{-1} f(x)$ (analog of S-regime), then at $t \to 0$ (24) is degenerated to the equation (25).

7. Prove by contradiction the non-negativity of solutions of the problems (32) and (33) (use the comparison theorem (5) and the fact, that at zero boundary conditions the solutions are identical to zero).

3 An Averaging Method

We will now consider a variant on the averaging method used to study the spatial-temporal dynamics of localized structures. We will formulate two approaches to an averaged description. Using them we shall achieve a classification of the processes of burning in thermal conducting media.

1. Localized structures in nonlinear media. The effect of the localization of perturbations studied in sections 1, 2 can be exhibited not only in the presence of external influences on the medium, given by means of corresponding boundary conditions with blow-up, but also due to its own nonlinear properties. Rather strong nonlinearity (see a simple example in section 6, Chapter I) generates blowing-up regimes, which in turn are the reason for the origin of structures – inhomogeneities localized in space.

We shall study the spatial-temporal behavior of the distribution of temperature in a thermal conducting substance with nonlinear heat sources. Energy is released as a result of combustion, and of chemical and other types of reactions. The medium is considered infinite (Cauchy problem), the process of combustion is one-dimensional. It is described by the equation

$$\frac{\partial T}{\partial t} = \frac{\partial}{\partial x}\left(k_0 T^\sigma \frac{\partial T}{\partial x}\right) + q_0 T^\beta , \tag{1}$$
$$q_0 > 0 , \qquad \beta > 1 , \qquad -\infty < x < \infty , \qquad t \leq t_0 ,$$

with initial function

$$T(x,\, t_0) = T_0(x) , \qquad -\infty < x < \infty . \tag{2}$$

Thermal physical characteristics of the substance, i.e. the thermal conductivity and strongly nonlinear ($\beta > 1$)) source of energy $q_0 T^\beta$ are power law functions of temperature, well approximating in certain cases the actual dependencies. The combustion is initiated by an initial distribution of temperature $T_0(x)) \not\equiv 0$ (otherwise the substance would remain cold).

We construct an example of *a localized structure of combustion* in the problem (1), (2). We seek its solution in separating variables $T(x,\, t) = U(t) f(x)$. Then from (1) we have

$$f\,\frac{dU}{dt} = U^{\sigma+1}\,\frac{d}{dx}\left(k_0 f^\sigma \frac{df}{dx}\right) + U^\beta\, q_0 f^\beta ,$$

whence it is seen that the variables are separated in the case, where $\beta = \sigma + 1$. And then

$$U^{-(\sigma+1)}\frac{dU}{dt} = \frac{1}{f}\,\frac{d}{dx}\left(k_0 f^\sigma \frac{df}{dx}\right) + q_0 f^\sigma = C , \tag{3}$$

where $C > 0$ (cases with temperature growing in time are sought). For $U(t)$ from (3) we obtain the expression

$$U(t) = (C_1 - \sigma C t)^{-1/\sigma} , \qquad C_1 > 0 , \tag{4}$$

having sense only at $t < C_1/(\sigma C)$ and turning to infinity in a finite instant. Thus, the temperature is blowing-up. Below, without losing generality we assume $C = 1/\sigma$ and denote $C_1 = t_f$.

To find $f(x)$, by a replacement $f^{\sigma+1} = y$ we transform (3), first, to an equation

$$\frac{\sigma k_0}{\sigma+1} y'' = y^{\frac{1}{\sigma+1}} - q_0 \sigma y ,$$

having the form an equation of oscillations of a ball on a spring with an appropriate external force (see subsection 1, section 4, Chapter I), and then, using the replacement $y' = \omega$, to the first order equation

$$\frac{\sigma k_0}{\sigma+1} \frac{d\omega}{dy} = \frac{y^{\frac{1}{\sigma+1}} - q_0 \sigma y}{\omega} .$$

Its integration gives the relation between ω and y:

$$\frac{\sigma k_0}{\sigma+1} \frac{\omega^2}{2} = \frac{\sigma+1}{\sigma+2} y^{\frac{\sigma+2}{\sigma+1}} - q_0 \sigma \frac{y^2}{2} + C_2 .$$

For the definition of the constant C_2 we have to take into account the fact that the function $\omega = y' = f^\sigma f'(\sigma+1)$ within a numerical factor represents the spatial part of the expression for the flux $W(x,\, t) = -k_0 T^\sigma \cdot \partial T/\partial x = -k_0 U^{\sigma+1} f^\sigma f'$. At the front of thermal structure both the temperature f, and the heat flux should turn to zero. Therefore, $\omega = 0$ at $y = 0$ $(f = 0)$ and from here $C_2 = 0$. Passing in the latter equation from y and ω back to f, we obtain (at $(C_2 = 0)$ the quadrature

$$dx = \pm \frac{(\sigma+1)\sigma k_0\, df}{f\sqrt{2\dfrac{\sigma+1}{\sigma+2} f^{-\sigma} - q_0 \sigma}} ,$$

which is integrated in an explicit form

$$f = A \cos^{2/\sigma} Bx , \quad A = \left[q_0 \frac{\sigma(\sigma+2)}{2(\sigma+1)} \right]^{-1/\sigma} , \quad B = \frac{\sigma}{2} \sqrt{\frac{q_0}{k_0(\sigma+1)}} . \tag{5}$$

Integrating (4) and (5), we come to a final form of the sought solution

$$T(x,\, t) = \begin{cases} [q_0(t_f - t)]^{-1/\sigma} \left\{ \dfrac{2(\sigma+1)}{\sigma(\sigma+2)} \cos^2 \dfrac{\pi x}{L_T} \right\}^{1/\sigma} , & |x| \le \dfrac{L_T}{2} , \\ 0 , & |x| > \dfrac{L_T}{2} , \end{cases} \tag{6}$$

where $L_T = 2\pi \sqrt{k_0/q_0} \sqrt{(\sigma+1)/\sigma^2}$.

The solution (6) describes a non-monotonous distribution of temperature (structure) with a motionless front and constant half-width – similar

S-regime of combustion in a nonlinear thermal conducting medium. Despite the infinite growth of the temperature in a structure at $t \to t_f$ (at $t \to -\infty$ the temperature tends to zero), the process of combustion is localized in a finite area $|x| \le L_T/2$, with scales determined by the parameters of substance k_0, q_0, σ. The localization means the absence of an influence of a burning structure on areas of the medium outside the region $|x| \le L_T/2$.

Thus, the combustion of a nonlinear medium can occur at arbitrary number of independent thermal structures (if their maxima are separated by a space, larger than L_T). The reason for the complication of the process, its disintegration on structures, is in the *openness* of the considered system, and the exchange of energy with an environment. In comparison with such systems, the thermodynamically closed processes do not permit the formation of structures. In a thermal conducting medium without sources, as it is seen from the properties of the solution of problem of an instantaneous point heat source (subsection 2, section 1) and the comparison theorems, the distribution of temperature at $t \to \infty$ becomes spatially homogeneous.

By analogy with similar boundary conditions with blow-up from section 1 for a medium without heat sources, the construction of similar LS- and HS-combustion regimes is also possible. They are sought as follows (exercise 1)

$$T(x, t) = [q_0(t_f - t)]^{-\frac{1}{\beta-1}} f(\xi),$$

$$\xi = \frac{x}{k_0^{1/2} q_0^{m-1}(t_f - t)^m}, \qquad m = \frac{\beta - (\sigma + 1)}{2(\beta - 1)}, \qquad (7)$$

and possess properties similar to their boundary analogs: in the case of a LS-regime ($\beta > \sigma + 1$) the half-width of a localized structure is reduced in time, and the temperature is limited from above by a limiting curve at all $t \le t_f$; in the case of HS-regime ($\beta < \sigma + 1$) the localization is absent.

But the literal repetition of the scheme used in sections 1, 2 is unacceptable for the study of thermal structures, for at least two reasons. Existence and properties of similar LS- and HS-combustion regimes are much more thinly revealed than by methods of sections 1,2 (because of the complexity of the corresponding equation; see exercise 2). Besides, the direct application of similar solutions and comparison theorems for the analysis of the problem (1), (2) in the general case does not give the final results. For example, it is rather easy to check that in the problem (1), (2) at $\beta = \sigma + 1$ any initial perturbations of temperature will develop with blow-up (exercise 3). But it is impossible to prove their localization using a regular comparison theorems even in this particular case.

Indeed, as distinct from the boundary conditions with blow-up, the various solutions of the problem (1), (2) can have various and previously unknown moments of blow-up $t_f^{(1)} \neq t_f^{(2)}$ (from (6) it is seen that at equal rest conditions t_f is less, the larger is the amplitude of the structure in a moment $t = t_0$). Therefore one of the compared solutions ceases to exist before the other, so that the further comparison becomes senseless (more complicated methods of comparison have to be used).

2. Various ways of averaging. For a simplified analysis of spatial-temporal characteristics of thermal structures we use the method of averaging. Certain variants of this method are based on a refusal to describe exactly the behavior of the solution both in space and in time, and on the passage to some average characteristic quantities estimated from more simple models.

Concerning the thermal structures as such quantities, one can select the "amplitude–half-width" or "amplitude–position of front" pairs.

Consider the first case. We adopt the initial function $T_0(x)$ finite, with a maximum at a point $x = 0$ and being close to a symmetrical function. Then the solution $T(x, t)$ of the problem (1), (2) also will be almost symmetrical. The approximate solution is sought in a way which is like to a similar (see (7))

$$T(x,\, t) = \psi(t)\, \theta(\xi)\,, \qquad \xi = \frac{|x|}{\varphi(t)}\,, \tag{8}$$

where $\psi(t)$ and $\varphi(t)$ are the time-dependent sought amplitude and the half-width of the structure, $\theta(\xi)$ is a fixed finite monotonously decreasing function, and $\theta(0) = 1$.

We require that (8) satisfies the integrated equalities (conservation laws)

$$\int_{-\infty}^{\infty} \frac{\partial T}{\partial t}\, dx - \int_{-\infty}^{\infty} \frac{\partial}{\partial x}\left(T^{\sigma}\, \frac{\partial T}{\partial x}\right) dx - \int_{-\infty}^{\infty} T^{\beta}\, dx = 0\,,$$

$$\int_{-\infty}^{\infty} \frac{\partial T}{\partial t}\, T\, dx - \int_{-\infty}^{\infty} \frac{\partial}{\partial x}\left(T^{\sigma}\, \frac{\partial T}{\partial x}\right) T\, dx - \int_{-\infty}^{\infty} T^{\beta+1}\, dx = 0\,,$$

The first among them – the energy conservation law, the second – the moment equation follows – from an integration (1), multiplied on $T(x, t)$ (compare with subsection 4, section 3, Chapter III). For simplicity in (1) we adopt $k_0 = q_0 = 1$, which does not limit the generality, as is equivalent to a replacement $t' = q_0 t$, $x' = x(q_0/k_0)^{1/2}$. By integrating two latter equalities by parts, taking into account that at $x \pm \infty$ the heat flux $-T^{\sigma}\, \partial T/\partial x$ is equal to zero, we obtain

$$\frac{d}{dt}\int\limits_{-\infty}^{\infty} T\,dx = \int\limits_{-\infty}^{\infty} T^{\beta}\,dx\,,$$
$$\frac{1}{2}\frac{d}{dt}\int\limits_{-\infty}^{\infty} T^2\,dx = -\int\limits_{-\infty}^{\infty} T^{\sigma}\left(\frac{\partial T}{\partial x}\right)^2 dx + \int\limits_{-\infty}^{\infty} T^{\beta+1}\,dx\,, \tag{9}$$

The substitution of (8) into (9) gives a system for $\psi(t)$, $\varphi(t)$

$$\frac{d}{dt}\left[\psi(t)\,\varphi(t)\right] = \nu_1\psi^{\beta}(t)\varphi(t)\,,$$
$$\frac{d}{dt}\left[\psi^2(t)\,\varphi(t)\right] = -\nu_2\psi^{\sigma+2}(t)\,\varphi^{-2}(t) + \nu_3\psi^{\beta+1}(t)\,\varphi(t)\,, \tag{10}$$

where ν_1, ν_2, ν_3 are positive constants,

$$\nu_1 = \int\limits_{-\infty}^{\infty} \theta^{\beta}\,d\xi\,,$$
$$\nu_2 = 2\int\limits_{-\infty}^{\infty} \theta^{\sigma}\left| ds\frac{d\theta}{d\xi}\right|^2 d\xi \Bigg/ \int\limits_{-\infty}^{\infty} \theta^2\,d\xi\,, \tag{11}$$
$$\nu_3 = 2\int\limits_{-\infty}^{\infty} \theta^{\beta+1}\,d\xi \Bigg/ \int\limits_{-\infty}^{\infty} \theta^2\,d\xi\,.$$

It is assumed that the function $\theta(\xi)$ is chosen in such a way, to make ν_1, ν_2, ν_3 sensible.

It is easy to solve the system (10) with respect to the derivatives

$$\frac{d\psi}{dt} = \frac{\psi^{\sigma+1}}{\psi^2}\left[(\nu_3-\nu_1)\psi^{\beta-(\sigma+1)}\varphi^2 - \nu_2\right],$$
$$\frac{d\varphi}{dt} = \frac{\psi^{\sigma}}{\psi}\left[(2\nu_1-\nu_3)\psi^{\beta-(\sigma+1)}\varphi^2 - \nu_2\right], \qquad t > t_0 = 0\,, \tag{12}$$

and from (12) – to pass to the equation

$$\frac{d\psi}{d\varphi} = -\frac{\psi}{\varphi}\,\frac{a\psi^{\beta-(\sigma+1)}\varphi^2 - 1}{b\psi^{\beta-(\sigma+1)}\phi^2 - 1}\,, \qquad \psi = 0, \quad \varphi > 0\,, \tag{13}$$

where $a = (\nu_3-\nu_1)\nu_2$, $b = (\nu_3 - 2\nu_1)/\nu_2$. We have to realize the condition $\nu_3 > 2\nu_1$, that is

$$a > 0\,, \qquad b > 0\,. \tag{14}$$

This is required for the system (12) to permit blowing-up regimes.

Thus, the analysis of the problem (1), (2) is reduced to the study of a first order ordinary differential equation. The same essential simplification is achieved for the averaging "amplitude–position of front". The solution is sought in the same form

$$T(x,\,t) = \psi(t)\,\theta(\xi)\,, \qquad \xi = \frac{|x|}{g(t)}\,, \tag{15}$$

where $\psi(t) > 0$ is the amplitude of the structure, while $g(t) > 0$ is not the half-width, but the position its moving front. The function $\theta(\xi)$ is chosen in such a way, that $\theta(\xi) > 0$, $0 < \xi < 1$ and $\theta(\xi) = 0$, $\xi \geq 1$, $\theta(0) = 1$, $\theta'(0) = 0$. As the first integral equation for ψ and g we adopt the energy conservation law, and similarly to (10) we have

$$\frac{d}{dt}\left[\psi(t)\,g(t)\right] = \nu_1 \psi^\beta\, g(t)\,, \qquad t > 0\,. \tag{16}$$

To derive the additional equation we take advantage of the fact that $g(t)$ is the front of a thermal wave, and consequently $W(g(t)) \equiv 0$, $T(g(t),\,t) \equiv 0$. Differentiating the second of these identities by time, t

$$\frac{\partial T}{\partial g}\frac{dg}{dt} + \frac{\partial T}{\partial t} \equiv 0\,,$$

using (1) and rewriting the derivatives as limits, we come to the equality

$$\frac{dg}{dt}\lim_{|x|\to g^-(t)}\left[\frac{T(g(t),\,t) - T(x,\,t)}{g(t) - x}\right] =$$

$$= \lim_{|x|\to g^-(t)}\left[\frac{W(g(t),\,t) - W(x,\,t)}{g(t) - x} - T^\beta(x,\,t)\right].$$

It becomes simpler if we take into account the fact that $g(t)$ is a point of the front and $T(g(t),\,t) = W(g(t),\,t) \equiv 0$:

$$\frac{dg}{dt}\lim_{|x|\to g^-(t)}\frac{T(x,\,t)}{g(t) - x} = \lim_{|x|\to g^-(t)}\left[\frac{W(x,\,t)}{g(t) - x} + T^\beta(x,\,t)\right].$$

The temperature near the front, in accordance with the assumption as in an inertial medium, has the following asymptotic representation: $T(x,\,t) \approx (g(t) - |x|)^{1/\sigma}$ (see in section 1 the solutions with moving front for the equation (1) without sources). Then it is easy to check that the second term on the right hand side of the latter equality is small and hence, can be neglected (this confirms the assumption regarding the small influence of energy sources on the structure of solution at the front). From here we obtain

$$\frac{dg}{dt} = \lim_{T\to 0} \frac{W}{T} = -\lim_{T\to 0} \left(T^{\sigma-1}\frac{\partial T}{\partial x}\right) \tag{17}$$

for the velocity dg/dt of the moving wavefront. The substitution of (15) into (17) gives

$$\frac{dg(t)}{dt} = \nu_4 \frac{\psi^\sigma(t)}{g(t)}, \qquad t > 0, \tag{18}$$

where $\nu_4 = -(\theta^\sigma)'(1)/\sigma > 0$.

Solving (16) relative $\psi'(t)$ we rewrite (16) and (18) as

$$\frac{d\psi}{dt} = \nu_1\psi^\beta - \nu_4\psi^{\sigma+1}g^{-2}, \qquad \frac{dg}{dt} = \nu_4\psi^\sigma g^{-1}, \tag{19}$$

and then we pass to the equation

$$\frac{d\psi}{dg} = \frac{\psi}{g}\left[\mu\psi^{\beta-(\sigma+1)}g^2 - 1\right], \qquad g > 0, \quad \mu = \frac{\nu_1}{\nu_4}. \tag{20}$$

As in the previous case, the evolution of average characteristics of a thermal structure is described by a much simpler model than the initial one.

3. A classification of combustion regimes of a thermal conducting medium. We start the analysis of the problem (1), (2) from the case $\beta = \sigma + 1$, using the first method of averaging. The equation (13) has a simple form

$$\frac{d\psi}{d\varphi} = -\frac{\psi}{\varphi}\frac{a\varphi^2 - 1}{b\varphi^2 - 1}, \qquad \psi > 0, \quad \varphi > 0, \tag{21}$$

and is easily integrated

$$C_0 = \psi^{-1}\varphi^{-1}\left[1 - \frac{\nu_3 - 2\nu_1}{\nu_2}\varphi^2\right]^{-\frac{\nu_1}{2(\nu_3-2\nu_1)}},$$

where $C_0 \geq 0$ is a constant determined by the initial values of $\varphi(0)$, $\psi(0)$. The character of evolution of the thermal structure is clearly seen from the behavior of phase trajectories of the equation (21) (Fig. 75). The bold line indicates the trajectory

$$\varphi \equiv \varphi_S = [\nu_2(\nu_3 - 2\nu_1)]^{1/2} = \frac{L_T}{2}, \tag{22}$$

corresponding to a similar S-regime of combustion process (6) (it is also the isocline of the infinity of the equation (21)), the dashed line is the isocline of zero $\varphi = a^{-1/2} < \varphi_S$. The validity of (22) follows from the fact, that in accordance with (6) the function $\theta(\xi)$ in (8) obviously, can be taken in a form

$$\theta(\xi) = \begin{cases} \cos^{2/\sigma} \dfrac{\pi\xi}{2}, & |\xi| < 1\,, \\ 0, & |\xi| \geq 1\,. \end{cases}$$

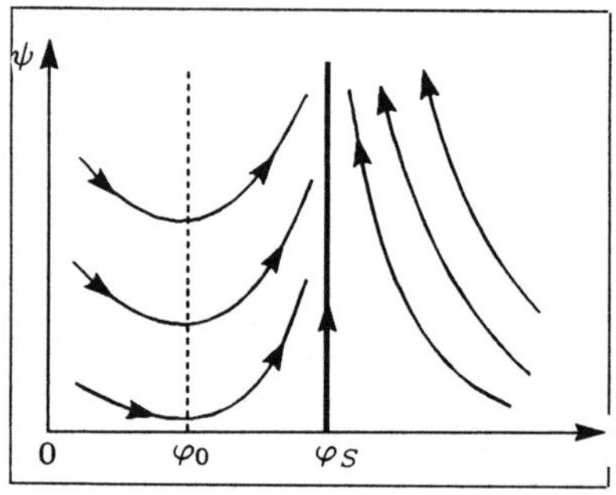

Fig.75.

Then from (6) we obtain the relation for the amplitude and half-width

$$\psi(t) = \left[\frac{2(\sigma+1)}{\sigma(\sigma+2)}\right]^{1/\sigma} (t_f - t)^{-1/\sigma}\,, \quad \varphi(t) = \frac{L_T}{2}\,,$$

and for coefficients ν_1, ν_2, ν_3 from (11) we obtain

$$\nu_1 = \frac{\sigma+2}{2(\sigma+1)}\,, \qquad \nu_2 = \frac{\pi^2}{\sigma(\sigma+2)}\,, \qquad \nu_3 = \frac{\sigma+4}{\sigma+2}\,.$$

The substitution of these expressions in (21) leads to an identity, i.e. the similar solution is precisely described by the method of averaging.

Concerning the non-similar solutions, we note that all of them, as it is clear from Fig. 75, develop in blowing-up regime, their trajectories tend to those of the similar solutions: $\varphi(t) \to \varphi_S$, $t \to t_f$. From here we have (exercise 4)

$$\psi(t) \simeq (\sigma\nu_1)^{-1/\sigma}(t_f - t)^{-1/\sigma}\,, \qquad t \to t_f\,, \tag{23}$$

i.e. at a developed stage combustion proceeds in accordance with a self-similar law. At an initial stage it can proceed in a more complicated way: at functions $T_0(x)$ with a half-width smaller than φ_0, the temperature in a structure first decreases and starts to grow only after reaching the half-width of its *critical* size $\varphi = \varphi_0$.

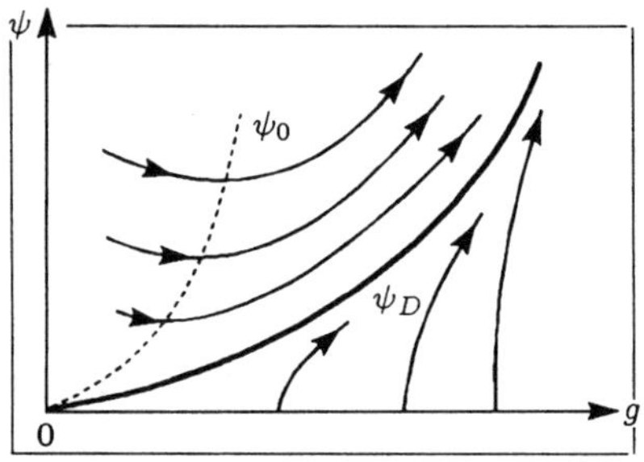

Fig.76.

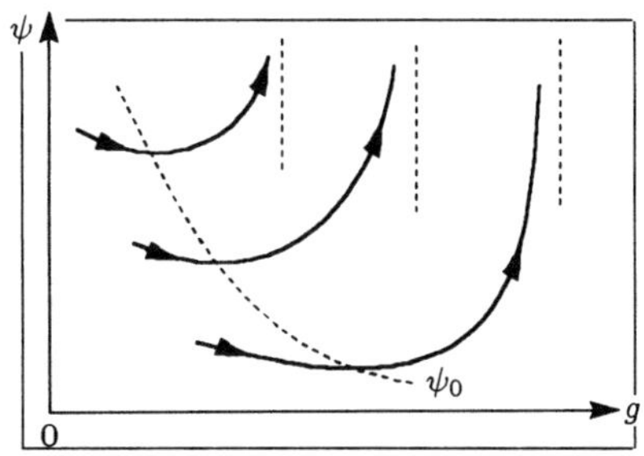

Fig.77.

We analyze the case $\beta \neq \sigma + 1$ with the help of equations (20). At $\beta \neq \sigma + 2$ a it has a general solution (exercise 5)

$$\left|\psi^{\beta-(\sigma+1)} g^2 - l_0\right|^{\frac{1}{\sigma+1-\beta}} \psi g = C_0 \,, \quad l_0 = \frac{1}{\mu} \frac{\beta-(\sigma+3)}{\beta-(\sigma+1)} \,, \tag{24}$$

where $C_0 \geq 0$ is determined by the initial data, while at $\beta = \sigma + 3$ has a solution

$$\psi^2 = [g^2(C_0 - 2\mu \ln g)]^{-1} \,, \qquad C_0 = \text{const} > 0 \,. \tag{25}$$

The behavior of the integral curves of the equation is different in ranges $\beta < \sigma + 1$, $\sigma + 1 < \beta < \sigma + 3$, $\beta > \sigma + 3$. In Fig. 76, 77, for the cases $\beta < \sigma + 1$, $\sigma + 1 < \beta < \sigma + 3$, the dashed lines indicate the isoclines ψ_0 of zero, the bold line indicates the special trajectory (separatrix)

$$\psi = \psi_D = l_0^{\frac{1}{\beta-(\sigma-1)}} g^{-\frac{2}{\beta-(\sigma+1)}} \,, \tag{26}$$

existing at $\beta < \sigma + 1$ (and $\beta > \sigma + 3$) and corresponding to a value $C_0 = 0$ in the general solution.

Consider the behavior of the average characteristics of thermal structures. In the case $\beta < \sigma + 1$ (Fig. 76) all trajectories converge to the separatrix (26), and the structure as follows from (19), tends asymptotically to a similar regime (see (7)):

$$\psi(t) \sim (t_f - t)^{\frac{1}{\beta-1}} \,, \qquad g(t) \sim (t_f - t)^{\frac{\beta-(\sigma+1)}{2(\beta-1)}} \,, \qquad t \to t_f \,,$$

so that $g(t) \to \infty$ at $t \to t_f$, i.e. heat localization is absent in the HS-regime.

The behavior of thermal structures of the LS-regime ($\beta > \sigma + 1$) is more diverse. At $\sigma + 1 < \beta < \sigma + 3$ (Fig. 77) any trajectory has a vertical asymptote with a coordinate

$$g_* = C_0^{\frac{\beta-(\sigma+1)}{\beta-(\sigma+3)}} \,,$$

i.e. $g(t) \to g_*$, $t \to t_f$, that implies a localization of combustion in area $|x| < g_*$. The amplitude of the structure grows by the self-similar law $\psi \sim (t_f - t)^{-1/(\beta-1)}$, $t \to t_f$ (the behavior of a half-width at the given way of averaging is obviously not described).

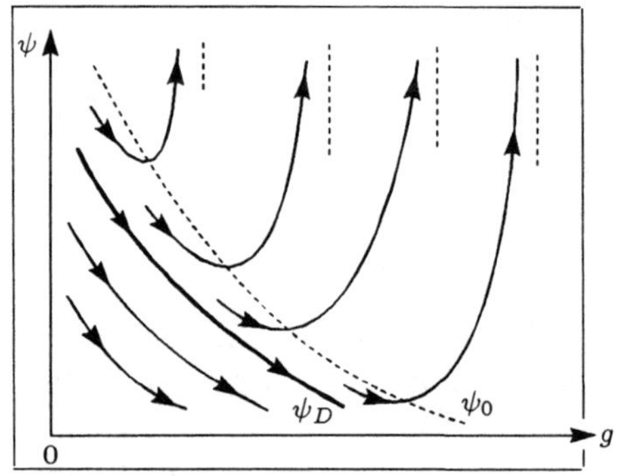

Fig.78.

The conclusion about the development of a blowing-up localized structure is valid for a part of trajectories also in the case $\beta \geq \sigma+3$, so that for $\beta = \sigma+3$ from (25) we obtain $g_* = e^{C_0/(3\mu)}$. However at $\beta > \sigma + 3$ (Fig. 78; the lines ψ_D and ψ_0 have the same content as in Fig. 76) on a phase plane of the LS-regime there is a separatrix (26) dividing the basically different classes of solutions. For initial functions $T_0(x)$, corresponding to areas higher than the line ψ_D, the investigated blowing-up regime does occur. If the initial perturbation has either a small amplitude, or a small size (the area lower than ψ_D), the blow-up does not develop. The temperature in a structure (see (24)) evolves by the law

$$\psi(g) \simeq F_0 g^{-1}, \qquad F_0 = C_0 l_0^{-\frac{1}{\sigma+1-\beta}}, \qquad g \to \infty .$$

From here and from (19) the following asymptotic evaluations are obtained

$$\psi(t) \sim t^{-\frac{1}{\sigma+2}}, \qquad g(t) \sim t^{\frac{1}{\sigma+2}}, \qquad t \to \infty .$$

Similar behavior is peculiar to self-similar solutions in a medium without release of energy, for example, for a solution of the problem on an instantaneous point heat source from subsection 2, section 1. Thus, in the case $\beta > \sigma+3$ where initiation is not sufficiently strong the process of combustion of the medium slows.

Note that in the HS-regime and in the blowing-up LS-regime there is a critical size of the structure $\varphi = \varphi_0$ (similarly to the S-regime); upon reaching it, the temperature starts to increase.

The method of averaging has enabled us to obtain quite a complete classification of the structures of combustion in a nonlinear medium. We will

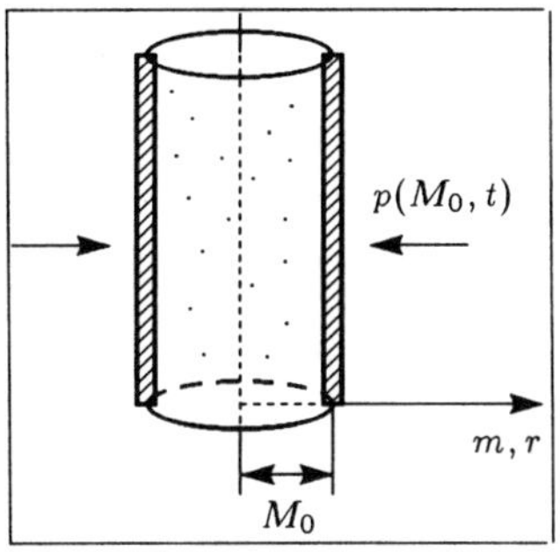

Fig.79.

now show, that their appearance is not the specific property of processes described by the parabolic equations. We construct a simple example of *localized gas dynamical structures.* Consider a continuous (without shock waves) compression of a final mass of gas $2M_0$, located within a cylindrical piston (Fig. 79). The one-dimensional process is described by a set of equations

$$\frac{\partial}{\partial t}\frac{1}{\rho} = \frac{\partial}{\partial m} rv\,, \qquad \frac{\partial v}{\partial t} = -r\,\frac{\partial p}{\partial m}\,,$$

$$p\rho^{-\gamma} = \varphi(m) \geq 0\,, \tag{27}$$

$$\frac{dr}{dt} = v\,, \qquad 0 < m < M_0\,, \qquad t_0 \leq t\,,$$

where t is the time; m is the mass coordinate counted from the axes of symmetry and related with Eulerian coordinate via expression $dm = \rho r\, dr$; ρ, v, p are the sought density, velocity and pressure of gas; $\gamma > 1$ is the adiabatic index; φ is the given entropy function. The system (27) can be derived easily (similarly to subsection 5, section 4, Chapter II) from the equations of motion in Eulerian form. Recall: the first two equations (27) are the laws conservation of mass and momentum, the third is the adiabatic integral (in the absence of shock waves the entropy of a fixed particle of gas does not vary in time), the fourth is the kinematic connection between r and v.

By analogy with thermal S-regimes we construct the solution (27) in separating variables

$$p(m,\, t) = p_1(t)\,\pi(m)\,, \qquad v(m,\, t) = u_1(t)\,u(m)\,,$$

$$r(m,\, t) = r_1(t)\,R(m)\,, \tag{28}$$

$$0 \leq m \leq M_0\,, \qquad t_0 \leq t \leq t_f\,,$$

so that the temporal part in (28) without any significant restriction of generality (exercise 6), can be readily taken in the form of power law functions

$$p = p^0(t_f - t)^n \pi(m)\,, \qquad v = u^0(t_f - t)^{n_1} u(m)\,,$$

$$r = r^0(t_f - t)^{n_2} R(m)\,, \tag{29}$$

$$0 \le m \le M_0\,, \qquad t_0 \le t < t_f < \infty\,.$$

To simplify the further calculations we adopt $p^0 = u^0 = r^0 = 1$.

Substituting these expressions into (27), we obtain

$$n = -2\,, \qquad n_1 = \frac{1-\gamma}{\gamma} < 0\,, \qquad n_2 = \frac{1}{\gamma} > 0\,. \tag{30}$$

As distinct from the flat case the index n in the law for pressure does not depend on γ (see the corresponding solution in subsection 3, section 1). To determine π, u, R we have the system

$$-\frac{2}{\gamma}\varphi(m)^{1/\gamma}\pi^{-1/\gamma} = \frac{d(Ru)}{dm}\,, \qquad \frac{1-\gamma}{\gamma}u = R\frac{d\pi}{dm}\,, \qquad u = -\frac{1}{\gamma}R\,,$$

$$0 < m \le M_0\,,$$

which can be solved easily. For example, the spatial profile of $\pi(m)$ is given by a simple formula

$$\pi(m) = \pi(M_0) + \frac{(\gamma-1)(m-M_0)}{\gamma^2}\,, \qquad 0 \le m \le M_0\,, \tag{31}$$

where $\pi(M_0)$ is determined from the law for the pressure on the piston. Note that by virtue of equation $u = -R/\gamma$ the velocity of gas on the axis of the cylinder $r = 0$ ($m = 0$) satisfies the natural condition of symmetry $v(0,\,t) = 0$.

From (29)–(31) we obtain the final form of the solution for pressure

$$p(m,\,t) = (t_f - t)^{-2}\left[\pi(M_0) + \frac{(\gamma-1)(m-M_0)}{\gamma^2}\right],$$

$$0 \le m \le M_0\,, \qquad t_0 \le t < t_f\,,$$

and (in view of the third equation (27)) for the density

$$\rho(m,\,t) = (t_f - t)^{-2/\gamma}\varphi(m)^{-1/\gamma}\left[\pi(M_0) + \frac{(\gamma-1)(m-M_0)}{\gamma^2}\right]^{1/\gamma},$$

$$0 \le m \le M_0\,, \qquad t_0 \le t < t_f\,.$$

In the latter formula we assume for simplicity that $\pi(M_0) = (\gamma - 1)\, M_0/\gamma^2$ and, hence, obtain

$$\rho(m,\, t) = \left(\frac{\gamma - 1}{\gamma^2}\right)^{1/\gamma} (t_f - t)^{-2/\gamma} \left[\frac{m}{\varphi(m)}\right]^{1/\gamma} \tag{32}$$

From (32) it is seen that the function $\rho(m,\, t)$ as distinct from $p(m,\, t)$, can have extremes. The degree of compression of a given region of the medium is determined by its entropy $\varphi(m)$ and pressure. Due to non-isoentropy at the monotonous profile of pressure in a wave of compression it is possible to reach greater density in areas with smaller pressure and to achieve gas dynamical structures (in the constructed example these are structures of any complexity; see, for example, exercise 7). They represent localized inhomogeneities of density connected with a fixed mass of gas and growing in blowing-up regime.

EXERCISES

1. Adopting formally in the problem (1), (2) $t_0 \to -\infty$ and $T(x,\, -\infty) \equiv 0$, obtain with the help of Π-theorem, the representation of its solution as (7).

2. By substituting (7) in (1) deduce for $f(\xi)$ a nonlinear ordinary differential second-order equation. Prove that it does not permit a similarity transformation of expansion-compression type (and consequently is not reduced to a first order equation).

3. Check that the solution of the problem (1), (2) at $\beta = \sigma + 1$ and arbitrary $T_0(x)) \not\equiv 0$ will tend to infinity at $t \to t_f < \infty$. For the proof, construct sequentially the functions bounding it from below, i.e. the solution of the problem of an instantaneous point source of heat and solution (6), and then apply the comparison theorem.

4. Derive the formula (23) by means of approximate solution of (12) at $\varphi \to \varphi_S$, $t \to t_f$.

5. Check that the equation (20) permits a similarity transformation of expansion-compression. Using the replacements of a form $\ln \psi = u$, $\ln g = v$, obtain its general solution (24), (25).

6. By substituting (28) into (27) check that the functions $p_1(t)$, $v_1(t)$, $r_1(t)$ are given by a quadrature from which at $t \to t_f < \infty$ the asymptotic time dependence (29) follows.

7. Select examples of such a distribution of entropy $\varphi_N(m) > 0$, $0 < m < M_0$ in (32), so that the density will have any given number of extremes N. Prove that the function $\varphi_N(m)$ can then be monotonous.

4 On Transition to Discrete Models

Now we offer some ideas on the simplest requirements of numerical methods and about basic concepts of the theory of difference schemes. Consider some typical approaches to the construction of discrete analogs of initial models, used for numerical research.

1. Necessity of numerical modeling, elementary concepts of the theory of difference schemes. However deep and varied the methods of qualitative analysis of mathematical models might be, the area of their applicability is rather limited. They are either simple, mainly linear models, or separate fragments of complicated models, including nonlinear ones. The only universal way of studying the models is by the application of numerical methods for determining the approximate solution of the posed problem by means of modern computing facilities and information science.

The computer algorithm, which is "understandable" to a computer, i.e. the sequence of operations (arithmetical, logical, etc.) leading to the solution, should satisfy the rather rigorous and occasionally contradictory requirements. These include, first of all, the need to obtain solutions of given accuracy by reasonable and whenever possible least number of operations, so far as the time of a single calculation should be measured by minutes and only in unique cases by hours. The amount of corresponding information cannot exceed the capacities of the machine's memory, during evaluations one should avoid using numbers which are too large or small and which are not accepted by the computer, the structure of the algorithm should be simple enough and has to take into account the computers' architecture and so on.

Only computing algorithms which fulfill these requirements, allow us to perform a multi-level numerical research of an initial model, to proceed with computing experiments, to conduct analysis in rather diverse situations and to obtain exhaustive information on the investigated object. Such understanding of mathematical modeling means not only a simple more precise definition of quantitative characteristics of the phenomena, but also the study of their basic qualitative properties. The latter is especially important for nonlinear objects, with diverse and unexpected behavior.

We stress that the problems of numerical modeling are not removed by themselves with the appearance of more and more powerful and cheap computers. It is connected with at least two reasons: more complex problems posed by the practical science, and the theory of the problems and the need to realize large number of series of computing experiments for an adequate study of object.

Therefore the development of effective computing algorithms always remains one of the key problems of mathematical modeling. To develop them, the methods, ideas and approaches used in the construction of initial mathematical models are widely used. This connection is well traced on an example of a rather broad class of models – those, which are reduced to differential equations. For them the process of creation of computing algorithms consists of two principal stages: first, the *discrete analogs* of initial models are created and their properties are studied, second, the discrete equations are solved numerically (an elementary example is represented in subsection 4, section 6, Chapter I).

Below we pay most attention to the first stage, by considering, first, the simplest boundary problem for a second-order equation on an interval

$$\frac{d}{dx}\left(\frac{du}{dx}\right) = -f(x)\,, \qquad 0 < x < l\,, \quad u(0) = u_1\,, \quad u(l) = u_2\,, \tag{1}$$

assuming here and further, that its solution (in appropriate sense) exists and is unique.

The passage from (1) to a discrete model is divided into two stages. Replace the continuous area $0 < x < l$ by a discrete one, i.e. a population of a finite number of points N. The most simple way is by the uniform split of the interval $[0, l]$ by the rule $x_i = ih$, $h = l/N$, $0 \le i \le N$. The set $\bar{\omega}_h = \omega_h \cup \{x = 0,\ x = l\}$, $\omega_h = \{x_i\}$, $i \ne 0,\ N$, of these points represents a (uniform) *difference grid with a step h*, the points x_i are the *knots* (see Fig. 80, where the bars denote the basic knots of the grid, and crosses denote the auxiliary knots $x_{i+1/2} = (x_{i+1} - x_i)/2$ with half-integer indices). All the functions appearing in (1) are considered now as functions not of a continuous argument x, but of a discrete argument x_i, (*grid functions*), for example, the analog of a solution $u(x)$, $0 \le x \le 1$, is *the approximate solution* $y(x_i)$, $0 \le i \le N$.

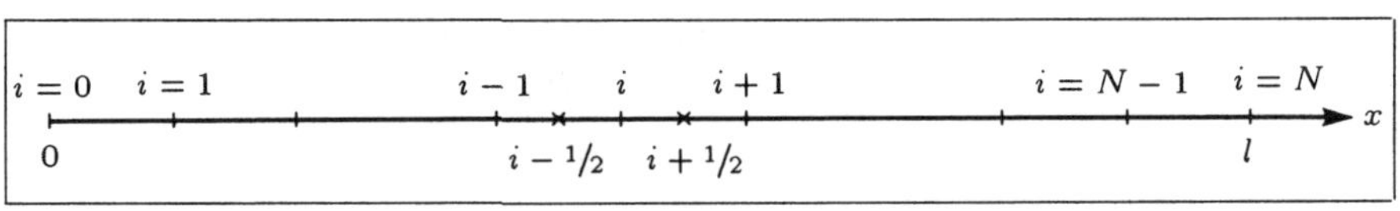

Fig.80.

On the second stage the discrete analogs of the differential equation (1) and input data are created. The most natural discretization of a differential operator is the replacement of derivatives by appropriate *finite differences*. We introduce the notations

$$u_x = \frac{u(x+h) - u(x)}{h}\,, \qquad u_{\bar{x}} = \frac{u(x) - u(x-h)}{h}\,, \qquad u_{\bar{x}x} = \frac{u_x - u_{\bar{x}}}{h}\,. \tag{2}$$

The first two expressions in (2) are the discrete approximation of a derivative du/dx; to obtain it, it is enough to use the value of function $u(x)$ only at two points (*the two-point stencil*). Decomposing $u(x)$ via Taylor series, it is easy to prove that

$$\frac{du}{dx}(x) = u_x + O(h)\,, \quad \frac{du}{dx}(x) = u_{\bar{x}} + O(h)\,,$$

$$\frac{du}{dx}\left(x + \frac{h}{2}\right) = u_x + O(h^2)\,, \quad \frac{du}{dx}\left(x - \frac{h}{2}\right) = u_{\bar{x}} + O(h^2)\,,$$

In other words, du/dx in the integer knots of a grid $x = x_i$ is approximated by expressions (2) in the first *order of approximation*, while in half-integer points $x = x_{i+1/2}$, $x = x_{i-1/2}$ (by virtue of symmetry) – with the second. To replace the second derivative of the function u (the third expression in (2)) a three-point stencil $x - h$, x, $x + h$ is required, so that

$$\frac{d}{dx}\left(\frac{du}{dx}\right) = u_{\bar{x}x} + O(h^2) = \frac{u(x+h) - 2u(x) + u(x-h)}{h^2} + O(h^2)\,,$$

i.e. the approximation has a second order (exercise 1).

The discrete approximation of input data in the considered case does not represent any difficulty and is carried out precisely: $\varphi_i = f(x_i) = f_i$, $i = 0, \ldots, N$; $y_0 = y(0) = u_1$, $y_N = y(l)) = u_2$.

Unifying all these considerations, we replace (1) by a set of $N-1$ *difference equations* for determination $N-1$ of unknown values of the approximate solution y_i in the knots x_i of the grid ω_h:

$$y_{\bar{x}x} = -\varphi\,, \qquad \text{or} \qquad \frac{y_{i+1} - 2y_i + y_{i-1}}{h^2} = -\varphi_i\,,$$
$$i = 1, 2, \ldots, N-1, \qquad y_0 = u_1\,, \qquad y_N = u_2\,. \tag{3}$$

The system (3) with boundary conditions is called a *difference scheme* and serves as a discrete analog of the model (1). Its solution is obtained relatively easily (see subsection 2).

Using (1)–(3), we illustrate the elementary concepts connected with the difference schemes. The grid function $z_i = y_i - u_i$, $i = 0, \ldots, N$, i.e. the difference between exact and approximate solutions in the knots of a grid $x = x_i$, is called an *error*. If $z_i = O(h^\alpha)$, $i = 1, \ldots, N$, $\alpha > 0$, then the difference scheme (3) *converges* (with an order α), and $z_i \to 0$, $h \to 0$ for all i: at a smaller grid cell y_i approximates the exact solution $u(x)$ in knots x_i as well as is required. Between the knots the sought solution, if necessary, can be determined further by means of a usual interpolation. In this case the construction of a difference scheme achieves its purpose.

To reveal the convergence of the scheme (3), we consider the grid function

$$\psi_i = u_i'' - u_{\bar{x}x_i}\,, \qquad i = 1, \ldots, N-1\,,$$

called *the error of approximation of a differential operator by difference one*, or *residual* (here the primes denote the derivation by x, $u_i'' = u''_{(x_i)}$). If $\psi_i = O(h^\beta)$, $i = 1, \ldots, N-1$; $\beta > 0$, then one has an approximates: the continuous operator approximates the discrete one with an order β (here, as was already shown above, $\beta = 2$), and $\psi_i \to 0$, $h \to 0$, $i = 1, \ldots, N-1$. The function $f_i - \varphi_i$, $i = 0, \ldots N$, is the error of approximation of the right hand side (in an investigated example it is identically equal to zero, as well as the error of approximation of the boundary conditions).

Subtracting in internal knots of a grid $i = 1, \ldots, N-1$ the equation (1) from (3), we obtain

$$y_{\bar{x}\,x_i} - u_i'' = -\varphi_i + f_i\,, \qquad i = 1, \ldots, N-1\,,$$

or, taking into account the equalities $z = y_i - u_i$, $\psi_i = u_i'' - u_{\bar{x}x_i}$, $\varphi_i = f_i$, we come to a system of difference equations for an error z_i, with zero boundary conditions

$$z_{\bar{x}x} = \psi\,, \qquad i = 1, \ldots, N-1, \qquad z(0) = z(N) = 0\,, \tag{4}$$

with the right hand side being the error of approximation ψ.

The given property allows us to reveal the basic connection between the already entered concepts of convergence and the approximation of the scheme (3) and the concept of its *stability*. The latter implies that for any admissible input data φ, u_1, u_2 the following inequality is fulfilled

$$||y|| = \max_i |y_i| \leq C||\varphi|| = C \max_i |\varphi_i|\,, \qquad i = 0, \ldots, N\,, \tag{5}$$

where $C > 0$ is a constant not depending on i and h (in this case the stability is with respect to the right hand side).

The scheme (4) is the particular case of (3). Therefore if (5) is fulfilled (see exercise 2) we readily have

$$|z| \leq C|\psi|\,, \quad \text{or} \quad z_i = O(h^\beta) \to 0\,, \quad h \to 0\,, \quad i = 0, \ldots, N\,.$$

i.e. from the properties of approximation ($\psi = O(h^\beta)$) and stability (5) of the difference schemes (3) follows its convergence (with the same order, as the order of approximation, $\alpha = \beta = 2$).

We explain with the help of qualitative considerations, that the revealed connection generally concerns any class of difference schemes. Let some general (abstract) problem

$$\begin{aligned} &L_u = -f(x)\,, \qquad x \in G\,, \qquad x = \{x_1, \ldots, x_n\}\,, \\ &u = \mu(x)\,, \qquad x \in \Gamma\,, \qquad \bar{G} = G \cup \Gamma\,, \end{aligned} \tag{6}$$

where L is a linear differential operator acting in an open area $x \in G$ (Γ is the boundary of closure $\bar{G} \cup \Gamma$ in area G), f and μ are given functions $x \in \mathrm{R}^n$ (the right hand side and the value of the sought solution on the boundary), be approximated by a difference scheme

$$L_h y_h = -\varphi_h\,, \qquad x \in \omega_h\,,$$
$$y_h(\gamma_h) = \mu_h\,, \qquad x \in \gamma_h\,, \qquad \bar{\omega}_h = \omega_h \cup \gamma_h\,, \tag{7}$$

where $\bar{\omega}_h$, ω_h, γ_h are the difference analogs for $\bar{G}$, G, Γ; L_h denotes the appropriate difference operator; and grid functions y_h, φ_h, μ_h are the analogs of an exact solution u and input data f, μ.

Introduce, as above, an error $z_h = y_h - u_h$, residual $\psi_h = (Lu)_h - L_h u_h$, and considering the approximation of the right hand side and the boundary conditions to be exact, taking account also (6) we obtain

$$L_h\, z_h = \psi_h\,, \quad x \in \omega_h\,; \qquad z_h = 0\,, \quad x \in \gamma_h\,. \tag{8}$$

Similar to case of scheme (3) the right hand side of the difference problem for z represents the error of approximation ψ. If the scheme (7) is stable, that is $||y_h||_{(1)} \le C||\varphi_h||_{(2)}$ (where the symbols $||\cdot||_{(1)}$ and $||\cdot||_{(2)}$ denote some, generally speaking, different norms of grid functions y_h and φ_h) then from (8) we have

$$||z_h||_{(1)} \le C||\psi_h||_{(2)}\,.$$

From here it is clear that the approximation ($||\psi_h|| \to 0$, $|h| \to 0$) and stability of the scheme (7) ensure its convergence ($||z_h|| \to 0$, $|h| \to 0$). The same statements are fair also for other problems for differential equations approximated by difference schemes, including also nonlinear ones (then the definition of stability is modified correspondingly).

Thus, there are at least two requirements for discrete models – the approximation of the initial model and its stability. Then for a sufficiently accurate numerical solution of difference equations (as a rule, they represent systems of linear and nonlinear algebraic equations of the order N, where N is the number of knots of a grid) and sufficiently small steps one has an adequately exact approximate solution. We stress: the number of knots cannot be too large (and the steps too small), in so far as the numerical solution has to be found in an acceptable number of operations, i.e. using actual grids. Consider some typical methods of constructing of discrete models.

2. Direct formal approximation. This historically first and, as follows from the title easily interpreted method is simple, clear and frequently gives high quality discrete models. We show this by constructing the difference approximation of the first boundary problem in an interval $[0,\, l]$ for the thermal conducting equation

$$\frac{\partial u}{\partial t} = \frac{\partial}{\partial x}\left(\frac{\partial u}{\partial x}\right), \qquad 0 < x < l, \quad 0 < t \le T,$$
$$u(0,\, t) = u_1(t), \quad u(0,\, l) = u_2(t), \quad 0 \le t \le T, \tag{9}$$
$$u(x,\, 0) = u_0(x), \quad 0 \le x \le l.$$

Its solution $u(x,\, t)$ is assumed to exist and is sought within $0 < x < l$ at all $0 < t \le T$ (Fig. 81).

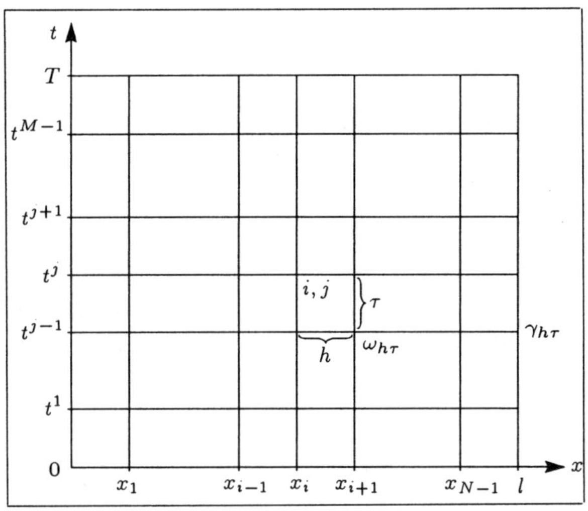

Fig.81.

We select the simplest discretization of the calculated area by splitting uniformly by $N-1$ vertical and M horizontal lines. The points of their intersection with each other and with intervals $[0,\, l]$, $[0,\, T]$, $[l,\, T]$ will give the knots of the grid $\bar{\omega}_{ht}$. Such a grid is said to be *uniform with a spatial step* $h = l/N$, $x_i = ih$, $0 \le i \le N$, and *time step* $\tau = T/M$, $t^j = j\tau$, $0 \le j \le M$, the set of knots with identical indices j is called as *a time layer* (Fig. 81). In boundary knots (belonging to intervals $[0,\, l]$, $[0,\, T]$, $[l,\, T]$) the function $u(x,\, t)$ is known from the boundary conditions (9), and the approximation is obvious: $y_0^j = u_1(t^j)$, $y_N^j = u_2(t^j)$, $j = 0, 1 \ldots, M$; $y_i^0 = u_0(x_i)$, $0 \le i \le N$.

The approximate solution y_i^j is necessary to find on the set of internal knots $\omega_{h\tau}$ of the grid $\bar{\omega}_{h\tau} = \omega_{h\tau} \cup \gamma_{h\tau}$. We conduct a natural approximation of the differential operator by difference, rewriting the latter uniformly in any point $(x_i,\, t^j)$ of grid $\omega_{h\tau}$. We replace the derivative by time by the first difference: $\partial u/\partial t \approx (y_i^{j+1} - y_i^j)/\tau$. Taking into account the fact that it contains the values of the difference solution from two time layers, we replace the second derivative by x with a sum of expressions (obtained in subsection 1) taken on $(j+1)$th and jth layers: $\partial^2 u/\partial x^2 \approx \sigma y_{\bar{x}x}^{j+1} + (1-\sigma) y_{\bar{x}x}^j$, where $0 \le \sigma \le 1$. As a result, instead of (9) we come to *the scheme with weights*

$$\frac{y_i^{j+1} - y_i^j}{\tau} = \sigma\, \frac{y_{i+1}^{j+1} - 2y_i^{j+1} + y_{i-1}^{j+1}}{h^2} + (1-\sigma)\, \frac{y_{i+1}^j - 2y_i^j + y_{i-1}^j}{h^2}\,,$$

$$(x_i,\, y^j) \in \omega_{h\tau}\,, \tag{10}$$

$$y_0^j = u_1(0,\, t^j)\,, \quad y_N^j = u_2(l,\, t^j)\,, \quad j = 0,\, 1,\, \ldots,\, M\,;$$

$$y_0^0 = u_0(x_i)\,, \qquad 0 \le i \le N\,,$$

representing a system of $(N-1)\,M$ linear algebraic equations for the determination of the same number of functions y_i^j. Each of equations (10) is rewritten at $\sigma \ne 0,\, 1$ on the six-point stencil using knots with indices $(i-1,\, j+1)$, $(i,\, j+1)$, $(i+1,\, j+1)$, $(i-1,\, j)$, $(i,\, j)$, $(i+1,\, j)$. The error of its approximation in the general case is the magnitude $O(\tau + h^2)$, i.e. is of the first order by time and second by space. For *a symmetrical scheme* with $\sigma = 1/2$ the order of approximation by time is increased to $O(\tau^2)$ (exercise 3).

At $\sigma = 0,\, 1$ from (10) we obtain a simpler purely *explicit* ($\sigma = 0$, four-point stencil $(i,\, j+1)$, $(i-1,\, j)$, $(i,\, j)$, $(i+1,\, j)$) and a purely *implicit* ($\sigma = 1$, four-point stencil $(i-1,\, j+1)$, $(i,\, j+1)$, $(i+1,\, j+1)$, $(i,\, j)$) schemes.

At $\sigma = 0$ each of the equations of the scheme (10) contains only one unknown quantity y_i^{j+1}. Therefore its solution is easily obtained by means of explicit formulae during the passage from the jth to $(j+1)$th layer using known values of the solution on the boundary (on the layer $j = 0$ the solution is known from the initial data).

In case of the implicit scheme (and all schemes with $\sigma \ne 0$) the difference equations contain three unknown quantities: y_{i-1}^{j+1}, y_i^{j+1}, y_{i+1}^{j+1}; on each time layer we obtain a problem of type (3). The similar problems for the three-point equations of the form

$$A_i y_{i-1} - C_i y_i + B_i y_{i+1} = -F_i\,, \qquad i = 1,\, \ldots,\, N-1\,,$$
$$y_0 = \kappa_1 y_1 + \mu_1\,, \qquad y_N = \kappa_2 y_{N-1} + \mu_2\,, \tag{11}$$

where $A_i \ne 0$, $B_i \ne 0$, $i = 1,\, \ldots,\, N-1$, at conditions $|C_i| \ge |A_i| + |B_i|$, $i = 1,\, \ldots,\, N-1$; $|\kappa_\alpha| \le 1$, $\alpha = 1,\, 2$; $|\kappa_1| + |\kappa_2| < 2$ (for (3) and (10) these conditions are fulfilled) are relatively easily solved by the sweep method. For the solution (11) the presence of a recurrent dependence of the following form is assumed

$$y_i = \alpha_{i+1}\, y_{i+1} + \beta_{i+1}\,. \tag{12}$$

Its substitution into (11) gives for the coefficients α_{i+1} and β_{i+1} the recurrent relations

$$\alpha_{i+1} = \frac{B_i}{C_i - \alpha_i A_i}\,, \qquad \beta_{i+1} = \frac{A_i \beta_i + F_i}{C_i - \alpha_i A_i}\,, \qquad i = 1, \ldots, N\,.$$

From these, with the help of the boundary condition at $i = 0$ values α_{i+1}, β_{i+1} are obtained in all knots of the grid (direct sweep). Further, with the help of conditions in a point $i = N$ at known α_{i+1}, β_{i+1} by the formula (12) we estimate the values y_N, y_{N-1}, $\ldots, 1, y_i$, y_0 (inverse sweep). Note: the simplicity of the solution of the algebraic system (10) is determined by the simple (three-diagonal) structure of its matrix.

Slightly more complicated considerations reveal the stability of the scheme (10), so that there is a qualitative difference between the cases $\sigma = 0$ and $\sigma = 1$. The purely implicit scheme is stable at any ratio between the steps h and τ (*unconditional stability*), while for the purely explicit scheme the realization of an inequality $\tau \leq Ch^2$, $C > 0$ is some constant (*conditional stability*) is necessary. The given requirement (typical of the explicit difference schemes generated by the parabolic equations) can pose too strict restrictions on the time step. Therefore, despite the simplicity, the explicit schemes for the solution of these problems are not practically used.

The difference schemes of type (10) are constructed in a similar way for boundary problems distinct from (9) and more general parabolic equations. For example, for the equation of nonlinear thermal conduction

$$\frac{\partial u}{\partial t} - \frac{\partial}{\partial x}\left(k(u)\frac{\partial u}{\partial x}\right)$$

one of the obvious approximations of the first boundary problem is as follows

$$\frac{y_i^{j+1} - y_i^j}{\tau} = \frac{1}{h}\left[k^j_{i+1/2}\frac{y^{j+1}_{i+1} - y^{j+1}_i}{h} - k^j_{i-1/2}\frac{y^{j+1}_i - y^{j+1}_{i-1}}{h}\right],$$

$$(x_i,\, y^j) \in \omega_{h\tau}\,,$$

$$y^j_0 = u_1(0,\, t^j)\,, \quad y^j_N = u_2(l,\, t^j)\,, \quad j + 0,\, 1,\, \ldots,\, M\,;$$

$$y^0_i = u_0(x_i)\,, \quad 0 \leq i \leq N\,,$$

where $k^j_{i+1/2}$ and $k^j_{i-1/2}$ are certain approximations of thermal conductivity coefficient in half-integer knots $i + 1/2$, $i - 1/2$, taken for simplicity on jth layer. The given implicit stable scheme is easily solved, as (10), by the sweep method. If the thermal conductivity is approximated on $(j+1)$th layer, it becomes nonlinear and is solved by means of appropriate methods of *sequential approximations (iterative procedures).*

Thus, based on direct approximation in certain cases the discrete models possessing the necessary qualities are obtained. At the same time its formal application can lead also to the construction of discrete analogs of the initial models, having nothing in common with their pre-images. Consider the rather natural approximation of the equation (9) – the difference equation

$$\frac{y_i^{j+1} - y_i^{j-1}}{2\tau} = \frac{y_{i+1}^j - 2y_i^j + y_{i-1}^j}{h^2}. \tag{13}$$

written on the five-dot stencil $(i,\ j+1)$, $(i+1,\ j)$, $(i,\ j)$, $(i-1,\ j)$, $(i,\ j-1)$.

Together with the appropriate boundary conditions this forms *the three-layer scheme* (as distinct from *the two-layer scheme* (10)), with an error of approximation $O(\tau^2 + h^2)$ and being easily solved by the explicit formulae. However the given scheme is unsuitable, as it is unstable at any h and τ (absolutely unstable). Its solution at limited boundary conditions can become as large as possible, when $j \to \infty$ $(t \to \infty)$, which contradicts the maximum principle for the equation (9).

We explain this, representing the solution of equation (13) with zero boundary conditions as a sum of partial solutions (harmonics); each of the latter ones has the form $y_{(k)}(x,\ t) = T_{(k)}(t)\, X^{(k)}(x)$, $k = 1, 2, \ldots, N-1$ (separation of variables). Then, substituting $y_{(k)}(x,\ t)$ into (13) and separating the variables, for any harmonics we have

$$\frac{T_{(k)}^{j+1} - T_{(k)}^{j-1}}{2\tau T_{(k)}^j} = \frac{X_{i+1}^{(k)} - 2X_i^{(k)} + X_{i-1}^{(k)}}{h^2 X_i^{(k)}} = -\lambda_k\,, \tag{14}$$

where $\lambda_k = 4/(h^2) \cdot \sin(\pi k h/2) > 0$ is the parameter of separation, or the eigenvalue for harmonics with number k (exercise 4).

The temporal part of the solution $y_{(k)}(x,\ t)$, as follows from (14), satisfies the equation

$$T_{(k)}^{j+1} - T_{(k)}^j = -\alpha_k T_{(k)}^j\,, \qquad \alpha_k = 2\tau\lambda_k > 0\,,$$

with a partial solution of a form $T_{(k)}^{j+1} = q_k T_{(k)}^j$ (from this connection follows $T_{(k)}^{j+1} = q_k^{j+1} T_{(k)}^0$). Then q_k should satisfy the equation $q_k^2 + \alpha_k q_k - 1 = 0$ with the real roots, one of them by module is more than one at any α_k. By virtue of this for a rather large value of j, harmonics can be present in a general solution of the equation (13) as big as is possible by their absolute value.

These examples do not exhaust the shortages of direct formal approximation. In more complicated situations for discrete models generated by it, fundamental properties intrinsic to initial objects can not be fulfilled.

We illustrate this by considering the following problem for a stationary thermal conduction equation

$$\frac{d}{dx}\left(k(x)\frac{du}{dx}\right) = 0\,, \qquad 0 < x < 1\,, \quad k(x) \geq C_1 > 0\,, \tag{15}$$
$$u(0) = 1\,, \qquad u(1) = 0\,.$$

The model (15) is one of the simplest in the theory of heat transfer, except that the thermal conductivity $k(x)$ can be a discontinuous function (thermal conducting material consists of different substances).

We expand the differential operator (15)

$$\frac{d}{dx}\left(k(x)\frac{du}{dx}\right) = k\,\frac{d^2u}{dx^2} + \frac{dk}{dx}\cdot\frac{du}{dx}\,,$$

and introduce absolutely natural, on the first sight, replacements (uniform grid)

$$\frac{d^2u}{dx^2} \approx u_{\bar{x}x}\,, \qquad \frac{dk}{dx} \approx \frac{k_{i+1} - k_{i-1}}{2h}\,, \qquad \frac{d^2u}{dx^2} \approx \frac{u_{u+1} - u_{i-1}}{2h}\,,$$

In view of boundary conditions (15) we come to the difference scheme

$$k_i\,\frac{y_{i+1} - 2y_i + y_{i-1}}{h^2} + \frac{k_{i+1} - k_i}{2h}\,\frac{y_{i+1} - y_{i-1}}{2h} = 0\,, \tag{16}$$
$$0 < i < N\,, \qquad y_0 = 1\,, \qquad y_N = 0\,,$$

approximation of an order $O(h^2)$.

Let $k(x)$ be a discontinuous step function

$$k(x) = \begin{cases} k_1\,, & 0 < x < \xi\,, \\ k_2\,, & \xi < x < 1\,, \end{cases} \tag{17}$$

where ξ is an irrational, $\xi = x_n + \theta h$, $x_n = nh$, $0 < \theta < 1$, $k_1 \neq k_2$.

At such function $k(x)$ the solution (15) obviously depends linearly on x, so that the declination of function $u(x)$ is different in areas with different values of thermal conductivity. The only solution is determined from conditions of conjugation in a point $x = \xi$, i.e. from a condition of temperature continuity $u^-(\xi) = u^+(\xi)$ and heat flux continuity $W^-(\xi) = W^+(\xi)$ $(W(x) = -k(x)du/dx)$:

$$u(x) = \begin{cases} 1 - \alpha_0 x\,, & 0 \le x \le \xi\,, \quad \alpha_0 = (\kappa + (1-\kappa)\xi)^{-1}\,, \\ \beta_0(1-x)\,, & \xi \le x \le 1\,, \quad \beta_0 = \kappa\alpha_0\,, \quad \kappa = k_1/k_2\,. \end{cases} \tag{18}$$

It is easy to find the solution of the difference problem (16) in case (17), in view of the fact that the equation (16) has a form $y_{i-1} - 2y_i + y_{i+1} = 0$ at $i \ne n$, $i \ne n+1$ and, therefore,

$$y_i = y(x_i) = \begin{cases} 1 - \alpha x_i\,, & 0 \le x \le x_n\,, \\ \beta(1 - x_i)\,, & x_{n+1} \le x \le 1\,. \end{cases} \tag{19}$$

The coefficients α, β are obtained from the equations (16) in points $i = n$, $i = n+1$ $(x_n = \xi - \theta h,\ x_{n+1} = \xi + (1-\theta)h)$:

$$\alpha = \frac{1}{\mu + (1-\mu)\,\xi + h(\lambda - \theta - (1-\theta)\,\mu)}\,,$$

$$\mu = \frac{3+\kappa}{5-\kappa}\,, \qquad \lambda = \frac{5\kappa - 1}{3\kappa + 1}\,, \qquad \beta = \mu\alpha\,.$$

From these formulae it is seen that $\alpha \to \bar{\alpha}_{\lambda_0} = (\mu + (1-\mu)\xi)^{-1}$, $\beta \to \bar{\beta}_0 = \mu\alpha_0$ at $h \to 0$. Therefore at $h \to 0$ the function $\tilde{y}(x,\,h)$ (a solution of (9), determined between knots of a grid with the help of linear interpolation) has a limit

$$\bar{u}(x) = \lim_{h\to 0} \tilde{y}(x,\,h) = \begin{cases} 1 - \bar{\alpha}_0 x\,, & 0 \le x \le \xi\,, \\ \bar{\beta}_0(1-x)\,, & \xi \le x \le 1\,. \end{cases} \tag{20}$$

The functions (20) and (19) coincide only in the case $\kappa = 1$ $(k_1 = k_2)$. Therefore, the solution of the difference problem tends at $h \to 0$ not to a solution of problem (13), but to a completely different function: the difference scheme (16) *diverges.*

The reason for its divergence is revealed by the analysis of the solution (20) in a point $x = \xi$: the temperature there is continuous $(u^-(\xi) = u^+(\xi))$, while the heat flux is discontinuous ($W^-(\xi) \ne W^+(\xi)$, exercise 5). The scheme (16) violates the conservation law (balance) of heat in the substance, i.e. the fundamental law, being at the basis of the deduction of all models of thermal conduction. Such discrete models are called *non-conservative* and, as a rule, cannot be used for a study of initial models.

3. The integro-interpolational method. From the examples mentioned it becomes clear that the transition to discrete models cannot be realized in a purely formal way. The discrete model should preserve whenever possible the greatest number of basic properties, which the models of initial objects possess.

Concerning equation (15), this means that for its difference scheme a realization of the discrete analog of the energy conservation law is necessary. The widely used approach to the construction of such schemes is based on a formulation of this law in the integrated form for meshes $x_{i-1} \leq x < x_i$, $x_i = ih$, $i = 1, \ldots, N$, of the difference grid with a consequent replacement of the obtained integrals and derivatives by approximate difference expressions (*a integro-interpolational method*).

For a stationary process of thermal conduction without absorption and release of energy, the equation of heat balance in the interval $x_{i-1/2} < x < x_{i+1/2}$ (half-integer indices are chosen) means the equality of fluxes on its boundaries

$$W_{i-1/2} - W_{i+1/2} = 0\,, \qquad i = 1, \ldots, N-1\,. \tag{21}$$

Integrating the equality $W(x) = -k(x)du/dx$ in an interval $x_{i-1} \leq x \leq x_i$

$$u_{i-1} - u_i = \int\limits_{x_{i-1}}^{x_i} \frac{W(x)}{k(x)}\, dx\,,$$

and assuming $W(x) = \tilde{W}_{i-1/2} = \text{const}$ at $x_{i-1} \leq x \leq x_i$ (simplest interpolation), we obtain

$$u_{i-1} - u_i \approx \tilde{W}_{i-1/2} \int\limits_{x_{i-1}}^{x_i} \frac{dx}{k(x)}\,,$$

whence the approximate value for $\tilde{W}_{i-1/2}$ is given by the formula

$$\tilde{W}_{i-1/2} = a_i \frac{u_i - u_{i-1}}{h}\,, \qquad a_i = \left(\frac{1}{h} \int\limits_{x_{i-1}}^{x_i} \frac{dx}{k(x)} \right)^{-1}.$$

By substituting it into (21), we come to a *conservative difference scheme* (compare with (16))

$$\frac{1}{h}\left(a_{i+1}\frac{y_{i+1}-y_i}{h}-a_i\frac{y_i-y_{i-1}}{h}\right)=0\,,\qquad 1\le i\le N-1\,,$$

$$a_i=\left[\frac{1}{h}\int\limits_{x_{i-1}}^{x_i}\frac{dx}{k(x)}\right]^{-1}=\left[\int\limits_{-1}^{0}\frac{ds}{k(x_i+sh)}\right]^{-1}, \tag{22}$$

for which the energy conservation law is fulfilled both for each mesh, and for the whole interval [0, 1] (exercise 6). The conservative discrete models for more complicated processes, including nonlinear and non-stationary processes of heat transfer, are constructed in the same way.

The applicability of the integro-interpolational and similar methods includes broad classes of models. It is particularly important for the problems of gas dynamics, when the solutions (see subsection 2 section 1) can be discontinuous. We will now construct conservative difference approximations for the equations of one-dimensional flow of gas using the mass coordinates (subsection 5, section 4, Chapter II) in divergence form

$$\frac{\partial\eta}{\partial t}=\frac{\partial v}{\partial m}\,,\quad \frac{\partial v}{\partial t}=-\frac{\partial p}{\partial m}\,,\quad \frac{\partial}{\partial t}\left(\varepsilon+\frac{v^2}{2}\right)=-\frac{\partial}{\partial m}(pv)\,, \tag{23}$$
$$0<t\le T\,,\qquad 0<m<M_0\,.$$

Here t is time, m is the mass coordinate; $\eta = 1/\rho$ is the specific volume (ρ is the density), v is the velocity, p is the pressure and $\varepsilon = \varepsilon(\eta,\, p)$ is the internal energy of gas.

Introduce a grid with steps which are uniform both in time and space

$$\bar\omega_n=\{m_i=ih,\ i=0,\,1,\,\dots,\,N\,;\ h=M_0/N\}\,,$$

$$\omega_\tau=\{t^j=j\tau\,,j=0,\,1,\,\dots,\,M\,;\ \tau=T/M\}\,,\qquad \bar\omega_{h\tau}=\bar\omega_h\cup\omega_\tau\,.$$

For the sake of simplicity for the grid analogs of gas dynamic quantities, we keep the same notations, referring the function v to the integer points of the grid ($m = m_i$), while p, η, ε refers to the half-integer ones ($m = m_{i+1/2}$).

We integrate the second equation (23) in a rectangle $m_{i-1/2} \le m \le m_{i+1/2}$, $t^j \le t \le t^{j+1}$:

$$\int\limits_{m_{i-1/2}}^{m_{i+1/2}}(v^{j+1}-v^j)\,dm+\int\limits_{t^j}^{t^{j+1}}(p_{i+1/2}-p_{i-1/2})\,dt=0\,,$$

and the rest in a rectangle $m_i \le m \le m_{i+1}$, $t^j \le t \le t^{j+1}$:

$$\int_{m_i}^{m_{i+1}} (\eta^{j+1} - \eta^j)\, dm - \int_{m_{t^j}}^{t^{j+1}} (v_{i+1} - v_i)\, dt = 0\,,$$

$$\int_{m_i}^{m_{i+1}} \left[\left(\varepsilon + \frac{v^2}{2}\right)^{j+1} - \left(\varepsilon + \frac{v^2}{2}\right)^{j} \right] dm + \int_{t^j}^{t^{j+1}} [(pv)_{i+1} - (pv)_i]\, dt = 0\,.$$

We replace time integrals included in these identities by expressions

$$\int_{t^j}^{t^{j+1}} p\, dt \approx p^{(\sigma_1)}\tau\,, \qquad \int_{t^j}^{t^{j+1}} v\, dt \approx v^{(\sigma_2)}\tau\,, \qquad \int_{t^j}^{t^{j+1}} (pv)_i\, dt = p_{*i}^{(\sigma_3)} v_i^{(\sigma_4)}\tau\,,$$

where $f^{(\sigma_\alpha)} = \sigma_\alpha f^{j+1} + (1 - \sigma_\alpha) f^j$, σ_α - weight ($\alpha = 1, 2, 3, 4$), $p_{*i} = 0.5(p_{i-1/2} + p_{i+1/2})$. We replace space integrals by an obvious rule, for example,

$$\int_{m_{i-1/2}}^{m_{i+1/2}} v\, dm \approx v_i h\,, \quad \int_{m_i}^{m_{i+1}} \eta\, dm \approx \eta_{i+1/2}\, h\,.$$

In sum we come to a four-parametric family of difference schemes

$$\begin{gathered} \frac{v_i^{j+1} - v_i^j}{\tau} = -\left(\frac{p_{i+1/2} - p_{i-1/2}}{h}\right)^{(\sigma_1)}, \\ \frac{\eta_{i+1/2}^{j+1} - \eta_{i+1/2}^j}{\tau} = \left(\frac{v_{i+1} - v_i}{h}\right)^{(\sigma_2)}, \\ \frac{E_{i+1/2}^{j+1} - E_{i+1/2}^j}{\tau} = \frac{p_{*i+1}^{(\sigma_3)} v_{i+1}^{(\sigma_4)} - p_{*i}^{(\sigma_3)} v_i^{(\sigma_4)}}{h}, \end{gathered} \tag{24}$$

where $E_{i+1/2} = \varepsilon_{i+1/2} + (v_i^2 + v_{i+1}^2)/2$ is the total energy of ith mesh of gas. They represent the conservative discrete analogs of the equations (23) at any σ_α, $\alpha = 1, \ldots, 4$. For them there are valid difference analogs of laws of conservation of mass, of momentum and total energy in any mesh of the grid (integrating the equations (24) on a grid $\bar{\omega}_{h\tau}$, it is easy to reveal the validity of conservation laws in a discrete form for the whole mass of gas $0 \le m \le M_0$).

At $\sigma_1 = 0$, $\sigma_2 = \sigma_3 = \sigma_4 = 1$ the system (24) can be solved via the explicit formulae: first, we obtain v_i^{j+1}, then $\eta_{i+1/2}^{j+1}$, and from the third equation (24) and the equations of state $p\eta = (\gamma - 1)\varepsilon$ (in case of ideal gas) the values $p_{i+1/2}$ for all $i = 0, 1, \ldots, N$ are determined by the sweep method using boundary conditions at $i = 0$, $i = N$. The implicit schemes (24) are solved be means of iterative methods.

As distinct from the parabolic type equations, the explicit schemes for hyperbolic equations are stable not at the condition $\tau \le Ch^2$, but at the realization of a weaker inequality $\tau \le Ch$, and hence, are frequently applied in practice. Note, that at practical evaluations in gas dynamics difference schemes *an artificial "viscosity"* is introduced, smoothing the strong discontinuities. It essentially facilitates the calculations, since it avoids the need to specially distinguish areas of discontinuities (*the homogeneous schemes*).

4. Principle of complete conservatism. The more essential difference of discrete models of gas dynamics from models of heat transfer is in the diversity of ways of mathematically representing conservation laws for the gas. In particular, the third equation (23) can be rewritten not for a total energy $E = \varepsilon + v^2/2$, but for the internal energy ε, and at least in two ways

$$\frac{\partial \varepsilon}{\partial t} = -p\,\frac{\partial v}{\partial m}\,, \qquad \frac{\partial \varepsilon}{\partial t} = -p\,\frac{\partial \eta}{\partial t}\,. \tag{25}$$

For an initial model these and other forms of representation are equivalent and turn to each other at the appropriate transformations of equations. In the case of discrete models the given property is by no means guaranteed. For example, the discrete analogs of equations (25) do not generally follow from the third equation of the conservative scheme (24). The opposite is true as well: for a scheme approximating the equation of internal energy, the conservation law of total energy is not necessarily fulfilled. A simple example is the non-conservative scheme "cross"

$$\begin{gathered}
\frac{v_i^{j+1} - v_i^j}{\tau} = -\frac{p_{i+1/2}^{j+1/2} - p_{i-1/2}^{j+1/2}}{h}\,, \\
\frac{\eta_{i+1/2}^{j+3/2} - \eta_{i+1/2}^{j+1/2}}{\tau} = \frac{v_{i+1}^{j+1} - v_i^{j+1}}{h}\,, \\
\frac{\varepsilon_{i+1/2}^{j+3/2} - \varepsilon_{i+1/2}^{j+1/2}}{\tau} = -p_{i+1/2}^{j+3/2}\frac{v_{i+1}^{j+1} - v_i^{j}}{h}\,,
\end{gathered} \tag{26}$$

when a "chess" is used (see Fig. 82); the points and crosses denote the knots of the used six-point stencil; the functions ε, p, η refers to the half-integer, while v refers to the integer knots) and is easily solved in case $p\eta = (\gamma - 1)\varepsilon$ by the explicit formulae.

A similar partial conservation of the discrete models can make them unsuitable for numerical modeling (conservation of total energy does not

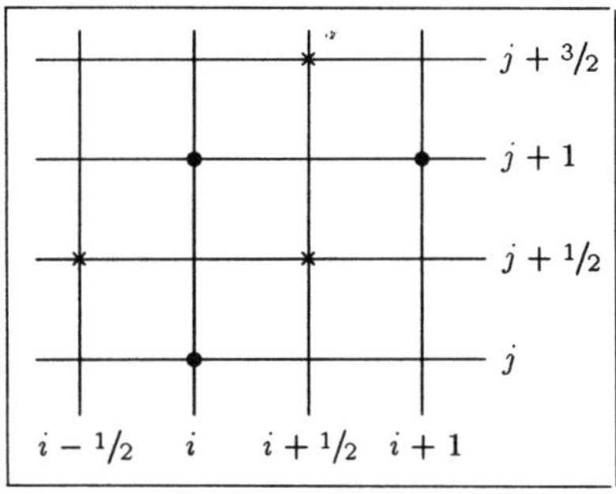

Fig.82.

mean, that its components – the kinetic and internal energies, and hence, the velocity and the temperature of gas – are obtained correctly). Therefore, in the construction of discrete models it is necessary to reflect with them as many fundamental properties of the initial object as possible.

Taking as an initial four-parametric set of the conservative schemes (24), we obtain difference approximations of equations (23) possessing the formulated property.

To simplify further calculations we introduce the notations

$$\bar{p}_i = p_{i+1/2}\,, \quad \bar{\eta} = \eta_{i+1/2}\,, \quad \bar{\varepsilon}_i = \varepsilon_{i+1/2}\,, \quad \bar{p} = \bar{p}_i^j\,, \quad v = v_i^j\,,$$

$$\frac{p_{i+1/2} - p_{i-1/2}}{h} = \bar{p}_{\bar{m}}\,, \qquad \frac{v_{i+1} - v_i}{h} = v_m\,,$$

below we omit the bar above p, η and ε, (except in some special cases). With the same purpose instead of the third equation (24), we consider the approximation of the first equation (25) for the internal energy. Besides, the following formula is required

$$f^{(\beta)} = f^{(\alpha)} + \tau(\beta - \alpha) f_t\,, \tag{27}$$

where α and β are arbitrary numbers, $f^{(\alpha)} = \alpha\hat{f} + (1-\alpha)f$, $\hat{f} = f^{j+1}$.

In view of the remarks made from (24) the four-parametric family of the schemes is obtained

$$v_t = -p_{\bar{m}}^{(\sigma_1)}\,, \qquad \eta_t = v_m^{(\sigma_2)}\,, \qquad \varepsilon_1 = -p^{(\sigma_3)} v_m^{(\sigma_4)}\,, \tag{28}$$

Here the *complete conservatism* is required, i.e. they have to approximate also the equation for total energy (the third equation (23)) and the second equation (25).

From the second equation (28), the equality $v_m^{(\sigma_4)} = v_m^{(\sigma_2)} - \tau(\sigma_4 - \sigma_2)v_{mt} = \eta_t + \tau(\sigma_4 - \sigma_2)\, v_{mt}$ and the third equation (28), it follows that

$$\varepsilon_t = -p^{(\sigma_3)}\,\eta_t + \delta_1 E\,,$$

where $\delta_1 E = -\tau(\sigma_4 - \sigma_2)\, p_{mt}^{(\sigma_3)}$. Thus, the scheme (28) approximates the second equation (25) only when equality $\sigma_2 = \sigma_4$ is fulfilled. If this equality is not fulfilled, there is an *imbalance* $\delta_1 E$ of internal energy, caused by the appearance in the discrete medium of additional (virtual) sources and losses of energy.

Now we obtain $\delta_2 E$ – the imbalance of the total energy. Multiplying the first equation (28) by $v^{(0.5)} = 0.5(v + \hat{v})$, we obtain the equation

$$\frac{v_t^2}{2} = -(v^{0.5}) \cdot p_{\bar{m}}^{(\sigma_1)},$$

then we combine it with the third equation (28)

$$\left(\varepsilon + \frac{v^2}{2}\right)_t = -p^{(\sigma_3)} v_m^{(\sigma_4)} - v^{0.5} p_{\bar{m}}^{(\sigma_1)}. \tag{29}$$

We transform the right hand side in (29) by means of formulae (27)

$$p^{(\sigma_3)} v_m^{(\sigma_4)} + v^{0.5} p_{\bar{m}}^{(\sigma_1)} =$$

$$= (p^{(\sigma_1)} + \tau(\sigma_3 - \sigma_1)\, p_t)(v_m^{(0.5)} + \tau(\sigma_4 - 0.5) v_{mt}) + v^{(0.5)} p_{\bar{m}}(\sigma_1) =$$

$$= (p_{(-1)}^{(\sigma_1)} v^{0.5})_m + \delta_2 E\,,$$

where the following notations are introduced

$$p_{(-1)} = \bar{p}_{i-1} = p_{i-1/2}\,,$$

$$\delta_2 E = \tau(\sigma_3 - \sigma_1) v_m^{(0.5)} p_t +$$

$$+\tau(\sigma_4 - 0.5) p_{mt}^{(\sigma_1)} + \tau^2(\sigma_3 - \sigma_1)(\sigma_4 - 0.5) p_t v_{mt}\,.$$

In sum, (29) takes the form

$$\left(\varepsilon + \frac{v^2}{2}\right)_t = -(p_{(-1)}^{(\sigma_1)} v^{(0.5)})_m - \delta_2 E\,.$$

Thus, the equation (29) does not approximate the equation for total energy (the same concerns scheme (26); see exercise 7). The imbalance $\delta_2 E$, as well as the imbalance $\delta_1 E$, has an artificial origin.

It is absent only in the case $\sigma_3 = \sigma_1$, $\sigma_4 = 0.5$ and, therefore, $\sigma_2 = 0.5$.

Unifying these results, instead of a four-parametric family of the schemes (28) obtained with the help of integro-interpolational method, we come to an one-parametric family

$$v_t = -p_{\bar{m}}^{(\sigma_1)}, \qquad \eta_t = v_m^{(0.5)}, \qquad \varepsilon_t = -p^{(\sigma_1)} v_m^{(0.5)} \tag{30}$$

of completely conservative ($\delta_1 E = \delta_2 E = 0$) discrete models of gas dynamics. Note that the third equation in (30) can be replaced by one of the following equations

$$\varepsilon_t = -p^{(\sigma_1)}\,\eta_t\,, \qquad \left(\varepsilon + \frac{v^2}{2}\right)_t = -(p_{(-1)}^{(\sigma_1)} v^{(0.5)})_m\,,$$

and the last equation can be replaced by one of the following equations

$$\left(\varepsilon + \frac{v_{(+1)}^2}{2}\right)_t = -(p^{(\sigma_1)} v^{(0.5)})_m\,, \qquad \left(\varepsilon + \frac{v^2 + v_{(+1)}^2}{4}\right)_t = -(p_*^{(\sigma_1)} v^{(0.5)})_m\,.$$

In the general case the scheme (30) has an approximation $O(\tau + h^2)$, and at $\sigma_1 = 0.5$ the approximation by time is of second order. At $\sigma_1 = 0$ the scheme (30) is explicit, and its solution is obtained by simple formulae.

From the structure of $\delta_1 E$ and $\delta_2 E$ it is clear that they are especially big when the characteristics of the gas strongly vary in space and time (discontinuous flows, strong inhomogeneity of medium, blowing-up regimes, etc.). In these situations the virtual energy sources become comparable with the actual quantities, and the results of numerical simulations of the initial object differ remarkably from its real behavior. Imbalance can be reduced by taking a smaller time step τ, which naturally leads to a corresponding increase in the number of calculating operations (the decrease of the step of the grid in space has no influence on the value $\delta_1 E$, $\delta_2 E$).

The principle of complete conservatism, used to solve many complex problems is one of the reliable approaches to the construction of discrete models with required properties.

5. Construction of difference schemes by means of variational principles. Variational principles, being one of basic methods of deriving mathematical models of various objects, are also widely used to construct their corresponding discrete analogs. This is due to their universality, the relative simplicity of application, their connection with conservation laws and symmetry properties. The most natural way in such an approach is to discretize the initial object (medium), to formulate the variational principle for the obtained object and to derive the corresponding connections (equations) for discrete quantities, i.e. to complete the construction of the sought discrete model.

Following this logic we construct the difference schemes for the equations (23) of one-dimensional gas dynamics in Lagrangian coordinates and compare them with the schemes of subsections 3, 4.

Consider the gas as a configuration of a close material "points" ("particles"), and split its mass M_0 for simplicity into N equal fractions of mass $m_{i+1/2} = 0/N$, $i = 0, 1, \ldots, N-1$ (the mass $m_{i+1/2}$ is the analog of step h in (24), (25)). In other words, as in subsection 3, we introduce a difference grid with knots at points $i = 0, 1, \ldots, N$ and a uniform mass of meshes. We describe coordinates and velocities of a particle (mesh) through Cartesian coordinates x_i, x_{i+1} and the velocity v_i, v_{i+1} of the corresponding knots. The rest parameters of the mass points are the density $\rho_{i+1/2}$, specific volume $\eta_{i+1/2}$, pressure $p_{i+1/2}$, internal energy $\varepsilon_{i+1/2}$, – we refer to the middle of a mesh (half-integer index).

We apply the Hamiltonian principle for the constructed discrete medium by selecting $x_i(t)$, $x_{i+1}(t)$ as generalized coordinates of the meshes (the quantities $v_i(t) = dx_i/dt = \dot{x}_i$, $v_{i+1}(t) = dx_{i+1}/dt = \dot{x}_{i+1}$ play a role of generalized velocities). We define the kinetic energy of the system by the rather obvious expression

$$T = \sum_{i=0}^{N-1} T_{i+1/2} = \sum_{i=0}^{N-1} m_{i+1/2} \frac{v_i^2 + v_{i+1}^2}{4}$$

as a sum of energies of each particle (other expressions are possible as well)

$$T_{i+1/2} = \frac{m_{i+1/2}}{2} \frac{v_i^2 + v_{i+1}^2}{2} .$$

We correlate the "potential" energy of particles with their internal energy $\varepsilon_{i+1/2} = \varepsilon_{i+1/2}(x_i, x_{i+1})$, in so far as its "release" enables us to perform a work, such as the circuit of the charged capacitor releasing the energy accumulated in it, transforming into a motion of charges (see subsection 4, section 2, Chapter III). Note that such definition of a potential energy for the considered system, as well as revealing the potentiality of its motion, requires detailed analysis (in subsection 3, section 2, Chapter III it is performed for the simpler case of small oscillations of a string). Therefore we confine ourselves to the above mentioned qualitative considerations.

The summarized potential energy of the set of meshes is

$$V = \sum_{i=0}^{N-1} V_{i+1/2} = \sum_{i=0}^{N-1} m_{i+1/2}\varepsilon_{i+3/2} ,$$

where $V_{i+1/2}$ is the energy of a separate particle.

"The Lagrangian" of a discrete gas medium is given by the expression

$$L = T - V = \sum_{i=0}^{N-1} m_{i+1/2} \left(\frac{v_i^2 + v_{i+1}^2}{4} - \varepsilon_{i+1/2} \right), \tag{31}$$

and the functional of action is given by the formula

$$Q = \int_{t_0}^{t_1} L(x(t),\, \dot{x}(t),\, t)\, dt =$$

$$= \int_{t_0}^{t_1} \sum_{i=0}^{N-1} m_{i+1/2} \left(\frac{v_i^2 + v_{i+1}^2}{4} - \varepsilon_{i+1/2}(x_i,\, x_{i+1}) \right) dt\,. \tag{32}$$

Performing the variation in accordance with the Hamiltonian principle (subsection 3, section 1, Chapter III) of the action (32) by all kinematically possible paths, we calculate the variations of its terms. In so far as $v_i = \dot{x}_i$,

$$\delta \left(\frac{v_i^2}{4} \right) = \frac{1}{4} \frac{\partial v_i^2}{\partial \dot{x}_i} = \frac{1}{2} v_i\, \delta \dot{x}_i\,.$$

The variation of the internal energy is given by the formula

$$\delta \varepsilon_{i+1/2} = \frac{\partial \varepsilon_{i+1/2}}{\partial x_i}\, \delta x_i + \frac{\partial \varepsilon_{i+1/2}}{\partial x_{i+1}}\, \delta x_{i+1}\,,$$

and for the variation of action we have

$$\delta Q = \int_{t_0}^{t_1} m_{i+1/2} \sum_{i=0}^{N-1} \left(\frac{1}{2} v_i\, \delta \dot{x}_i + \frac{1}{2} v_{i+1}\, \delta \dot{x}_{i+1} - \right.$$

$$\left. - \frac{\partial \varepsilon_{i+1/2}}{\partial x_i}\, \delta x_i - \frac{\partial \varepsilon_{i+1/2}}{\partial x_{i+1}}\, \delta x_{i+1} \right) \delta t\,.$$

Integrating in the latter equality terms $v_i\, \delta \dot{x}_i$ and $v_{i+1}\, \delta \dot{x}_{i+1}$ by parts, we take into account the commutation of operations of variation and differentiation by time ($\delta \dot{x}_i = d(\delta x_i)/dt$) and the condition $\delta x_i = 0$, $i = 0, \ldots, N-1$, at $t = t_0$, $t = t_1$. We obtain an expression indicating that by virtue of independence of variations δx_i for δQ we have $\delta Q = 0$ only when the coefficients are equal to zero at any δx_i, i.e. when the following equations are fulfilled

$$\frac{dv_i}{dt} = -\frac{\partial \varepsilon_{i+1/2}}{\partial x_i} - \frac{\partial \varepsilon_{i-1/2}}{\partial x_i}\,, \qquad i = 0, \ldots, N-1\,.$$

The acceleration of gas meshes is written on the left hand sides. To reveal the content of the right hand sides, recall that for the considered adiabatic flows the equalities $d\varepsilon = -p\,d\eta$ and $\varepsilon(p,\,\eta) = \varepsilon(\eta)$ (see formulae (27), (29) from section 4, Chapter II) are valid. Then, for a discrete medium we obtain $d\varepsilon_{i+1/2} = -p_{i+1/2}\,d\eta_{i+1/2}$, $\varepsilon_{i+1/2} = \varepsilon(\eta_{i+1/2}) = \varepsilon[(x_{i+1} - x_i)/m_{i+1/2}]$. Conducting the transformations

$$\frac{\partial\varepsilon_{i+1/2}}{\partial x_i} = \frac{\partial\varepsilon_{i+1/2}}{\partial\eta_{i+1/2}}\cdot\frac{\partial\eta_{i+1/2}}{\partial x_i} = \frac{p_{i+1/2}}{m_{i+1/2}}\,,$$

$$\frac{\partial\varepsilon_{i-1/2}}{\partial x_i} = \frac{\partial\varepsilon_{i-1/2}}{\partial\eta_{i-1/2}}\cdot\frac{\partial\eta_{i-1/2}}{\partial x_i} = -\frac{p_{i-1/2}}{m_{i-1/2}}\,,$$

we come to a final form of the equations of motion of gas "particles" (on a homogeneous grid) which were obtained using the Hamiltonian principle

$$\frac{dv_i}{dt} = -\frac{p_{i+1/2} - p_{i-1/2}}{(m_{i+1/2} + m_{i-1/2})/2}\,, \qquad i = 0,\,\ldots,\,N-1\,. \tag{33}$$

Join to (33) the equations

$$\frac{d\eta_{i+1/2}}{dt} = \frac{v_{i+1} - v_i}{m_{i+1/2}}\,, \qquad i = 0,\,\ldots,\,N-1\,, \tag{34}$$

the following from equality $\eta_{i+1/2} = (x_{i+1} - x_i)/m_{i+1/2}$ (the conservation law of a mesh mass), and also following from equality $d\varepsilon_{i+1/2} = -p_{i+1/2}\,d\eta_{i+1/2}$, and from (34), the equation

$$\frac{d\varepsilon_{i+1/2}}{dt} = -p_{i+1/2}\frac{d\eta_{i+1/2}}{dt} = -p_{i+1/2}\frac{v_{i+1} - v_i}{m_{i+1/2}}\,, i = 0,\,\ldots,\,N-1\,, \tag{35}$$

expressing the law of variation of the internal energy.

Together with the given equation of state $\varepsilon_{i+1/2} = \varepsilon(\eta_{i+1/2})$ the system (33)–(35) represents the *semi-discrete* model of gas dynamics adequate to all required conservation laws – momentum, mass, energy (note, that the conservation law of total energy $E = T + V$ readily follows from the invariance of the Lagrangian (31) with respect to the time shift).

Replacing the derivatives by time with finite differences in (33)–(35), passing to non-index notations and introducing the weights σ_1, σ_2, σ_3, σ_4, we obtain the discrete model of gas dynamics

$$v_t = -p^{(\sigma_1)}_{\bar{m}}\,, \qquad \eta_t = v^{(\sigma_2)}_m\,, \qquad \varepsilon_t = -p^{(\sigma_3)}\,v^{(\sigma_4)}_m\,,$$

coinciding with the difference scheme (28) constructed in subsection 3 by means of the integro-interpolational method (at an appropriate choice of weights it becomes completely conservative).

The applicability of the demonstrated approach is not limited by the above considered elementary case (see, for example, exercise 8). Variational principles are effectively used to derive discrete models in rather difficult situations (many-dimensional processes, grid of complicated structure and so on).

6. Use of the hierarchical approach in derivation of discrete models. The basic idea of the given approach is the use of knowledge about the position of the model to be discretized within the hierarchy of models of the investigated object. If the hierarchy is constructed by the principle "from above downwards", then, by conducting a partial or complete discretization of a general model, we pass to a lower level discrete model. Obviously, the method of passage should correspond to one adopted for the initial model. As a result of realizing of such a procedure in the constructed discrete model certain additional features of higher level models are introduced, and this can increase the adequacy of the model approximating the initial object.

Specify these considerations by considering equations describing one-dimensional flows of non-viscous non-thermal conducting gas (rewriting in divergence form the one-dimensional equations (4), (10), (15) from section 4, Chapter II)

$$\frac{\partial \rho}{\partial t} + \frac{\partial(\rho V)}{\partial x} = 0\,,$$

$$\frac{\partial(\rho V)}{\partial t} + \frac{\partial}{\partial x}\left(\rho V^2 + p\right) = 0\,, \qquad (36)$$

$$\frac{\partial E}{\partial t} + \frac{\partial}{\partial x}[V(E+p)] = 0\,.$$

Here $E = \rho(V^2/2 + 3RT/2)$ is the sum of kinetic and internal energies of a unit volume of gas, i.e. its total energy, $T = p/(\rho R)$ is the temperature, $R = k/m$ is the gas constant, k is the Boltzmann constant, m is the mass of an atom or molecule of gas.

Recall that in subsections 4, 5, section 3, Chapter III, the equations (36) were obtained from the Boltzmann kinetic equation as a result of its "averaging" provided that the distribution function $f(x, v, t)$ is locally Maxwellian (formula (5), section 3, Chapter III)

$$f^{(0)}(V) = n\left(\frac{m}{2\pi kT}\right)^{3/2} \exp\left[-\frac{m}{2kT}(v-V)^2\right], \qquad (37)$$

where v is the velocity of random motion of gas particles (at Maxwellian distribution there are no viscous stresses and heat fluxes in a medium).

The discrete analogs of the equations (36) can be built using any of the methods described in the previous sections. The hierarchical approach in this case (and for certain generalizations of (36)) yields the following.

1) the distribution function f (solution of equation (2) in section 3, Chapter III) and macrocharacteristics of the gas are considered as a function of a discrete argument t^j, $j = 0, 1, \ldots,$, i.e. the discretization is performed by time of an initial model – Boltzmann kinetic equation;

2) the quantity $\tau = t^{j+1} - t^j$ (step of discretization) is taken sufficiently small to satisfy the condition $\tau \le l/|\langle v \rangle|$, where l is the characteristic length of the free path, $|\langle v \rangle|$ is the module of characteristic velocity of particles. Therefore during τ the number of collisions between particles is insignificant and can be neglected (see the deduction of the Boltzmann equation);

3) the distribution function $f^{j+1} = f(x, v, t^{j+1})$ varies (see the deduction of the Boltzmann equation) in comparison with function $f^j = f(x, v, t^j)$ only due to the modification by particles of their phase volume, hence the following simple relation is valid

$$f(x, v, t^{j+1}) = f(x - v\tau, v, t^j), \tag{38}$$

being a solution of the Boltzmann equation in the absence of collisions;

4) the connection (38) is simplified via expansion of f^{j+1} in Taylor series by the parameter $v\tau$.

$$f^{j+1} = f^j - \tau \frac{\partial f^j}{\partial x} v + \frac{\tau^2}{2} \frac{\partial^2 f^j}{\partial x^2} v^2 + \ldots \tag{39}$$

up to terms of the third order; since the distribution function is decreasing exponentially with the growth of v^2, the contribution of particles with large velocities can be neglected and the corresponding terms in (39) can be omitted;

5) similar to the derivation of the equations for moments in subsection 4, section 3, Chapter III, the relations (39) are sequentially multiplied by functions $\Psi(v)$, respectively equal to 1, mv, $mv^2/2$ (*summary invariants*), and are integrated over all velocities v; as a result the relation between f^{j+1} and f^j will be transformed into connections between the average of such hydrodynamic parameters of gas as the density, flux of mass, total energy

$$\rho = \int mf\, dv, \qquad \rho V = \int mvf\, dv, \qquad E = \int \frac{mv^2}{2} f\, dv,$$

and other quantities taken at the moments t^j, t^{j+1} (in other words we obtain a hydrodynamic model of a medium which is *discrete in time*);

6) after corresponding discretization in space, the final discrete (both in time and in space) models of gas are constructed, taking into account

some features of the initial kinetic equation (*kinetically consistent difference schemes*); they cannot generally be obtained by approximating hydrodynamic models (for example, equations (36)), in so far as the properties of the higher level models are used to construct them.

The simplest variant of the above described procedures corresponds to the case of locally Maxwellian distribution function (37) and corresponding (for one-dimensional flows) equations (36); their semi-discrete analogs are as follows

$$\frac{\hat{p}-p}{\tau}+\frac{\partial}{\partial x}(\rho V)=\frac{\tau}{2}\frac{\partial^2}{\partial x^2}(\rho V^2+p),$$

$$\frac{\widehat{(\rho V)}-\rho V}{\tau}+\frac{\partial}{\partial x^2}(\rho V^2+p)=\frac{\tau}{2}\frac{\partial^2}{\partial x^2}(\rho V^3+3pV), \tag{40}$$

$$\frac{\hat{E}-E}{\tau}+\frac{\partial}{\partial x}[V(E+p)]=\frac{\tau}{2}\frac{\partial^2}{\partial x^2}\left[V^2(E+2p)+\frac{p}{\rho}(E+p)\right],$$

where for simplicity we have denoted $g^{j+1}=\hat{g}$, $g^j=g$. In their deduction the integrals obtained on the right hand side (39) are represented (in view of the properties of distribution function $f^{(0)}$) as sums and products of the moments of function e^{-z^2} of various orders. The odd moments are equal to zero, while evens are determined by the formula

$$\int_{-\infty}^{\infty} e^{-z^2} z^{2r}\,dz=\frac{\pi^{1/2}}{2^r}(2r-1)\,!!.$$

Upon realizing the discretization of the space operators, from (40) we obtain completely discrete analogs of equations (36), which can be numerically solved by means of the explicit formulae (one can also easily construct the implicit variants of the model (40), by mutual replacement in (38) of f^{j+1} and f^j). The terms on the right hand side of (40) of the first order of smallness by τ represent the terms which are additional to the usual discrete models for (36), and which carry additional information about the origin of the approximate model.

Not representing the derivation of the whole system (40), we confine ourselves only by the equation of continuity. Representing (39) as

$$m\,\frac{f^{j+1}-f^j}{\tau}+m\,\frac{\partial f^j}{\partial x}\,v=m\,\frac{\tau}{2}\,\frac{\partial^2 f^j}{\partial x^2}\,v^2,$$

and multiplying this expression by $\Psi(v)=1$ and and integrating over v, we obtain

$$\frac{\rho^{j+1}-\rho^j}{\tau}+\int m\,\frac{\partial f^j}{\partial x}\,v\,dv=\frac{\tau}{2}\int m\,\frac{\partial^2 f^j}{\partial x^2}\,v^2\,dv,$$

or, extracting the sign of derivation outside the integral (see formulae (10), section 3, Chapter III),

$$\frac{\rho^{j+1}-\rho^j}{\tau}+\frac{\partial}{\partial x}\int m\,f^j\,v\,dv=\frac{\tau}{2}\frac{\partial^2}{\partial x^2}\int m\,f^j\,v^2\,dv\,.$$

Transforming the integrals in the latter equality in accordance with the definitions of functions V and p (see subsection 4, section 3, Chapter III), we come to an equation

$$\frac{\hat{\rho}-\rho}{\tau}+\frac{\partial}{\partial x}(\rho V)=\frac{\tau}{2}\frac{\partial^2}{\partial x^2}(\rho V^2+p)\,,$$

i.e. to the first of the equations (36).

The approach described above is generalized for many-dimensional analogs of equations (36) and for more complicated models of gas flows, for example, on the Navier-Stokes equations (see subsection 5, section 3, Chapter III). In the latter case, instead of the Maxwellian distribution function $f=f^{(0)}$, the following approximation – the function $f=f^{(0)}+f^{(1)}$ – is used, corresponding to a medium with non-zero viscous stresses and heat fluxes (see subsection 4, section 3, Chapter III).

We have to explain: the boundary conditions for models of type (40) cannot be taken exactly the same as those in the case of the Eulerian equations (or of Navier-Stokes equations) already due to a loss of conservativeness of the discrete model at such automatic transition. Therefore, not only the equations of model (40), but also the boundary conditions have to be matched with the boundary conditions of the kinetic equation of a higher hierarchical level.

Due to their construction the kinetically consistent difference schemes possess series of properties making them rather effective for the numerical modeling of many complicated gas currents.

EXERCISES

1. Decomposing the function $u(x)$ in the Taylor series in the vicinity of a point x and keeping a sufficient number of terms, check that the difference derivative $u_{\bar{x}x}$ approximates the derivative d^2u/dx^2 with the second order.

2. Let in the problem (3) with $y_0=y_N=0$ the knot $i_0\neq 0, N$ be such, that $|y_{i_0}|=\max|y_i|$, $0\le i\le N$. Rewriting the equation (3) in a point i_0, show the validity of an inequality $|y_{i_0}|\le|\varphi_{i_0}|/2$ and, hence, of inequality (5), where $C=1/2$.

3. Decomposing the function $u(x,t)$ in the vicinity of a point $x=x_i$, $t=(t^{j+1}+t^j)/2$ in a Taylor series, prove that at $\sigma=1/2$ the error of approximation of the scheme (10) equals $O(\tau^2+h^2)$.

4. Check by direct substitution that the eigenvalue problem $X_{\bar{x}x}+\lambda X=0$, $0<x<1$; $X(0)=X(1)$, $X(x))\not\equiv 0$ has nontrivial solutions – eigenfunctions of a form $X^{(k)}=\sqrt{2}\sin\pi kx$.

5. Check that for the solution (20) the difference between $W^{-}(\xi)$ and $W^{+}(\xi)$ is equal to $q = -\alpha_0(\mu - \kappa)\,k_2$ and that $q \to \pm 0$ at $x \to \pm 0$, i.e. the power of the virtual heat source can be as large as possible.

6. Summarizing equation (22) over $i = 1, \ldots, N-1$, obtain the difference heat conservation law in all grid area $\bar{W}_{1/2} - \bar{W}_{N-1/2} = 0$, where $\bar{W}_{i-1/2} = -a_i(y_i - y_{i-1})/h$.

7. Introduce notations $p^{j+1/2}_{i+1/2} = \bar{p}^j_i = \bar{p}$, $\eta^{j+3/2}_{i+1/2} = \bar{\eta}^{j+1}_i = \hat{\bar{\eta}}$, etc. Rewrite the scheme "cross" (26) as

$$v_t = -\bar{p}_{\bar{m}}\,, \qquad \bar{\eta}_t = \hat{v}_m\,, \qquad \bar{\varepsilon}_t = -\hat{\bar{p}}\bar{v}_m\,.$$

Multiplying $v_t = -\bar{p}_{\bar{m}}$ on $v^{(0.5)}$ and repeating the considerations of the derivation of (30), represent the third equation (26) as

$$\left(\bar{\varepsilon} + \frac{v^2}{2}\right)_t = -(\bar{p}_{-1}v^{(0.5)})_m - \delta E\,.$$

where the imbalance of the total energy $\delta E = \tau\bar{p}_t v^{(0.5)}_m + 0.5\,\tau p v_{mt} + 0.5\,\tau^2 p_t v_{mt}$ can not be removed, as distinct from the case of the scheme (24).

8. Check, repeating for the case of non-uniform by mass difference grid constructions in subsection 5, that the equations (34), (35) of the model (33)–(35) remain unchanged, while the equation (33) is somewhat modified, but conserves its physical content.

Bibliography for Chapter V: [2, 6, 8, 12, 15, 17, 18, 21, 22, 24, 26, 29–31, 33, 34, 46, 51, 59, 61–65, 67, 68, 70, 71, 74, 75, 80, 87-89].

Chapter VI

MATHEMATICAL MODELING OF COMPLEX OBJECTS

1 Problems of Technology and Ecology

We will now show the need for broad application of computing experiment for the analysis and prognosis of big technological and ecological projects. We will represent concrete examples illustrating the tight correlation of problems of technology and ecology, and the inevitability of using mathematical modeling for their solution.

1. Physically "safe" nuclear reactor. Nuclear energy is one of the bases of industry in many developed countries, in some of them it is the foremost method of producing electricity. The problems of its safety are complicated and diverse: the burial of the nuclear waste and the reduction of the need to extract uranium, effective utilization of uranium which has been used and liberated at nuclear disarmament, and, certainly, the prevention of major accidents at reactors.

In any type of reactors the defining physical processes are the neutron-nuclear reactions leading to a release of energy in its active zone, and the heat transfer from this zone is later used to produce electrical energy. The working reactor is kept in a critical state, when the number of released neutrons is such that the power created does not practically depend on time. In a subcritical state fewer neutrons appear than are lost, and the reaction of decay weakens rapidly. In a supercritical state, on the contrary, the neutron's output is too big, and it can lead to a heating and "explosion" of the active zone.

In regular reactors the critical state is physically unstable, and is supported artificially with the help of a very complicated control system. Without such a system either the subcritical or supercritical conditions would be realized. The reactor is made with a reserve of reactivity (supercritical), which is compensated by inserting special rods into the active zone which swallow the "extra" neutrons. If in the course of the burn-out of the fuel the reactivity decreases, the controlling rods are partially removed from the system and the neutron flux grows to the magnitude necessary for the scheduled work of the reactor.

The characteristic time of deviation from the critical state is determined mainly by the time scale of delayed neutrons, i.e. of neutrons appearing from products of decay, some time after disintegration. This time period is less than one minute, which presents rather strong requirements to the control system. Just for this short time it should "find out" and realize an appropriate solution for any unexpected urgent situations.

The principal idea of a physically safe reactor is that the fuel components should be selected in such a way that firstly its characteristic time should be noticeably more than one minute and, secondly, that elements of self-regulation have to appear in the active condition.

This can be achieved, if the following chain of transformations will be rather noticeable in the reactor

$$^{238}\mathrm{U} + \mathrm{n} \to {}^{239}\mathrm{U} \overset{\beta^-}{\to} {}^{239}\mathrm{Np} \overset{\beta^-}{\to} {}^{239}\mathrm{Pu}\,,$$

where ^{238}U, ^{239}U, ^{239}Np, ^{239}Pu denote, correspondingly, the isotopes of uranium, neptunium and plutonium, n denotes neutron, the symbol β^- denotes beta-decay (emission of an electron by the nucleus). In this case the formed plutonium is the basic and immediately used fuel. The characteristic time scale of such a reaction – the time of two beta-decays – is approximately equal to 2.5 days, i.e. about four orders of magnitude more, than for the delayed neutrons.

The effect of self-regulation is connected with the fact that the increase (for whatever reason) of the flux of neutrons will lead to a fast decay of plutonium, to a decrease of its concentration and accordingly, a decrease of the flux of the neutrons (the formation of the new nuclei ^{239}Pu will proceed at the former rate approximately within 2.5 days). If, on the contrary, the neutron flux due to some external interference sharply decreases, the decay rate will decrease and hence, will increase the rate of running time of plutonium, with consequent magnification of number of neutrons released in the reactor during approximately the same time scale (several days).

The transformations described also take place in conventional reactors, however there they play a minor role in energy release, as they are used mainly for the accumulation of plutonium. The answer to the problem of whether there is such a mixture of U and Pu, when the given reaction becomes dominating, can only be obtained with the help of mathematical modeling of this complex system.

A rather complete mathematical model of the active zone of the reactor should include models of non-stationary spatially three-dimensional processes of neutron transfer in a hardly inhomogeneous medium, burn-out of fuels and reactor kinetics, and also a model of heat transfer.

However to check the reality of the proposed physical idea with good accuracy one can use a more simple model. The first simplification is the separate analysis of neutron-nuclear processes and the process of heat transfer (this is justified for large periods of regulation). It is quite possible to study purely neutron processes not in three-dimensional, but in one-dimensional geometry considering them in addition, in diffusion and one-group approximation (the latter means, that the spectral characteristics of neutrons are averaged in an appropriate way).

The equations of neutron transfer obtained with these simplifications are in some sense close to the equations of heat transfer (section 2, Chapter II). They are solved together with the equations of reactor kinetics for six groups of the predecessors of delayed neutrons (equation of the type of radioactive decay, but with the "source" of neutrons) and the equations of burn-out for almost twenty types of isotopes of U, Pu, Np and of other elements. The separation of these three models is not possible, as the majority of unknown quantities appear in all equations. The method of a solution is the semi-discrete (spatial) approximation of an initial model with consequent numerical integration by time of the obtained complex system of ordinary high dimension differential equations.

The typical computing experiment consists of the representation of initial critical assembly (i.e. of a mixture of substances and geometry of the active zone such that the obtained reactor is critical) and the study of evolution in time of basic characteristics of the reactor. Note: the initial assembly is obtained by means of a special preliminary computer experiment.

One of the variants of assembly in cylindrical geometry is as follows:

Zone 1	Zone 2
^{239}Pu – 8% ^{238}U – 92% Steel Na	Depleted natural uranium (^{235}U – 0.33%, ^{238}U – 99,77%) Steel Na
$r_1 = 89.48$ cm	$r_2 = 200$ cm

The assembly is divided into two zones of different sizes with different plutonium content and different isotopes of uranium (the presence of sodium in an homogeneous mixture corresponds to the presence of a heat-carrier, the steel corresponds to the constructional elements). In Fig. 83 the dynamics (measured in 10^{-6} s^{-1}) of $\lambda(t)$ which is a characteristic of time at which the considered two-zone reactor is no longer critical (at $\lambda > 0$ the number of released neutrons is proportional to $e^{\lambda t}$) is represented. It is seen that during almost nine months of work for a reactor without control $|\lambda| \leq 10^{-6}$ s^{-1}, i.e. the time of its deviation from critical is approximately 10 days (in this variant, if no external control is introduced, the reactor damps in nine months). Thus, the basic idea of a physically safe reactor is convincingly confirmed.

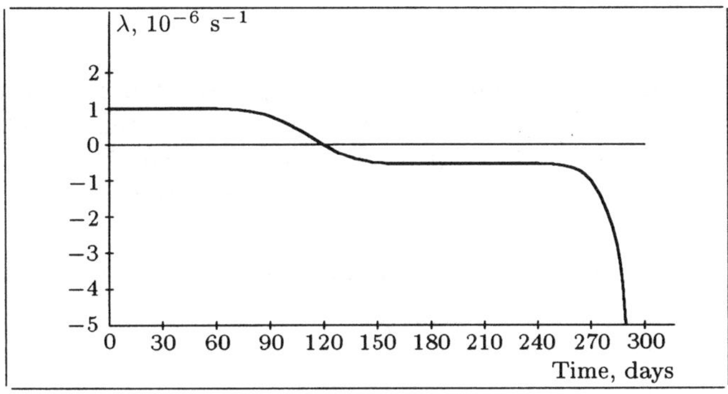

Fig.83.

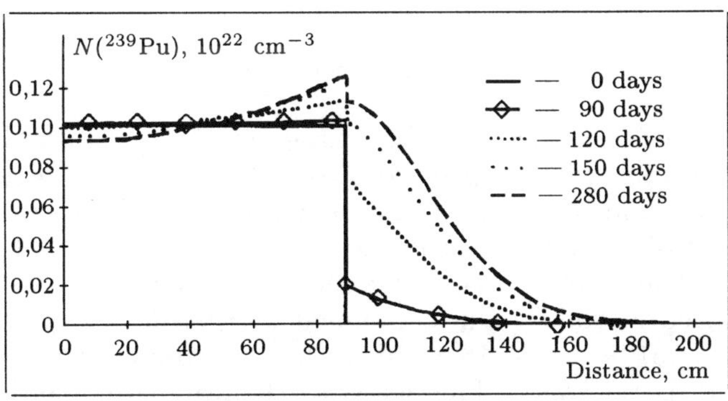

Fig.84.

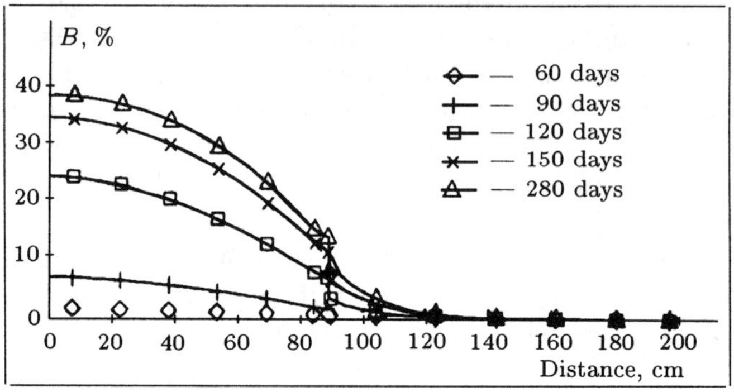

Fig.85.

The computing experiment, as well as a general conclusion, also provides a series of very important details about the investigated system. In Fig. 84 the spatial concentration of plutonium $N(^{239}\text{Pu})$ (in 10^{22} cm $^{-3}$), and in Fig. 85 – the degree of fuel use at various instants is shown. As distinct from the usual reactors, a high degree of fuel use (theoretically, without engineering restrictions, 40–50% instead of standard 4-5% is reached here).

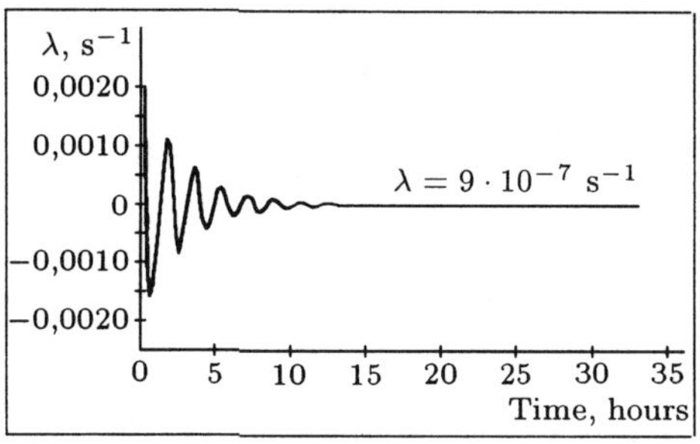

Fig.86.

The experiments model also the answer to the question about the stability of the reactor with respect to strong perturbations (accidents). For example, an ejection of a part of the heat-carrier (sodium) from the active zone as a result of a spontaneous or violent break in the pipeline, is possible. In Fig. 86 the dependence of λ on time after a similar accident is represented. It was simulated as follows: in a moment $t = 63$ days from the first zone within an interval $0 \le r \le 15.4$ cm during 15 minutes with constant rate all the sodium is removed. This leads to the appearance of oscillations of the characteristic time scale (then its minimum value is 8 minutes, which is quite enough for the actual correction of the work of the reactor). The perturbation appearing afterwards is damped by the reactor itself without any external action (self-regulation) and in 32 hours its characteristic time scale is stabilized, becoming equal, as before, to approximately 10 days. The experiments with a model also justify some other advantages of the considered reactor, for example there is no need to add new portions of plutonium to the active zone instead of the burnt-out plutonium (therefore there is no need to transport and manipulate with one of the principal components of nuclear weapon).

Thus, mathematical modeling and computing experiments allow us to study not only qualitatively, but also quantitatively one of the ways to improve the safety of nuclear plants.

Fig.87.

2. A hydrological "barrier" against the contamination of underground waters. The supply of large cities and industrial centers with good-quality drinking water and water for technical needs has long been a crucial ecological problem. To solve it, besides using open and therefore easily polluted sources (rivers, lakes, reservoirs) the underground waters of corresponding layers are intensely used. They are influenced less by anthropogeneous actions, however the problems associated with inevitable contamination remain actual. One of them is the localization of a dangerous impurity penetrating into a part of a layer, so that the water in other parts remains pure and suitable for consumption.

This purpose can be reached using part of the underground waters to create a kind of hydrological barrier on a path of propagation of contaminations. Its general scheme is shown in Fig. 87: between the sources of contamination (asterisks) and the water well, special slits pump up rather pure water into the layer and raise its level (barrier). The pumping creates a forced motion of the ground waters to the right and to the left from the barrier (arrows). The current filtered to the right pushes back the incoming water containing impurities, preventing the further propagation of contamination along the layer.

The mathematical model realizing this scheme contains the equations of

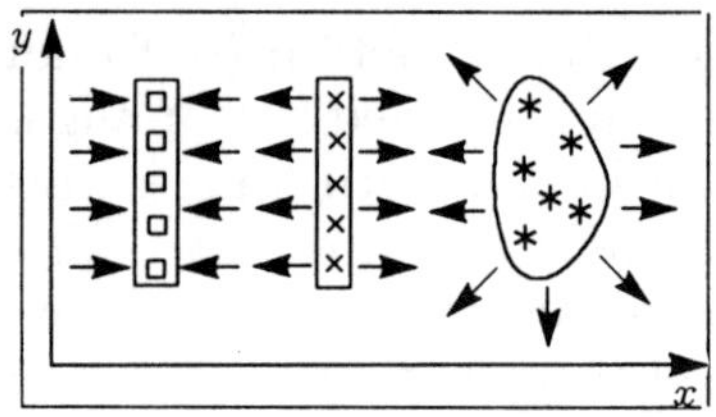

Fig.88.

filtration of underground waters and the equations of propagation of pollution, supplemented by appropriate input data (properties of the ground, water and impurities, information on the geometry of the considered area, boundary conditions, etc.). For a relatively small amount of impurity in the water there is no noticeable influence on its motion, and consequently the filtration can be considered separately from the dynamics of propagation of contamination. For some assumptions, the main one being the sufficient extension of the layer, the motion of the water is described within the framework of the Bussinesque model (section 1, Chapter II). In this case all characteristics of the process are functions of the variables x, y, t (the top view of the layer is represented in Fig. 88, where the crosses denote the pumping slits, the small squares denote the water wells, and the asterisks denote the sources of contaminations).

The model of propagation of impurity is obtained (as well as the Bussinesque equation) from a conservation law (balance) of the mass of impurity within the element of the ground. In the absence of diffusion of contamination it represents the usual continuity equation and in an elementary variant has the form

$$\frac{\partial}{\partial t}[C(H+h)] + \frac{\partial}{\partial x}[C(H+h)u] + \frac{\partial}{\partial y}[C(H+h)\,v] = Q(z,\, y,\, t)\,,$$

where $C(x,\, t)$ is the sought concentration of impurities, $Q(x,\, y,\, t)$ is the known intensity of sources of contaminations (the other notations are the same, as in section 1, Chapter II). This equation can also be treated at known functions h, u, v as a two-dimensional equation of concentration C transfer. In so far as by the Darsi law

$$u = -\nu\,\frac{\partial h}{\partial x}\,, \qquad v = -\nu\,\frac{\partial h}{\partial y}\,, \qquad \nu = \mu\rho g\,,$$

then the given equation can be rewritten in a form

$$\frac{\partial}{\partial t}[C(H+h)] - \nu\,\frac{\partial}{\partial x}\left[C(H+h)\,\frac{\partial h}{\partial x}\right] - \nu\,\frac{\partial}{\partial y}\left[C(H+h)\,\frac{\partial h}{\partial y}\right] = \\ = Q\,(x,\, y,\, t)\,.$$

If one takes into account the noticeable random motion of impurities, the latter equation becomes more complicated

$$\frac{\partial}{\partial t}[C(H+h)] - \nu\frac{\partial}{\partial x}\left[C(H+h)\frac{\partial h}{\partial x}\right] - \nu\frac{\partial}{\partial y}\left[C(H+h)\frac{\partial h}{\partial y}\right] =$$

$$= \frac{\partial}{\partial x}\left[D(H+h)\frac{\partial C}{\partial x}\right] + \frac{\partial}{\partial y}\left[D(H+h)\frac{\partial C}{\partial y}\right] + Q\,(x,\,y,\,t)\,,$$

becoming a second-order equation (of parabolic type) with respect to the function C. The additional terms on its right hand side correspond to the increase or decrease in the mass of an element of the ground due to the presence of flows estimated by Fick's law

$$W = -D\,\mathrm{drag}\,C\,,$$

where $D > 0$ is the coefficient of hydrodynamic dispersion (analog of thermal conductivity coefficient in Fourier's law; see Table 1).

One of the natural ways of making the transition to a discrete model is to approximate a system of two parabolic equations: the Bussinesque equations from section 1, Chapter II, for h and the above mentioned equation for C, via a difference scheme constructed, for example, by the integro-interpolational method (this is especially convenient in the case of uniform grids). Note that the equation for h does not contain the quantity C, and this property of the model considerably facilitates its numerical study. It is reasonable to use strictly stable implicit schemes permitting us to perform calculations with a sufficiently large time steps.

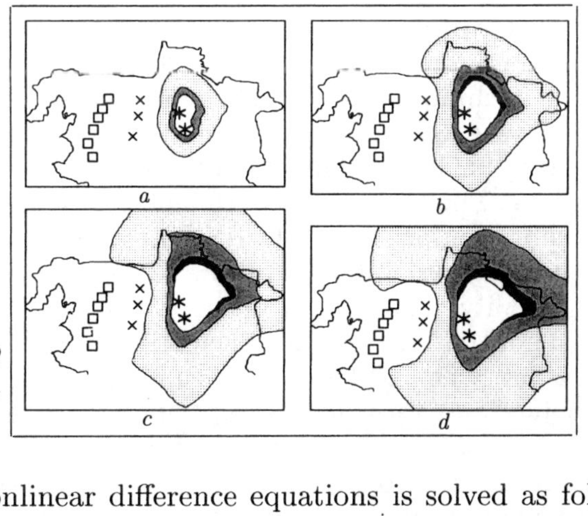

Fig.89. *a – t = 2 years; b – t = 7 years; c – t = 13 years; d – t = 33 years*

The resulting system of nonlinear difference equations is solved as follows. First from the difference Bussinesque equation for h_i^j (not containing function C_i^j) the values h_i^j are obtained. Then they are used in difference equations for C_i^j, from which the latter are determined. For the solution of discrete analogs of the two-dimensional parabolic equations for h and C, the standard method of variable directions (and iterative procedures) of a successive solution by running a set of one-dimensional problems (first in direction x, then in direction y, and vice versa) is applied.

In Fig. 89, a–d we represent the results of a demonstrating computing experiment via the above described technique simulating the dynamics of the phenomenon (closed lines correspond to constant values of the concentration of impurities in some conditional units). For simplicity, the properties of the ground were considered as constant, the quantities h_i^j and C_i^j were considered to be equal to zero at large distances from a source of contamination, the calculated steps in time and in space yield about one month and one kilometer, respectively. The characteristic scales of the whole process are tens of years and hundreds of kilometers (the size of a small central European country).

From Fig. 89 it is clearly seen that the impurities from a source of contaminations are spread in all directions, except for the direction of the hydrological barrier protecting the wells from low quality water.

Obviously, the final answer regarding the efficiency of the hydrological barrier can be obtained only after additional study by simulations of the investigated processes, including taking into account the properties of concrete grounds, various ways of disposition of wells, economic aspects, etc. Obviously, certain solutions of the considered problem should be based on the mathematical modeling and computer experiments, in so far as the scale of the phenomenon precludes full-scale experiments, while laboratory tests provide only limited information due to the absence of similarity.

3. Complex regimes of gas flow around body. This situation, similar to the one in subsections 1,2, is characteristic for problems of a modern flight technique design. The aerodynamic tubes, in which the fragments of the future apparatus are "tested", are very expensive and complicated engineering facilities. The cost of an hour's use can reach several thousands of dollars. The obtained experimental data are not complete (since it is impossible to conduct trials at all admissible velocities, angles of attack, etc.) and require correct interpretation. Moreover they concern only certain parts of the facility. This last problem cannot be eliminated due to the basic reason. The flow pattern by gas of a small material duplicate of a real object located in a tube, has a little in common with processes happening at actual scales (meters and tens of meters), since there is no detailed similarity between them.

It is easy to prove this by rewriting the two basic dimensionless parameters of similarity (see section 1, Chapter V) for a considered class of phenomena – the Mach number ($\mathbf{M} = u/c$ is the ratio of flow velocity to the velocity of sound) and Reynold's number ($\mathbf{Re} = uL\rho/\mu$, where μ is the coefficient of gas viscosity, L is the characteristic size of the facility). From the form of the numbers **M** and **Re** it follows that keeping the same number **M** for various L, it is impossible to also maintain the value $\mathbf{Re} = \mathbf{M}cL\rho/\mu$, describing the ratio of forces of pressure to the forces of viscosity.

Therefore the basic information on the behavior of designing flight vehicles is obtained via computing experiments, conducted for all the range of admissible parameters. Real tests are performed mainly as "standard" experiments: by comparing their results with those of mathematical modeling, the adequacy of the models used becomes clear, and the accuracy of the computing algorithms, if necessary, can be developed further.

We will now demonstrate the mutual relations of computing and real experiments on an example of mathematical modeling of certain complicated flows using a technique based on kinetically consistent difference schemes for Eulerian and the Navier-Stokes equations (described in subsection 6, section 4, Chapter V). In the examples considered below, the processes of viscous transfer are essential only within narrow area of a boundary layer near the boundary of streamlined bodies, therefore the Euler equations are solved.

One of the solved standard problems is the description of non-stationary two-dimensional flows in a "knee " of a flat rectangular channel, for which rather reliable experimental data are available. In the bottom of the channel (Figs. 90, 91), occupied initially by a motionless gas ($\gamma = 1.4$), a supersonic ($\mathbf{M} = 1,\ 2$) flows enters creating a series of complicated gas dynamic configurations upon the propagation. Grids with the number of knots $N_x \cdot N_y \cong 3\,000$ and step in space $h_x = h_y = 1/27$ and in time $\tau = 0.2$ were used.

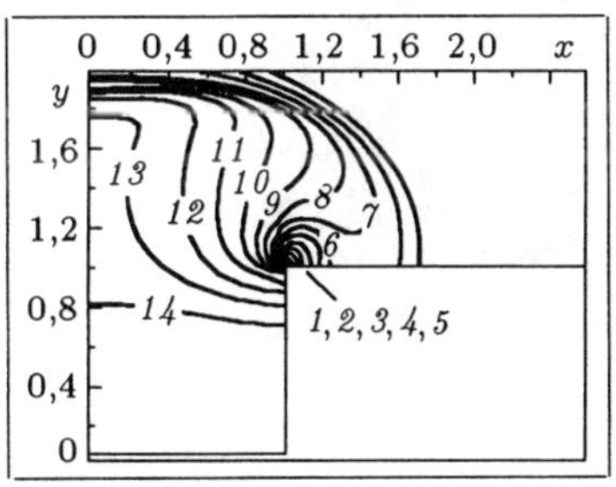

Fig.90.

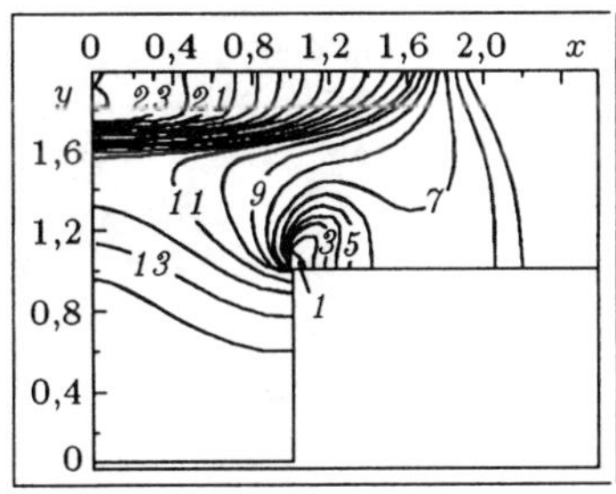

Fig.91.

In Figs. 90, 91 for instants $t = 30$, $t = 40$ we represent the equidistant with a step $\delta\rho = 0,083$ isoclines of gas density in a channel of a unit cross section ($\rho_1 = 0.747$, $\rho_{14} = 1.826$ for Fig. 90 and $\rho_{24} = 2.656$ for Fig. 91, the units of measurement are conditional). In Fig. 90 the zone of compression

– the shock wave formed at the stream motion near the rectangular cusp (the lines are condensed) – and the rarefaction zone after shock wave are clearly seen. At the upper wall of the channel the gas is decelerated and compressed. As a result of this deceleration, a shock wave reflected from a wall does appear (Fig. 91, the density of gas in the compression area is decreasing by distance from the left hand boundary, which is obvious).

The further picture becomes even more complicated: the shock waves, reflected from the boundaries, are propagating over the stream, interacting among themselves are again reflecting from the walls and so on. Their dynamics are traced in detail in a computing experiment. The adequacy of mathematical modeling for the investigated phenomena is confirmed by comparison of Fig. 91 and corresponding Fig. 92, where the interferogram of density distribution in a stream obtained at direct real experiment is shown. The good qualitative and quantitative coincidence of the results of computing and actual experiments is visible.

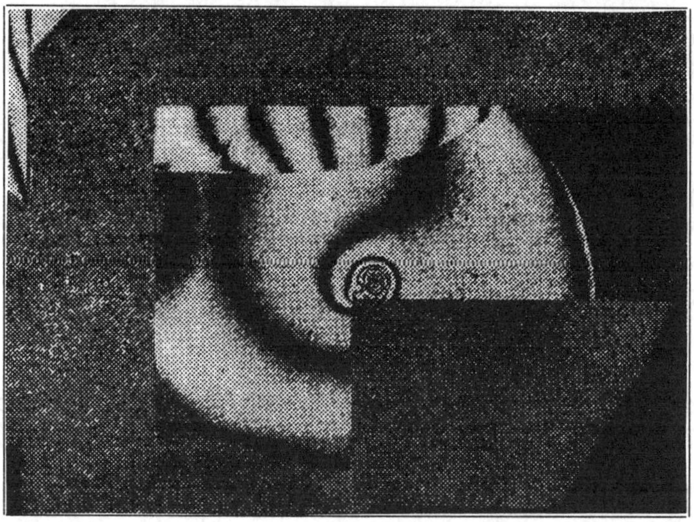

Fig.92. *Interferogram of distribution of density in full-scale experiment*

Thus, the numerical model used is quite suitable for describing flows of complex structure, and it enables to study also other processes (including those for which there are not reliable experiments). In Figs. 93 and 94 the results of numerical modeling of the flow by gas of the front part of a cylindrical body with a "needle" – a long thin cusp forwarded to the gas inflowing from the right (along the axis z). The parameters of gas in the flow: $\mathbf{M} = 3.5$, $\rho = 1$, $\gamma = 1.4$; dimensions of the body: the diameter of the main cylinder $D = 50$, the diameter of the needle $d = 2$, its length $l = 46$. The calculations were performed in coordinates r, z on a grid containing approximately 3 000 knots. The equidistant isoclines of pressure are shown (in Fig. 93 $p_1 = 1.44$, $p_{10} = 11.47$, in Fig. 94 $p_1 = 1.9$, $p_{10} = 24.44$), the bold line with points is the sound line with $\mathbf{M} = 1$.

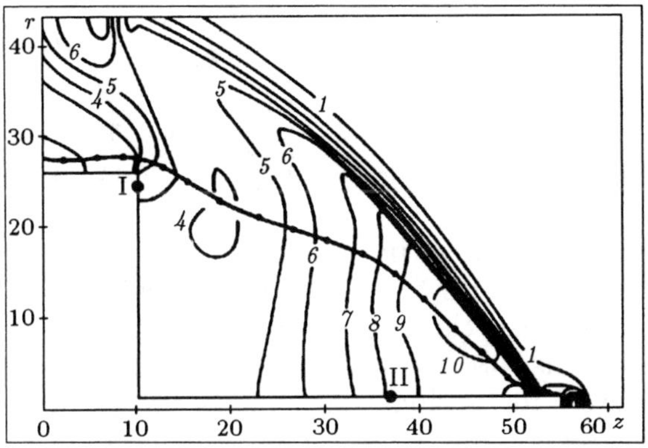

Fig.93.

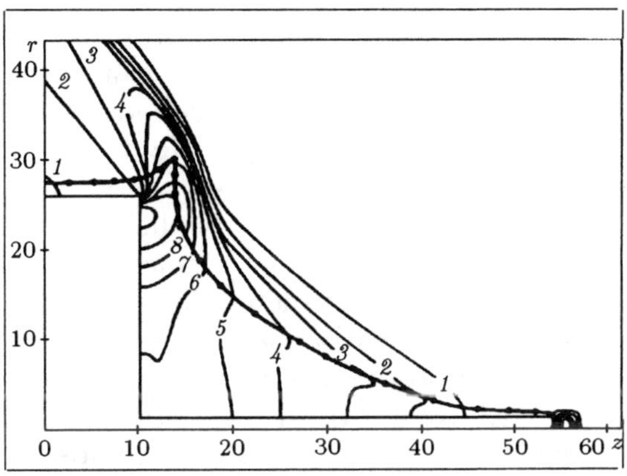

Fig.94.

The comparison of Fig. 93 ($t = 53$) and Fig. 94 ($t = 83$) shows that the flow is not steady-state, and has an essentially non-stationary character (so that the oscillations of quantities can be rather significant). The essence of these almost periodic oscillations at numerical modeling lies in the increase and decrease of the zone of the current between the shock wave (frequent isoclines) and the body. The shock wave resulting at the deceleration of the inflowing gas moves in time along almost the whole length of the needle, which leads to noticeable change of parameters of flow in the neighborhood of the needle. In Fig. 95 the temporal dynamics of pressure in two characteristic points of the body (bold points in Fig. 93) are shown. Line 1 indicates the pressure at point I, line 2 indicates the pressure at point II, line 3 corresponds to the characteristic size (by axis z) of the zone between the shock wave and the body.

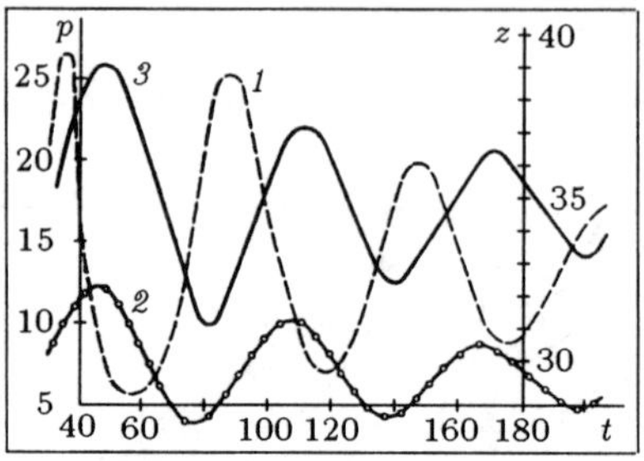

Fig.95.

The oscillation amplitude is rather big (2–3 times more than the average value), the same as the frequency, equal approximately to l/c (where c is the velocity of sound in the area between the shock wave and the body). Therefore the appearance of oscillations (observed also in real experiments) and their quantitative characteristics for various parameters of inflowing gas and the configurations of the bodies has to be taken into account in the design of corresponding flight vehicles. We stress that the constantly growing complexity of similar problems requires not only the use of more and more powerful computers, but also (in no lesser degree) the permanent development of discrete models and computing algorithms.

4. Ecologically acceptable technologies for burning hydrocarbon fuels. The majority of energy used in industry and everyday life is produced in the burning facilities – in thermal electric plants, gas turbines, internal combustion engines, etc. The chemical energy of hydrocarbon burned in air (petroleum, kerosene, black oil, methane) is either realized immediately for heating or is transformed to mechanical or electrical energy. The common property of all these processes is that the air contains not only oxygen (oxidizer), but also a noticeable fraction of an "extra" substance – nitrogen. Due to this fact during combustion junctions are produced which, as distinct from the neutral nitrogen, are chemically active (most of them are the oxides of nitrogen NO and NO_2). The harm caused by these oxides in atmosphere is diverse: they are the main source of acid rains, essentially promote the dissipation of the ozone atmospheric layer, they are toxic for breathing and so on.

Two methods of decreasing emissions of nitrogen oxides are usually used. The first is the cleaning of the combustion products. The corresponding equipment is rather effective, but simultaneously is very expensive. The cost of modern filters reaches up to 10–30% of the cost of the plant itself, in the same proportion increase also the exploitation expenses. The main idea of the second method is the organization of the combustion process in such a way that when the formation of nitrogen oxides is minimal at the same power characteristics of the installation. The realization of this approach would mean a noticeable decrease of expenses for gas cleaning of products of combustion from NO and NO_2 (it is impossible to completely suppress their formation).

The processes occurring in burning devices are very complicated. Plenty of chemical transformations of a large number of various substances, release and absorption of energy, gas dynamical motion of mixtures, turbulent mixing of components of fuel with air and combustion products, etc. do occur there. The dynamics of this phenomenon strongly depend on the condition of the inflow of the fuel, the configuration of the boilers, the location of the heaters and other characteristics. The search for ecologically sound conditions cannot be based either on purely theoretical representations, or on real experiments, which are expensive, time-consuming and unsafe.

The satisfactory mathematical model of the work of a boiler in a plant includes two connected parts. The first of them is the local description of chemical kinetics for 32 components of mixture, in particular of those reactions, which play the main role in the formation of toxic contaminants. The second part is the non-stationary spatially two-dimensional model of large-scale processes of thermal and mass transfer (diffusion of substances, heat transfer, convective motion and so on). It is complicated also by the need to accounting for the actual geometry of the boiler (Π–shaped form), by the presence of several rows of heaters serving to inject fuel and air, etc. From the computing experiments of such a model all necessary characteristics of combustion at various operational modes of the boiler are obtained, and then the optimal ones are selected.

We will now characterize one of the fragments of mathematical modeling of the described phenomena, confining ourselves by the first part – the study of the kinetics of formation of NO and NO_2 at combustion of CH_4 (methane) mixed with air (the mixture is considered isothermal and spatially homogeneous). The complete model includes 196 direct and 196 inverse reactions for the 32 substances participating in the combustion: O_2, N_2, CH_4, O, N, NO, NO_2, etc. From a mathematical point of view this represents a system of $196 \times 2 = 392$ ordinary differential equations describing the profit and the loss of components as a result of corresponding direct and inverse reactions. The integration by time gives the detailed dynamics of the process at a known initial concentration of the substances. Thus, not only the total concentration of each substance is traced, but also all paths of its formation, which is especially important for the analysis of mechanisms of emerging toxic impurities (one of the effective methodical ways facilitating the calculations: by the results of test evaluations the reactions making a minor contribution to the process are distinguished and hence are not taken into account in the model).

We will now describe one of the typical computing experiments performed

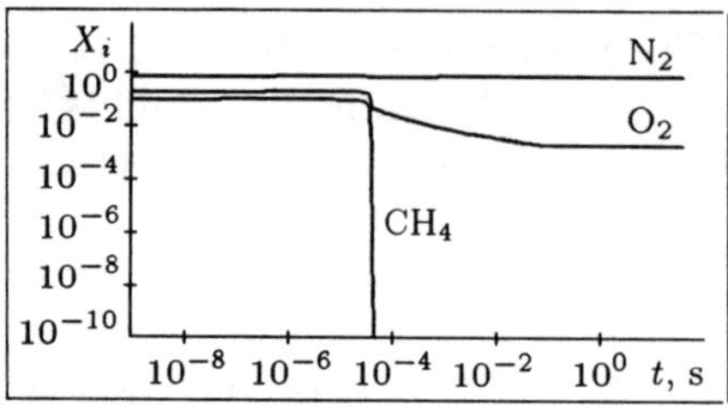

Fig.96.

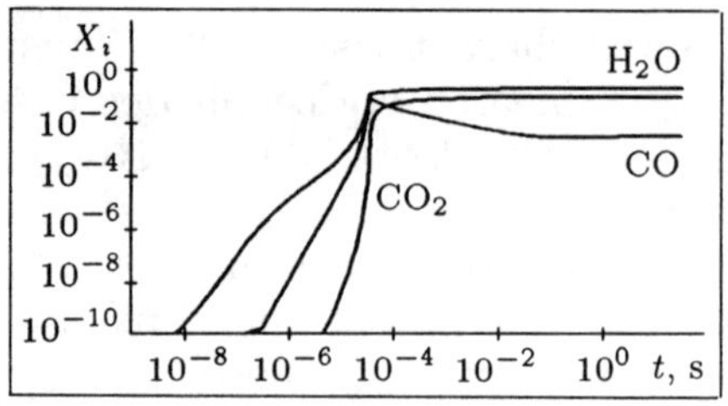

Fig.97.

in conditions close to the actual ones. The mixture is at atmospheric pressure, its temperature is 2 000 K, and the initial content at $t = 0$ is given in the following fractions: $[N_2] \approx 0.7$, $[O_2] \approx 0.2$, $[CH_4] \approx 0.1$. The calculations were performed before the relaxation to the equilibrium, i.e. until the period when the concentration of any of components ceases to vary.

Fig.98.

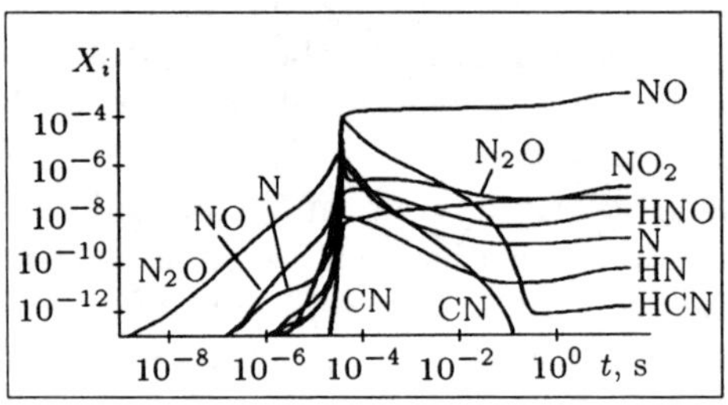

In Fig. 96 the dependence on time of the molar (volume) concentration of components of an initial mixture, in Fig. 97 – for H_2 O, CO_2, CO, in Fig. 98 – fractions of nitrogen containing substances including NO, NO_2 and N_2O (nitrous oxide), are shown. The first conclusion is that the combustion of methane occurs very quickly and its content near the moment $t = 4.1 \cdot 10^{-5}$ decreases practically to zero. The process of pure combustion ceases after this.

We shall now consider kinetics of transformations of substances nitrogen containing. First of all, from Fig. 98, it is clear that the total concentration of NO exceeds by many orders of magnitude the concentration of other harmful impurities. Therefore the basic attention should be directed towards diminishing just this component. It is also essential, that the content of NO is monotonously increased by time. Thus, the problem of diminishing of NO is easier to solve by suppression its origin (if NO was already formed, then it is much more difficult to get rid of it).

A more important result follows from comparison of dynamics of the variation on a content CH_4 and NO. For NO the experiment gives at the moments $t = 4 \cdot 10^{-5}$ s, $t = 0.1$ s, $t = 20$ s values of concentration equal, respectively to [NO] $\approx 8 \cdot 10^{-5}$, [NO] $\approx 2 \cdot 10^{-4}$, [NO] $\approx 6 \cdot 10^{-4}$. In other words, up to the moment when the combustion of methane is practically finished, the mixture contains only 13% of the total amount of the finally formed NO (the main part of the harmful impurities arise after the useful reactions of burning methane are finished).

Strong temporally different scaling of these two processes indicates of the principal possibilities of an essential diminution of NO in the products of combustion. It is necessary to remove the products of combustion from the zone of the flame as soon as possible (ideally immediately after the burn-out of methane) and to cool them quickly. As a result the further chemical transformations, including the formation of NO, will be stopped, but without loss in the power of the installation.

The modeling clearly indicates several other important properties of the process, for example, the output of NO monotonously decreases at a decrease of temperature of combustion, the exit of nitrous oxide N_2O (the ozone depleting substance) is significant only at intermediate stages of the process, while it later turns to harmless N_2, etc. These conclusions are very useful for the development of recommendations on designing ecologically acceptable power installations.

2 Fundamental Problems of Natural Science

Consider the application of the methodology of mathematical modeling for the solution of some fundamental problems from various areas of natural science. We show, that it allows us not only to determine the quantitative characteristics of the investigated processes, but also to discover qualitatively new phenomena.

1. Nonlinear effects in laser thermonuclear plasma. The actual perspective of the solution of future energy problems is connected with the controlled thermonuclear synthesis (CTS) of isotopes of hydrogen, first of all of deuterium (D) and tritium (T)

$$\mathrm{D} + \mathrm{T} =^4 \mathrm{He} + n + 17.6 \quad \mathrm{MeV}\,.$$

In this elementary reaction, D + T nucleus of helium (α – particle) and neutron are formed with total kinetic energy 17.6 MeV, with main part (14 MeV) being contained in neutrons (1 MeV is equal $1.602 \cdot 10^{-13}$). In a gram of DT-mixture a huge energy is hidden equivalent to the energy released in the burning of 15 tons coal. Doubtless advantages of CTS are the practically unlimited storage of "fuel" in the ocean possessing enormous energy, along with ecological cleanness.

To initiate the reaction D + T to heat the mixture must be heated to several tens of millions degrees and compressed up to density comparable with the density of DT-ice (of cooled DT-mixture) equal to 0.2 g/cm^3. If it will be possible to keep it further in this condition during a time scale sufficient for the burn out of a noticeable part of the "fuel", the released energy will be comparable to the energy spent on the heating and compressing plasma. The realization of this scheme first in the laboratory, and then in the industrial conditions would mean the practical solution of one of the oldest fundamental problems of physics.

We have the basic possibility of realizing thermonuclear synthesis known since the middle of twentieth century: in 1940s it was established, that the energy of Solar radiation is mainly due to internal reactions of merging isotopes of hydrogen. The first explosions of hydrogen bombs performed in the 1950s were example of a "hand-made" thermonuclear reaction.

However both these "ways " of releasing of energy are unsuitable for the purposes of CTS. On the Sun the plasma is kept in the necessary condition due to powerful gravitational forces, while the energy accompanying the explosion even of a miniature hydrogen bomb, many times exceeds those needed for peaceful use.

There are some basic physical ideas of realization of CTS. Historically, the first of them is the magnetic trap, and its most developed design – the tokamak, i.e. the toroidal chamber in which the external and generated by plasma currents magnetic fields do not allow scattering and cooling of the heated plasma "doughnut". The inertial synthesis competes with this relatively stationary way of obtaining of thermonuclear energy. Its idea is in the fast heating of a drop of a fuel, which due to the inertiality of motion of matter will have no time to scatter and cool until it will not possess the necessary thermonuclear energy. Periodically occurring microexplosions will give a constant flux of neutrons and α – particles, used outside the working chamber.

The most convenient source of such fast heating are lasers. In laser thermonuclear synthesis (LTS) the sequence of events is as follows. The radiation directed on a spherical target is absorbed by its outside layers, heats them up and evaporates. The "corona" which appears will expand with high velocity, compressing and heating the core of the target by its jet pressure. The intense thermonuclear combustion and release of energy occurs, then the target is blown away and cooled down.

The first test evaluations have shown that for realization of LTS, lasers capable of releasing energy 10^8–10^9 J during several nanoseconds (1 nsec = 10^{-9} s) are required, focusing it in a tiny area of several tens of microns. At present, lasers with such parameters do not exist, nor their appearance is anticipated in visible future.

Noticeably reducing the energy threshold of LTS is possible due to use various nonlinear effects intrinsic to laser plasma. These are studied both experimentally, and theoretically with the help of computer experiments with mathematical models of laser microexplosions. Not characterizing them explicitly, we shall explain that the basis of mathematical description of the processes in the central part of a target is given by the equations of heat transfer and dynamics of gas (sections 2 and 4, Chapter II). For their discrete approximation the completely conservative difference schemes were successfully used, as well as the schemes obtained from the discrete analogs of variational principles (subsections 4, 5, section 4, Chapter V).

One of the effects is connected with the typical gas dynamical nonlinearity. At compression of the core of the target by virtue of the "gradient catastrophe" (subsection 7, section 4, Chapter II) shock waves appear, which heat up the central part already at the initial stage of the process and prevent its further compression. As a result the densities reached are much smaller than it would be possible to achieve with adiabatic, non-shock compression (and at the same energy of the laser). This in turn leads to a sharp decrease in the release of thermonuclear energy, in so far as the rate of reaction of the synthesis is proportional to the square of the fuel density.

However compression process can be organized in such a way, that it will proceed in a non-shock manner, when all the generated gas dynamical perturbations during the contracting kernel by the "piston" arrive at the center of the target simultaneously and shock waves do not arise until the moment of its maximal compression. One simple analytical solution of the equation of gas dynamics demonstrating the basic realizability of such process, is represented in subsection 3, section 3, Chapter V.

The detailed mathematical modeling of conditions of non-shock supercompression of targets has convincingly shown that it can really be realized and the achieved density yields tens and hundreds of grams in a cubic centimeter. Then the energy threshold necessary for the realization of the "critical" experiment (the energy contribution into the target and its release are equal) is reduced by several orders of magnitude and is equal to 10^5–10^6 J. The laser pulse ensuring supercompression, should be "shaped", i.e. its power should vary in time by the law $G(t) = G_0(t_f - t)^{-2}$ (blow-up: see sections 1–3, Chapter V).

A similar result can be achieved if one replaces the temporal profile of the laser light by a spatial (hydrodynamic) profile. This is achieved by a noticeable complex design of the target (see Fig. 99), but the laser pulse as a function of time has a conventional "hat" form.

Both described concepts, checked in detail in computer experiments, have a basic role in the development of LTS, in so far as they provide one of the key ways of realizing controlled synthesis. They cannot be checked yet in direct real experiments, and therefore the adequacy of the mathematical models of LTS and the reliability of conclusions obtained on their basis, is determined by thorough comparison with existing experiments on heating and compression of laser targets.

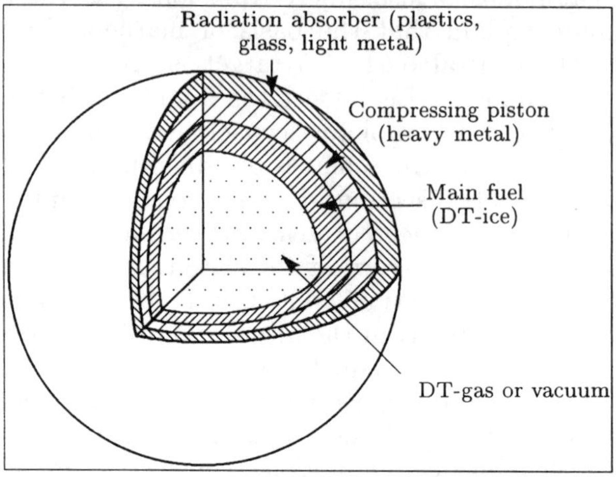

Fig.99.

Thus, for the targets of complex design one of the central problems is ensuring the symmetry and stability of compression. The thickness of shells the DT-gas, amounts to several percent of the initial size of the target, so that the compression process can be compared with the transformation of an empty egg shell to a core the size of a bean. Therefore the sometimes occurring non-coincidence of the results of one-dimensional spherical symmetrical calculations of compression with actual experiments is not surprising. In each such case the reason has to be revealed. In Fig. 100 a, b we represent the calculated form of the typical experimental target at the moment of maximal compression (a – general view, b – central part; the condensation of lines indicates the increase of density). It is visible that the center is nonsymmetric and, most importantly, is shifted by 50 microns relative to its initial position. As a result, the actual neutron release has decreased by 100 times in comparison with the calculated one in the one-dimensional model. The computing experiments have revealed the exact reason for the shift – the asymmetry of the laser radiation on the surface of the target, enabling us to obtain appropriate practical recommendations.

A no less important reason for the loss of one-dimensionality of compression is the hydrodynamic instability on the boundary of light and heavy fluids undergoing acceleration (due to this, mercury poured onto the surface of water in a vessel, undergoing the action of gravity, is inevitably mixed with the water, but this does not occur in the opposite case, when the water is poured onto the surface of mercury). Exactly this situation corresponds to a final stage of compression by a heavy shell with a less dense center: the

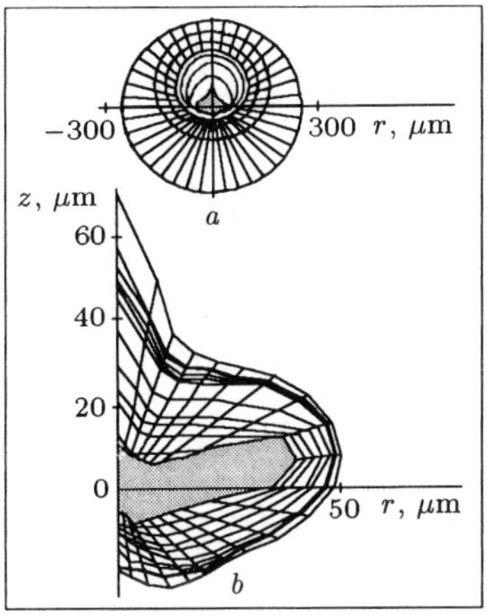

Fig.100.

shell is braked and hence, acceleration is induced by the forces of inertia, is directed from the heavy to the light substance. If the instability has time to develop strongly enough, the symmetry of compression will be violated and moreover, the inflow of a part of substance of the shell into the zone of combustion will immediately stop the reaction. It is impossible to observe this type of instability directly in experiments with targets. The universality of mathematical models is therefore used. First we discuss the results of modeling with real experiments in shock tubes – devices where the instability is induced by an artificial shock wave in a medium of non-uniform density (this is reached by separation of different substances by a diaphragm). Then the conclusions for laser targets are drawn.

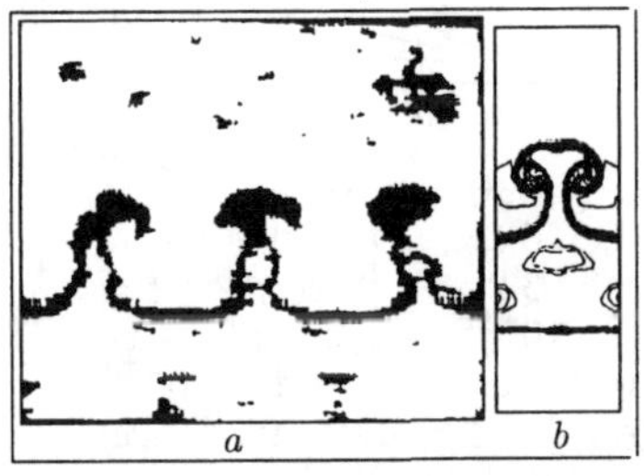

Fig.101.

In Fig. 101 a, b, the form of the boundary between heavy (xenon) and light (argon) gases at a moment $t = 100$ msec (the wavelength of initial perturbation is $\lambda = 24mm$) is represented; a are the experimental data, b is the result of the calculations. It is visible that numerical modeling provides not only a typical qualitative picture ("mushrooms"), but also quite precisely describes the quantitative characteristics (the amplitude of mushrooms, etc.). By virtue of this the computer experiment becomes a reliable tool for studying hydrodynamic instability, in particular for the definition of tolerances in accuracies of creating the targets.

The adequacy of models of laser synthesis allows us to consider the possibility of using one more characteristic of plasma-type nonlinearity, due to which localized structures of combustion appear in the medium (see section 3, Chapter V). If the plasma is rather dense (several tens of grams per cubic centimeter), the α-particles are absorbed at the place when they are released. Let in addition, in the first approximation the hydrodynamic motion be neglected in comparison with processes of heat transfer and combustion. Then the propagation of the released energy occurs due to heat transfer and is described by the equation

$$\frac{\partial T}{\partial t} = \frac{\partial}{\partial x}\left(k_0 T^{\sigma}\frac{\partial T}{\partial x}\right) + \frac{q_0\, T^{\beta}}{1 + BT^{b}},$$

where $k_0 > 0$, $q_0 > 0$ are constants at thermal conductivity of plasma and of energy source from thermonuclear reactions. For hydrogen plasma $\sigma = 2.5$, the constants $\beta = 5.2$, $b = 3.6$ and B in the source are such, that they can be approximated by a power law function of the form $q_0 T^{5.2}$ (at $1 < T < 3-4$ keV), $q_0 T^{3.5}$ (at $T \approx 4-5$ keV) and $q_0 T^{1.6}$ ($T > 5$ keV).

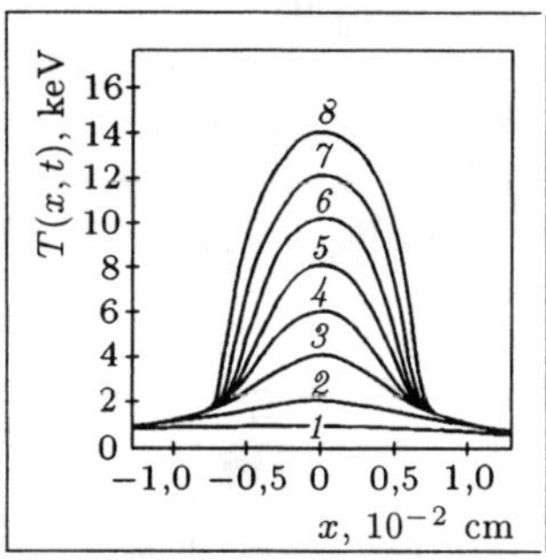

Fig.102. *1* – $t = 0.0$ *sec*; *2* – $t = 3.2 \cdot 10^{-10}$ *sec*; *2* – $t = 3.8 \cdot 10^{-10}$ *sec*; *4* – $t = 4.1 \cdot 10^{-10}$ *sec*; *5* – $t = 4.3 \cdot 10^{-10}$ *sec*; *6* – $t = 4.5 \cdot 10^{-10}$ *sec*; *7* – $t = 4.7 \cdot 10^{-10}$ *sec*

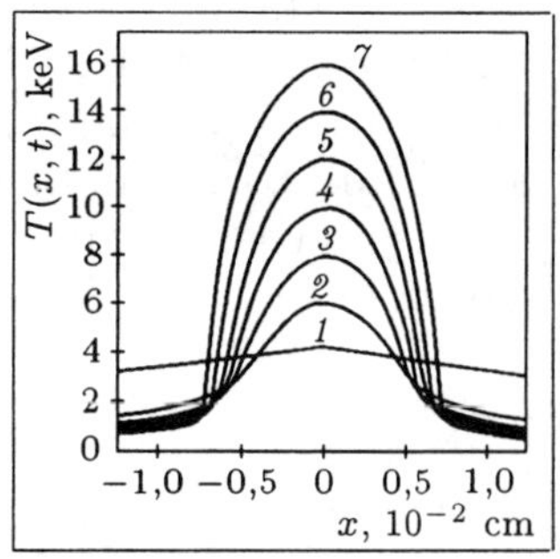

Fig.103. *1* – $t = 0.0$ *sec*; *2* – $t = 3.538 \cdot 10^{-8}$ *sec*; *3* – $t = 3.715 \cdot 10^{-8}$ *sec*; *4* – $t = 3.728 \cdot 10^{-8}$ *sec*; *5* – $t = 3.732 \cdot 10^{-8}$ *sec*; *6* – $t = 3.735 \cdot 10^{-8}$ *sec*; *7* – $t = 3.737 \cdot 10^{-8}$ *sec*; *8* – $t = 3.738 \cdot 10^{-8}$ *sec*

Therefore, up to temperatures of approximately $5 \cdot 10^6$ K the nonlinearity of medium implies the localized LS- and S-combustion regimes. In Fig. 102 the temporal dynamics of a combustion structure in plasma of density 20 g/cm^3 (numerical calculation), in Fig. 103 hold the same calculation, but in the presence of the plasma's own radiation. The radiation even improves somewhat the localization of the combustion, which is not propagated in the medium at temperatures higher than 5 keV. Therefore at the correct initiation of a thermonuclear reaction, the area of self-localized combustion, formed in the center of the target, can serve as a reliable "fuel" for burning the remaining mass of the combustible.

The cost of modern experimental installations for CTS – the prototypes of future power stations – amounts to hundreds of millions of dollars, the experiments on them are complicated, labor-intensive and time consuming. It is natural, that at the solution of the problem of CTS any physical or technological idea is not considered seriously without being carefully studied by methods of mathematical modeling and computer experiments.

2. Mathematical restoration of the Tunguska phenomenon. The brightest of known examples of large-scale collision of a celestial body with the Earth's atmosphere is the Tunguska phenomenon observed at about 7 a.m. on June 30, 1908 by many inhabitants of an extensive region in East Siberia centered near settlement Vanavara (the river Podkamennaya Tunguska). A huge cosmic fireball (angular size 0.5° on distance 100 kms, i.e. with a cross-sectional size of about 800 m) was moving in the clear sky under some angle to the horizon with a velocity of more than 1 km/sec. It disappeared behind a forest, then a bright flash occurred and the repeated acoustic waves broke the windows in an area with diameter more than 100 km. The witnesses also felt a noticeable thermal radiation wave and observed shadows caused by the flash. The geophysical and seismic stations of Russia and world wide registered air and seismic waves turning several times over the Earth. Consequent expeditions have discovered general destruction in taiga: destroyed forest and traces of light damage of trees over an area of about 2 000 km^2 around the epicenter of the event. Material remnants of the celestial body were not found.

All these data testify that a powerful air explosion had occurred (without forming a crater on the surface of the Earth) with an energy not less than 10^6 tons of trotil equivalent.

Thus, nature itself has performed a unique large-scale experiment. Studying it, one can obtain important information concerning astronomy, celestial mechanics, theories of comets and meteoroids (i.e. the meteorites before they fall onto the Earth). Note that meteorites (apart from examples of lunar rocks obtained on space expeditions) are the only samples of matter of the Universe arriving on the Earth.

The principal property of Tunguska phenomenon was the explosive disintegration of a body above the Earth and the absence of any considerable amount of its matter on the Earth's surface, which indicates that it could not have been a dense stone or iron meteoroid. It was either a rare type of stone meteoroid with an enhanced content of ice, carbon and hydrocarbon, or a fragment of comet core – the conglomerate of pieces of ice, gas and dust. All these substances evaporate easily or burn down into the atmosphere, not leaving any traces. The cometary hypothesis explains some properties of the phenomenon more completely. And if this is correct, then the Tunguska phenomenon is the only authentic example of collision of a comet with the Earth (though the probability of such an event is significantly less than of a collision with an meteoroid).

The general hypothesis does not solve the problem of the basic characteristics of the cosmic body – the mass, velocity, fall angle, energy liberated at explosion, and so on. Its refutation or confirmation can be obtained only by mathematical modeling of the phenomenon and by comparing the results of computing experiments with the available observations of destruction.

This very complicated inverse problem of mathematical restoration of the event in general form is formulated as follows: in an instant $t = 0$ in the atmosphere on height $z = z_0$ we consider the motion of a body of velocity v_0 under an angle θ_0, linear size L_0, density ρ_0, temperature T_0, heat of vaporization i_0 and characteristic stress of disruption σ_0.

The initial state of the atmosphere can be accurately described in an isothermal approximation (the temperature is constant) with density and pressure varying with the height by exponential law: $\rho = \rho^0 \exp(-z/H_0)$, $p = p^0 \exp(-z/H_0)$, where H_0 is some normalization constant.

The basis of mathematical description of the phenomenon are taken to be the non-stationary Navier-Stokes equations for a compressed viscous thermal conducting gas (subsection 5, section 3, Chapter III) in spatially three-dimensional geometry. In so far as the phenomenon is characterized by high temperatures and radiation, the hydrodynamic equations are supplemented also by three-dimensional equation of radiation transfer (section 3, Chapter II). They should be solved jointly at various sets of the above listed parameters varied in rather broad bands. To derive the solution, obviously, it is necessary to set the properties of the medium – coefficients of viscosity, thermal conductivity, absorption of light, equation of state, etc. The problem contains in total more than ten parameters defining the solution (the application of similarity methods – see section 1, Chapter V – allows us to reduce their number and somehow to simplify the analysis of the results). The numerical realization of the described models was carried out with the help of appropriate finite difference schemes.

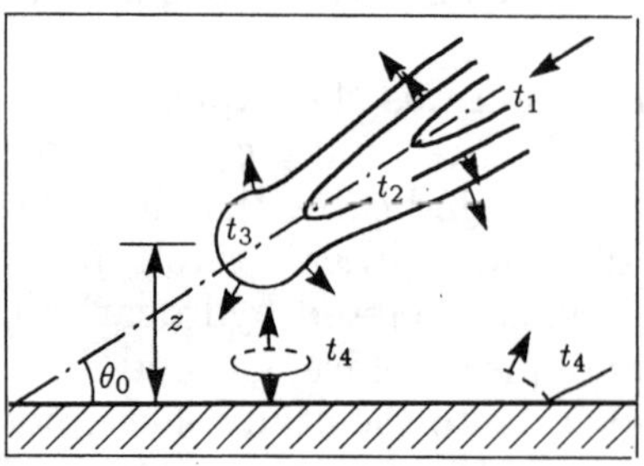

Fig.104.

The schematic chronology of events obtained from the computing experiment is represented in Fig. 104 (here $\theta_0 = 35°$, the body represents a hemispherical front part of radius 70m with a continued cylinder of thickness 140m). The dashed line denotes the trajectory of the body, the continuous line denotes the shock waves generated by its motion in successive instants. At moment t_1, t_2 the shock wave is ballistic, similar to that occurring around an object flying with supersonic velocity, for example of a jet plane. At a moment t_3 corresponding to the height of the body $z = 7$ km, the configuration of the shock wave becomes more complicated. By this period the body is braking and heated, its matter starts to expand explosively, creating a strong spherical shock wave. The further dynamics of the process, including the character of destruction, is determined by both waves. They arrive almost simultaneously (moment t_4) the Earth's surface (this is true only for the right part, lying on a surface under the trajectory) and are reflected from it. As the trajectory is inclined, the picture cannot be symmetrical relative to the center of the spherical shock wave considered as the epicenter of the explosion.

Indeed, in order for the propagating anti-clockwise ballistic shock wave to arrive at the Earth's surface to the left of the epicenter, it should cover a greater distance, than its lower part. Therefore it not only arrives at the surface later, but is also weaker and produces less significant destruction.

The conclusion about the asymmetry of destruction (on an almost plain taiga in the place of catastrophe) strongly agrees with real measurements. In Fig. 105 the results of analyzing characteristics of fallen wood in that region are shown; the circular lines imply equal values of force of destroying factors, the radial ones show the direction of the fallen trees. The picture is essentially nonsymmetrical, it has a "butterfly" form instead of concentric circles, which would have appeared if the body had fallen vertically or if there had been a point-like explosion of a small space vehicle, not producing a strong ballistic shock wave (a hypothesis about the artificial origin of the Tunguska body has also been made).

Fig. 106 shows similar data from the computing experiment (the arrows represent the direction of motion of the air over a surface, the dashed lines indicate the position of the front of the shock wave, the numbers indicate the instants in seconds), coinciding not only qualitatively, but for an appropriate selection of parameters of the body, also quantitatively with the consequences of the phenomenon.

Modeling the actions of the thermal radiation wave caused by the body is a relatively independent way of confirming the initial characteristics of the body. The comparison is represented in Fig. 107. The points denote the data of real observations: *1* is a weak burn, *2* is a moderate burn, *3* is a strong burn (charring). The solid line indicates the calculated value of the radiation $I = 16$ kal/cm ^{2} during 2 sec, necessary to burn the trees. It

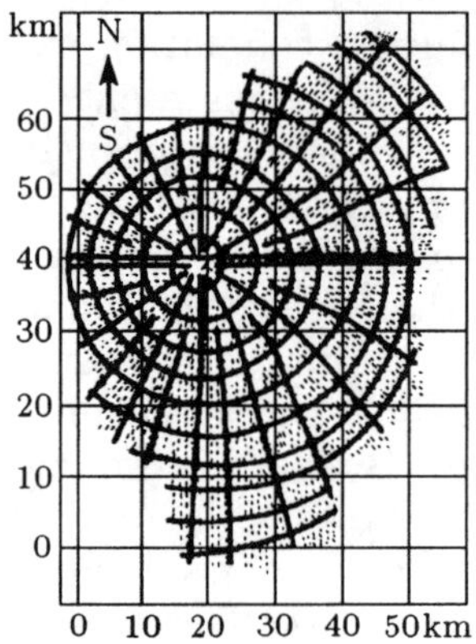

Fig.105.

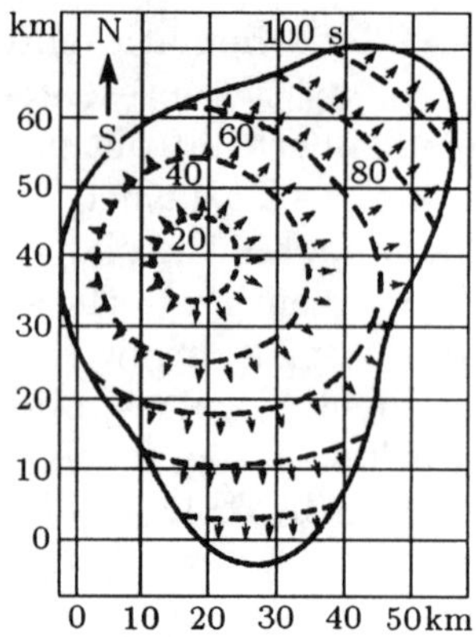

Fig.106.

precisely coincides with the actual boundary of the burn zone.

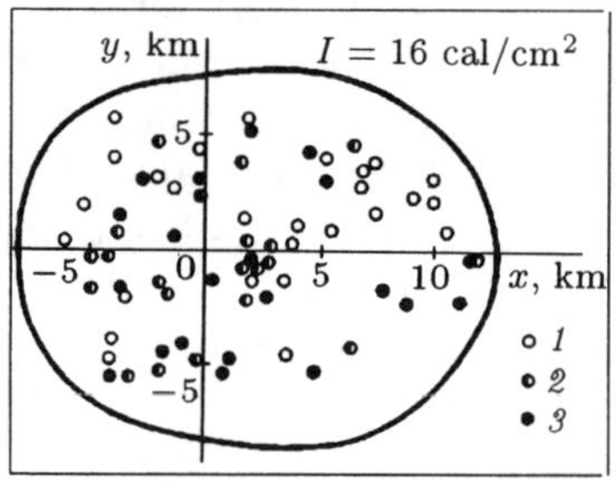

Fig.107.

The reproduction based on mathematical modeling of Tunguska phenomenon leads to the following quite convincing main conclusions, which cannot be obtained by other methods: the body of total mass 10^5 tons had penetrated the atmosphere under an angle $35°$ with a velocity of about 40 km/sec, was destroyed and braked sharply at height ~ 6.5 km, the shock waves have destroyed an area of wood, and the radiation from the heated up to $10 - 12 \cdot 10^3$ K body fragments, burned the trees, the power of the explosion was approximately 15 Megatons.

3. Climatic consequences of a nuclear conflict. Rather accurate long-term prediction of weather and changes of climate (caused first of all by anthropogeneous reasons) are extremely necessary for thorough scheduling of economic, technological, ecological and other forms of activity of humanity both a regional, and global scales. The optimal location of industrial centers, the best use of various forms of raw and energy resources, preferable choice of competing technologies, correct accents in agroindustrial policy – all these problems are in tight interdependence with the condition of the atmosphere, the oceans and the surface of the Earth.

The ongoing geophysical processes are very diverse and complicated. They include the hydrodynamic motion of atmospheric air and the waters of the seas and oceans, thermal and mass exchange in the "ocean-atmosphere" system, absorption, scattering and reflection of solar radiation (different in different seasons), seasonal variations of an underlying surface and many other phenomena. Their complexity is also connected with the inhomogeneity of the terrestrial surface and external non-stationarity of Earth's rotation around its own axes and around the Sun.

It is not surprising that the weather and climatic phenomena differ greatly by large-scale properties in time and in space. For example, the velocity of wind at a given point of Earth's surface can essentially differ from that at altitudes above this point. Besides, the atmospheric and oceanic currents are hardly turbulent, i.e. the quantities describing them undergo random fluctuations, with scales of hundreds of kilometers. Finally, all the mentioned processes are essentially nonlinear and their reaction to the variation of any parameters is hardly predictable.

Therefore the weather prognosis cannot be guaranteed. This is impossible not so much due to limited possibilities of computer facilities used for computations of a geometeofields, or lack of necessary data obtained with fixed and mobile stations, but due to basic reasons connected with the scale and complexity of the object. The precise prognosis of weather for several days is a quite good outcome.

The difficulties described have to be taken into account in the evaluation of possible consequences of numerous frequently proposed projects affecting the climatic processes, such as the transportation of several northern Russian rivers to dry southern districts, the closure of the Bering strait, so that the climate to the south of it will become warmer and so on. Similar experiments with a unique system can be performed only once, their results are irreversible and should be known with scientifically justified accuracy. Therefore the basic means of analysis and prognosis of these objects are computer experiments with their mathematical models.

Some problems of mathematical modeling of these phenomena are facilitated, when one deals with long-term weather (months) and climate (years, decades) changes. The average values in rather long time interval in depend to a smaller degree on small-scale fluctuations and instabilities. The climatic models of the "atmosphere–ocean" system include a series of interconnected units: three-dimensional non-stationary equations of motion of compressible and incompressible fluids taking into account the viscosities and thermal conduction (of Navier-Stokes type equations), the equation of radiative transfer in the atmosphere, etc. They are solved jointly taking into account initial conditions of the system (including the condition of an underlying layer) and known dynamics of solar radiation and other external phenomena. Their discrete analogs generally represent difference schemes. In so far as the computing algorithms for the solution of these problems should be economic and possess good resolution (accuracy), in the construction of difference schemes the approaches described (e.g. in Chapter V) are essentially used.

The climatic computing experiments can be conditionally divided into two types. The first includes those required to certify the adequacy of models via comparison of their outcomes with currently available reliable data based on real observations. The second type of experiments are directed to the prognosis of long-term climatic changes caused by natural or artificial phenomena.

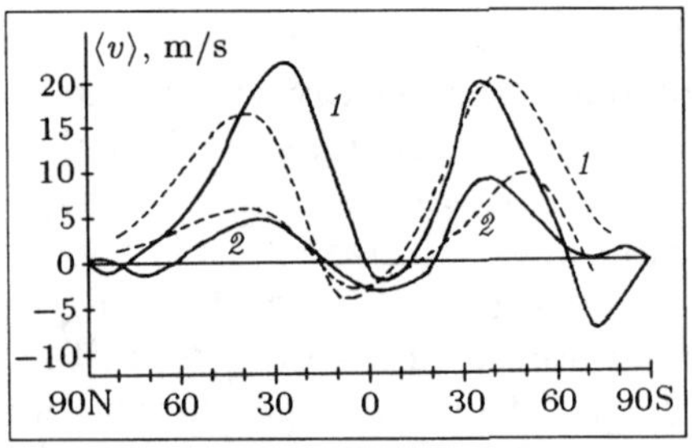

Fig.108.

Fig. 108 represents the comparison of "gauge" computing experiments on one of the most complete models of climate with observational data. The measured quantity $\langle v \rangle$ is the average for one year velocity of wind (in m/sec) for any possible latitudes on height adequate to pressures 400 mbar and 800 mbar (curves *1* and *2*). Dashed lines indicate the results of the observations. The calculations were performed as follows: the external actions typical for the given season were given and the calculations were conducted up to the obtained quasistationary state corresponding to the selected month, then the quantities were averaged in time and in space. The correspondence of results of computing experiments and observations appeared to be quite satisfactory for such a complex object, especially in the equatorial zone (trade-winds).

The established adequacy of the climatic model enables us to perform prognostic experiments connected, for example, with the "greenhouse effect". Large amount of CO_2 (carbon dioxide) of anthropogeneous origin is ejected into the atmosphere and prevents the escape of the Earth's own radiation into space, leading to the increase of the mean temperature of the atmosphere. This in its turn can lead to intense ice melting, raising the level of the oceans, and to other negative global consequences. In Table 4 some data of mathematical modeling of the "atmosphere—ocean" system are represented for various concentrations of CO_2 in the air. The first column includes the present concentration, the second and third include double and quadruple concentrations of CO_2, respectively.

Table 4

	1	2	3
Mean temperature of atmosphere, °C	-19.2	-17.54	-17.0
Temperature of the air at the underlying layer, °C	13.9	15.3	15.8
Temperature of the underlying layer, °C	12.5	13.6	14.3
Flux of short wavelength radiation on the underlying layer, Wt/m^2	255.3	254.3	254.8
Precipitations, mm/day	2.04	2.15	2.17

The computing experiment for these two hypothetical situations was carried out up to the coming to a new, different from the present quasi-stationarity in the system with consequent averaging of the results. It is seen that the mean temperature both of the air, and the underlying layer is noticeably increased (a variation of global temperature of 1–2°C is considered significant), the average of precipitations grows as well. The experiment also demonstrates the nonlinearity of the object. The increase in mean temperature at the transition from a double to a quadruple concentration of CO_2 is much less than for a doubling of the present concentration. The response of the "atmosphere–ocean" system to external (in this case anthropogeneous) action is not proportional to the magnitude of that action (in the considered situation the system softens the consequences of increasing human industrial activity).

The mathematical modeling allows us to evaluate the results not only of smooth, but also of sharp external interferences in the system. One of them can be a nuclear conflict between the struggling powers. The absolute unacceptability of a global nuclear war for civilization was recognized a long time ago. However the possibility of limited interchange of impacts ("attack on cities") using a small part of weapons has been considered. What climatic consequences can limited nuclear war have?

The experience of intense bombardments of large cities during the Second World War testifies to the inevitability of huge fires. Their intensity is such, that not only inflammable materials (trees, plastic) burn down, but also materials which are non-combustible in usual conditions – asphalt, concrete, brick. As distinct from the relatively pure combustion of forests, the powerful urban fires will be accompanied by the ejection into the atmosphere of a huge amount of soot – in some evaluations approximately 1 ton of soot per 1 ton of trotil equivalent fuel. This means, that a nuclear attack on cities with total power of 100 megatons (approximately 1% of the total reserves of nuclear powers) will lead to an immediate release into the atmosphere of 10^8 tons of soot.

Such a degree of "smoke" will reduce the solar light flux onto the underlying surface by tens of times. The computing experiments imitated just this scenario: in the models, corresponding characteristics of the atmosphere above the most probable regions of a conflict were instantly varied, and the temporal dynamics of climatic parameters were traced.

The main effect is fast and extremely strong cooling of the air above the continents: even in the case where only 1% of available weapons were used, the average temperature of the underlying surface will fall by 15°C in a week. The average temperature of higher atmospheric layers, on the contrary, will be increased by approximately the same magnitude (since they absorb all the solar radiation). The formed temperature inversion is extremely stable ("cold" – below, "warm" – above) and will be preserved for many months.

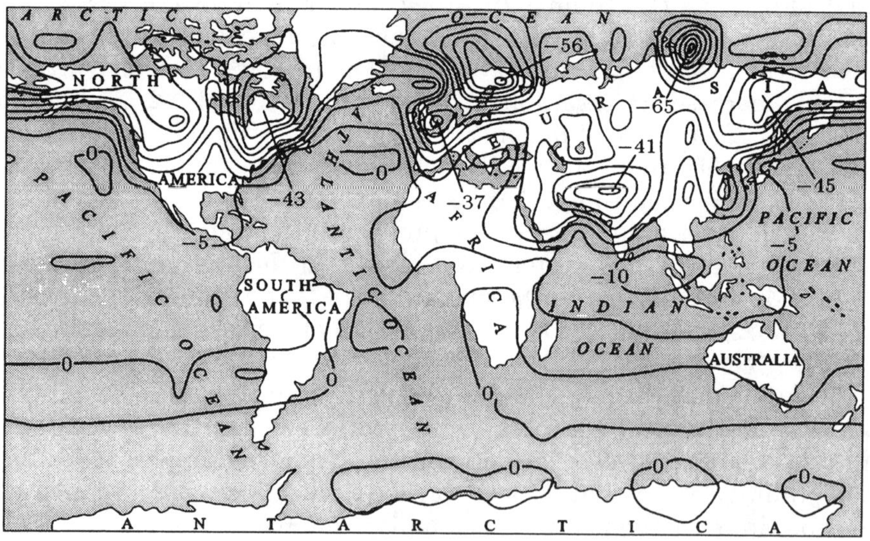

Fig.109.

This scenario is nothing other than a global climatic catastrophe. Thus the average climatic parameters do not give about its complete image. In Fig. 109 we show the isoclines of the air temperature at the Earth's surface at 30–40 days after a "100-megaton conflict". The temperature will fall below the norm on 56°C in northern Europe, on 65°C in northern Siberia, on 43°C in Northern America and on 41 °C in the southern Asia, etc. At the altitude of mountain glaciers the temperature will become much higher the normal, leading to intense ice melting. The huge masses of water flooding the supercool plains will cover them with ice. The ocean, because of large thermal capacity, will cool down much more slowly, and the contrast of temperatures between the water and land will generate hurricanes of unprecedented force in extensive coastal areas.

The catastrophe is really global, since the smoke released in a certain region will spread over the whole planet and will cause a "nuclear winter" at any point, not excluding also the point from which the sudden one-sided impact was performed.

Previously it was believed that the basic consequences of a nuclear weapon are the penetrating radiation and shock waves. Mathematical modeling convincingly testifies that apart from these relatively local consequences, the nuclear conflict will be accompanied by disastrous global change of climate, and hence, it is unacceptable even on a limited scale.

4. Magnetohydrodynamic "dynamo" of the Sun. Twenty two year cycles of solar activity have an important impact on terrestrial life. The noticeable increase in amplitude of a magnetic field generated by the Sun, influences on a number of diseases, conditions of agricultural production, operation of wireless and telecommunications.

The qualitative explanation and quantitative description of the solar "magnetic sinusoid" is one of the fundamental scientific problems containing, in addition, paradoxical riddles. For example, as testified by the measurements of annual increase of a tree's mass (the thickness of the rings on cross sections through a tree's trunk), several centuries ago the oscillations of the solar magnetic field had ceased over 50 years.

Various explanations for the nature of this phenomenon were made, including the influence of Jupiter having a period of rotation around the Sun equal to 11.9 years. In the mid 1950s, the first concept claiming scientific reliability – the theory of a solar MHD-dynamo – was formulated.

This term implies a complex sequence of processes causing periodic generation of magnetic fields of the Sun. They include the convective and turbulent hydrodynamic motions of solar plasma on its surfaces and in internal layers. The moving charged particles (currents) create a magnetic field, transforming into it a part of their kinetic energy. In its turn the magnetic field influences the motion of electrons and ions and hence, results in a system of periodic processes.

The validity of the MHD-dynamo theory can only be checked by computing experiments with mathematical models of the generation of solar magnetic fields and by comparing (where possible) their results with observational data. The basic equations of these models are deduced for corresponding assumptions from the system of Maxwell equations and have the form

$$\frac{\partial \vec{B}}{\partial t} = \mathrm{rot}\,[\vec{B} \times [\vec{r} \times \vec{\omega}]] + \mathrm{rot}\,\alpha\vec{B} - \mathrm{rot}\,\beta\,\mathrm{rot}\,\frac{\vec{B}}{\mu}$$

inside of Solar surface and the form

$$\mathrm{div}\,\vec{B} = 0$$

outside it.

Here $\vec{B} = \vec{B}(\vec{r},\, t)$ is the sought magnetic induction vector, $\vec{r}$ and t are the radius vector and time, respectively, $\vec{\omega} = \vec{\omega}(\vec{r})$ is the given angular-velocity vector depending on spatial coordinates (since the Sun, not being a rigid body, does not rotate as a unit), $\alpha = \alpha(\vec{r})$, $\beta = \beta(\vec{r})$, $\mu = \mu(\vec{r})$ is the given average characteristics of turbulent convective motion, the turbulent conductivity and magnetic permeability of plasma. Note that the input data $\vec{\omega}$, α, β, μ is not precisely known, and hence the adequacy of the results of modeling should be checked with particular care.

The solution of the problem is somewhat facilitated since the basic 22-year component of the large-scale varying magnetic field of the Sun is axisymmetrical (does not depend on an azimuthal meridional angle). Hence, instead of the initial three-dimensional problem, one can confine oneself to the analysis two-dimensional problem.

However, even with such noticeable simplification the model still remains rather complicated. Varying the input data, we have to solve a numerically large number of non-stationary problems for a system of two quasilinear parabolic equations at $r < R$, where R is the solar radius (internal problems), and as many problems for an elliptical equation at $r > R$, where $\beta = \infty$ (external problems). Obviously, their solutions should satisfy some interface conditions at $r = R$. The existence of two zones in area $r < R$ further complicates the solution of internal problems: the radiation core $0 < r < R_0 < R$ (where the quantity β is small, but cannot be neglected) and a convective zone $R_0 < r < R$, on its boundary with the core strongly varies not only β, but also the function μ.

The discrete analog of the considered model is based on a difference approximation of the differential equations for an internal problem and integral equation for an external problem (in area $r > R$ it is more convenient to solve numerically the integral equation equivalent to the initial elliptic one). For the numerical integration of the system of parabolic equations the implicit absolutely stable difference schemes of variable directions being solved by various variants of sweep method have been applied. Knowing the values of grid functions on the boundary $r = R$, it is rather easy, using the interface conditions to find the solution of a difference integral equation at $r > R$, to completely solve the problem. In the construction of discrete approximations the possibility of sharp spatial variations of parameters of equations (see subsection 3, section 4, Chapter V) was taken into account. The property of symmetry of the numerical solution (even or odd) relative to the solar equator, intrinsic to the solution of an initial problem at symmetrical initial data, was also taken into account.

The computing experiments have shown that the qualitative and in many respects the quantitative character of the process depends on the value of the basic dimensionless parameter of the problem – $D = R^3\alpha_0\omega_0/\beta_0^2$, where α_0, ω_0, β_0 are some characteristic values of functions α, ω, β. At $D \approx D_{\rm cr}$ the oscillations have a constant period and amplitude. At $D > D_{\rm cr}$ ($D < D_{\rm cr}$) the oscillations have a slightly different period and are growing (damping).

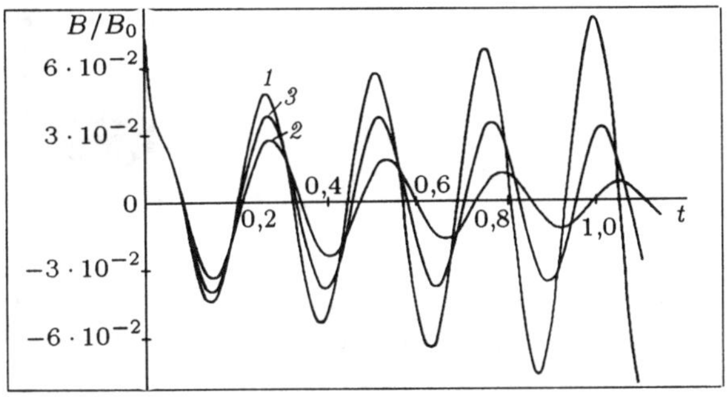

Fig.110.

The elementary interpretation of this result follows from the analysis of the structure of number D. Let, for example, the quantities R, α_0, β_0 be constant, and let ω_0 vary. At small ω_0 (the number D is small as well) the energy of solar rotation is not sufficient to maintain the periodic process and, starting, it damps in time. For a large value of ω_0 the solar "dynamo" works too intensely, generating a field, growing by amplitude. Finally, if the combination of characteristic parameters implies the critical value $D_{\rm cr}$, the oscillations have an observable regular character.

In Fig. 110 we represent (in dimensionless units) the results of computing experiments for the described cases (the curve *1* indicates the case $D = 2.75 \cdot 10^4 > D_{\rm cr}$, the curve *2* – $D = 2.3 \cdot 10^4 < D_{\rm cr}$, the curve *3* corresponds to $D = D_{\rm cr} = 2.56 \cdot 10^4$ and to regular oscillations with a period $P = 0.248$). The values of the period and amplitude of oscillations in steady-state do not depend on the adopted initial distributions, and with good quantitative accuracy are in agreement with the observational data.

This conclusion is not connected with the "tuning" of uncertain functions $\vec{\omega}$, α, β, μ for deriving from the model the previously known result. This is confirmed by comparison of more thin effects accompanying the solar activity. In Fig. 111 the time butterfly diagrams are presented, i.e. the con-

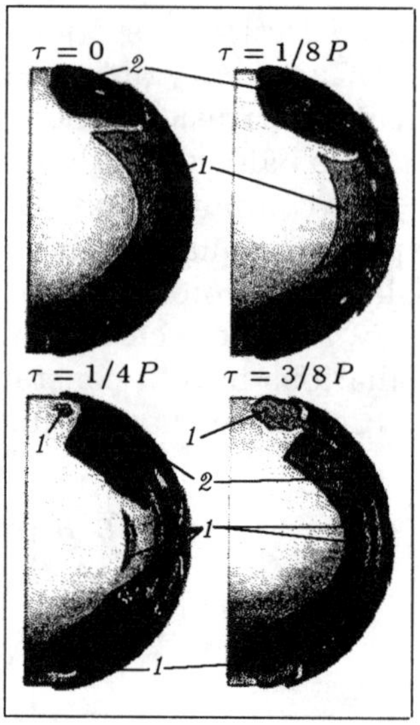

Fig.111.

figuration of curves of equal level of the magnetic field on the solar surface (calculated results for a half period). The zones with high magnitude magnetic field corresponding to solar "spots" (number *2*) in time shift from polar areas to the equator, and their place is occupied by zones with weaker magnetic field (number *1*) and vice versa. The process is repeated periodically, in good agreement with the actual picture. One more way of confirming the adequacy of the mathematical solar "dynamo" model is moving to the analysis of three-dimensional processes. They describe dynamics of non-axisymmetrical components of the solar magnetic field, coinciding in many features with the actual picture.

The mathematical modeling shows, that the Sun is a nonlinear MHD-generator with unexpected operational modes (the pauses in its activity are just being linked with nonlinearity). Therefore knowledge of this is important not only for fundamental science, but also for practical purposes.

3 Computing Experiment with Models of Hardly Formalizable Objects

We will now demonstrate the universality of mathematical modeling in examples studying hardly formalizable objects, for which there are no precisely formulated laws. We show that its application gives various possibilities for more deep understanding of their basic properties.

1. Dissipative biological structures. In the "predator-victim" type biological models considered in subsection 1, section 3, Chapter IV, and the possibility of the spatial inhomogeneous distribution of the population, are ignored. Such models serve only as a first approximation to a reality. In reality the living conditions of a population are never identical in various areas. Besides, even for a spatially homogeneous medium purely biological reasons of clustering or rarefaction of representatives of the population are always crucial: instinctive behavioral motives for gatherings in herds, seasonal variations in nature (for example, the approach of the mating season or growth of nestlings) and so on.

Therefore a more detailed mathematical description of populations should take into account the spatial phenomena. One such typical biological model is

$$\frac{dN}{dt} = (\alpha - cM)\, N + D_N \frac{\partial^2 N}{\partial x^2}$$

$$\frac{dM}{dt} = (-\beta + \gamma M)\, M + D_M \frac{\partial^2 M}{\partial x^2}\,,$$

where t is the time, x is the spatial coordinate (for a simplicity the process is considered to be one-dimensional), $N(x,\, t)$ and $M(x,\, t)$ are the "densities" of victims and predators, respectively, $\alpha > 0$, $c > 0$, $\beta > 0$, $\gamma > 0$, $D_N > 0$, $D_M > 0$ are constants describing the intrinsic properties of the populations.

The given model differs from the Lotki-Volterra equations by the presence in its right hand side of "diffusion" terms (D_N, D_M are coefficients of "diffusion") and represents a system of two equations of parabolic type relative to variables N and M. The origin these of "dissipative" terms is justified by the same assumptions as those made in subsection 1, section 1, Chapter IV at deducing the model of dynamics of clusters of amoebae: the velocity of variation of the number of population is influenced by the "chaotic" motion of amoeba in space forming a flux from more "populated" to less "populated" areas (it is considered to be proportional to the gradient of their densities).

The behavior of the population in spatial models can drastically differ from the picture described by point models. Consider, for example, the effect of the appearance in space of "waves of pursuit" of the predator after the victim. In Fig. 112 the results of demonstration computing experiment of the described model are shown (units of measurement are conditional). The problem was considered in an infinite area (Cauchy problem), the initial density of the victims decreases exponentially with the growth of the coordinate x. It was considered that no migration of victims occurs, i.e. $D_N = 0$ – this is the case in frequently occurring actual situations, when the mobility of victims is essentially less than the mobility of predators.

On the "front" of waves of pursuit a peak of density both of predators

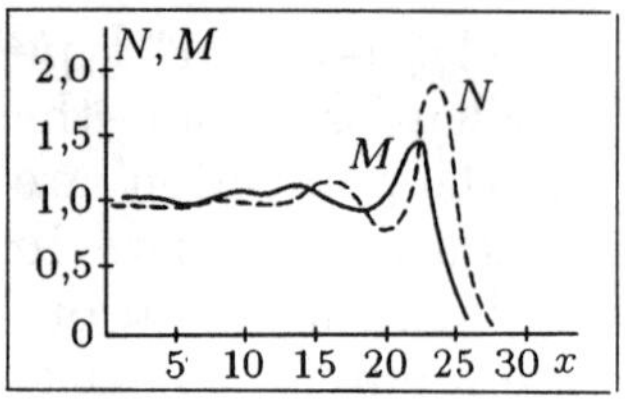

Fig.112.

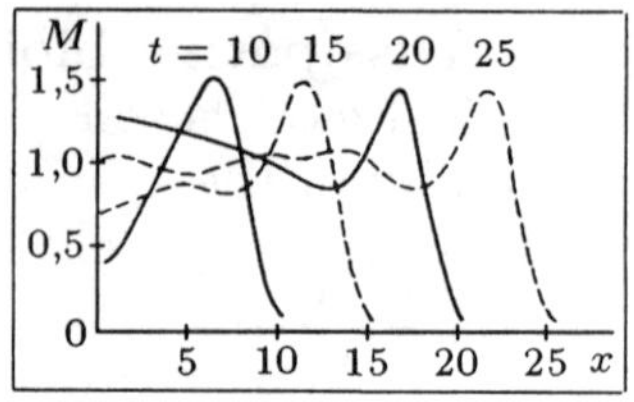

Fig.113.

and victims is formed (see Fig. 112, where the profiles of functions $M(x, t)$ and $N(x, t)$ in some instant are represented). Behind the wavefront a quasi-equilibrium is established in the system and the values of variables are close to constants. Temporal dynamics of formation of the profile of density of predators (Fig. 113) testifies the proximity of the process within the considered time interval to a self-similar case. The picture practically without modifications is reproduced in different instants in various areas of space, and the velocity of transition of characteristic points of the structure poorly depends on time (amplitude of the maxima somewhat decreases in time, in so far as the wave is moving along the decreasing "background" of density of victims). Such development corresponds to a similar traveling wave (see subsection 2, section 1, Chapter V), when all quantities depend on the combination $\xi = x - Dt$, $D > 0$ (for some particular values of parameters the self-similar solution can be obtained analytically).

The study of distributed biological systems illustrates well other relations between the models of various hierarchical levels – point and spatial. Consider the frequently used modification of the above mentioned equations

$$\frac{dN}{dt} = \alpha N^2 \frac{N_0 - N}{N_0} - cM N + D_N \frac{\partial^2 N}{\partial x^2},$$

$$\frac{dM}{dt} = (-\beta + \gamma N)\, M + D_M \frac{\partial^2 M}{\partial x^2},$$

where $N_0 > 0$. It differs from the classical Lotki-Volterra model by a form of terms describing the dynamics of a victim in the absence of a predator:

1) for small population densities (for types being multiplied by sexual reproduction) the growth rate of the number is proportional to the frequency of contacts between members, i.e. to the square of its density (compare with section 6, Chapter I);

2) a stable equilibrium density of population of victims $N = N_0$ exists, defined by the level of available resources (compare with the logistic model in subsection 5, section 1, Chapter I).

The point ($D_N = D_M = 0$) analog of the given model represents a nonlinear system of two ordinary differential equations. In an appropriate range of parameters it has a limiting cycle – a configuration of integral curves in a phase plane N, M, when at $t \to \infty$ the trajectories are "winded" over a limiting closed curve (auto-oscillation, the process is qualitatively similar to that described in subsection 1, section 3, Chapter IV).

The obtained from computing experiment, carried out for the same range of parameters, behavior of number of populations essentially depends on the initial conditions of the system. The experiment corresponds to the conditions of a ring area (the vicinity of coast lines of the closed reservoirs, levels of constant height in mountainous areas, etc.), and hence, the boundary conditions for functions $N(x,\, t)$, $M(x,\, t)$ are periodic by x. It was also considered that the mobility of the predators exceeds that of the victims ($D_N \ll D_M$).

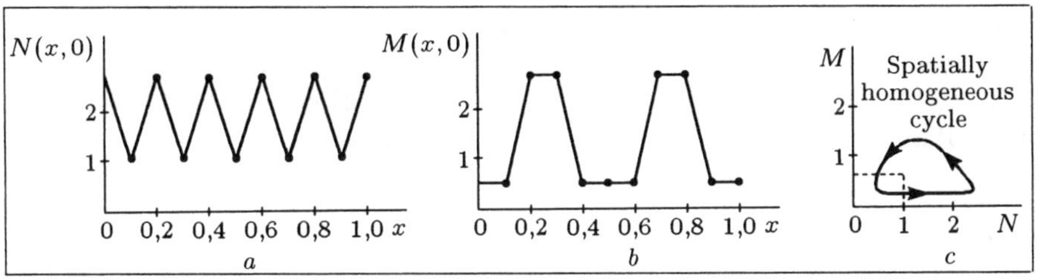

Fig.114.

Fig. 114 in dimensionless units represents the forms of several initial distributions of the victim (a) and the predator (b) leading to the establishing by time of a spatially homogeneous auto-oscillating process corresponding to a stable limiting cycle in the point model (c). The biological interpretation of this result is as follows: in a homogeneous area ("forest") of small size ("hares" and "wolves" have enough time to run over it many times during their life time) the populations can interact only in a manner which is auto-oscillating, varying in with a shift of phases manner.

If at other equal conditions "the hares" live permanently in the same places, and "the wolves" are actively migrating over the forest searching for food, then apart from homogeneous auto-oscillations, spatially inhomogeneous (but stationary in time) population distributions are possible.

Fig. 115 a–c shows some of the initial profiles of number of population of victims adequate to this condition, not described by a point model, (Fig. 116). There the number of "wolves" is uniform over all the forest, and the "hares" essentially dominate one edge of the forest – being approximately as many, as at the peak of homogeneous auto-oscillations, while on the other edge there are fewer of them.

Despite identical living conditions over the area, there appears a so-

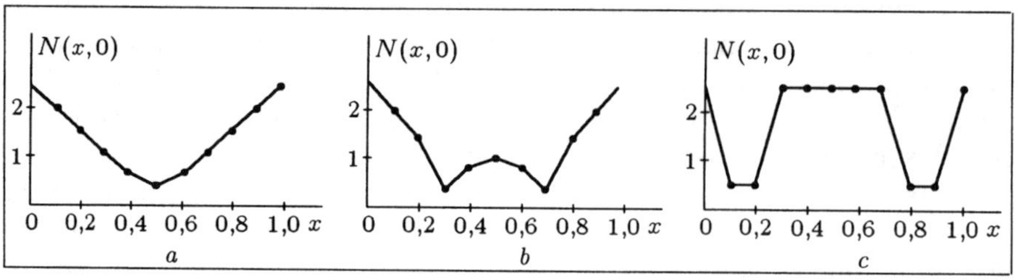

Fig.115.

called stationary dissipative structure (SDS). Note that, as distinct from the structures of combustion (section 3, Chapter V), the origin of SDS is stipulated not by blowing-up and localization of diffusion processes, but by competition of sources and losses of energy (in terms of the theory of heat transfer). The inhomogeneities of SDS correspond to really observed "spots" of populations in homogeneous territories with competing biological forms.

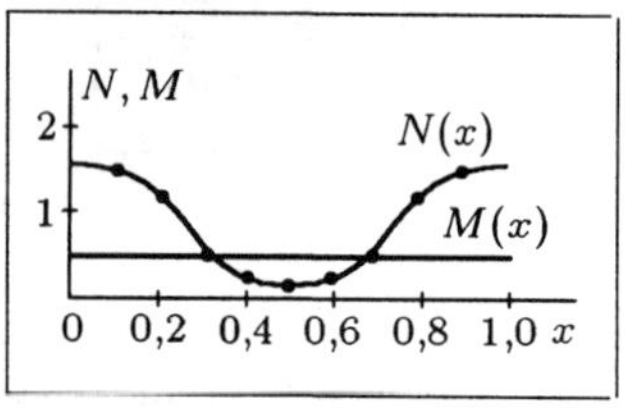

Fig.116.

2. Processes in transition economy. The transformation of an economic system with prevalence of state property and centralized planning to a competitive market economy with dominance of private property (or vice versa) is accompanied by complex transition processes. It is not surprising, as although both these systems do not exist in "pure" form, the difference between them has a fundamental character: regulation with the help of financial tools, reacting to the variations of economic parameters, or by means of orders following from the analysis of appeared deficits; the aim to maximize the profit of the firm or to execute the plan; almost complete self-sufficiency of working individuals or the guarantee, as far as possible, of minimum of state social security for all the population, etc.

The mathematical models of a market economy have been developed for a long time and have been investigated rather well, which is not the case for the models of planned and, especially, transition economies. The latter cannot be (even in principal sense) reduced to models of a classical type, for example, to those considered in section 2, Chapter IV, as they should reflect in themselves the basic features of both economic systems. The effective methodological approach to the construction of models possessing this synthetic property, is that, first, the models of balance of material and financial fluxes are being created, which in a certain sense are universal, i.e. are suitable for describing any type of economy. They are "deliberately" unclosed, and the mode of their closure depends directly on the behavior of the economic agents, policy of the state, etc. For various given forms of economic relations (scenarios) and hence, for different ways of closure, models for different types of economics are obtained.

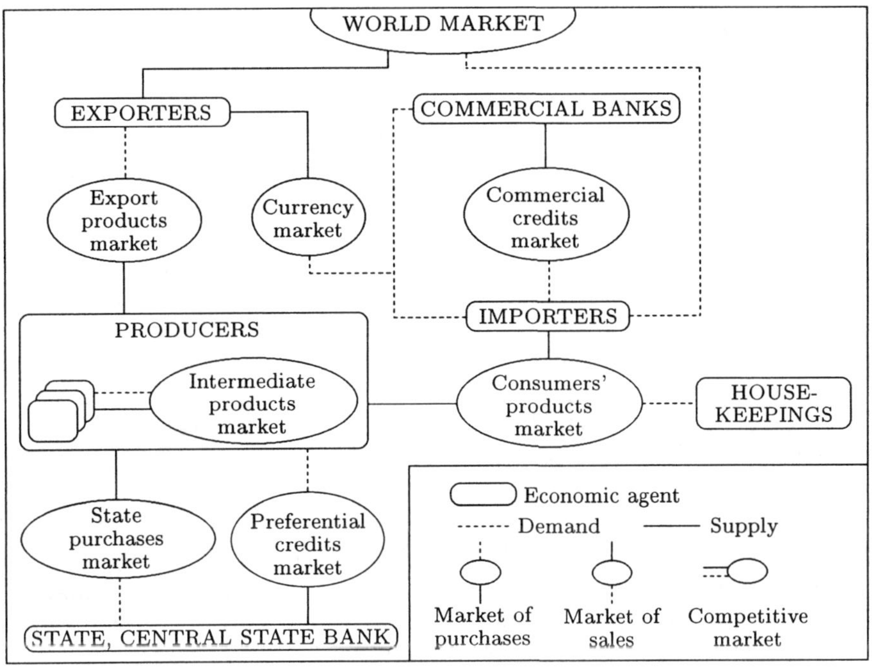

Fig.117.

The scheme of one of them is shown in Fig. 117. It represents on the macrolevel the rather complicated mutual relations of the economic partners governing the production, exchange and distribution of goods and services, in the Russian economy in the early 1990s. It is visible that the model represents a mixed transition economy: apart from the state (basic agent of the planned system), there are, for example, business banks working competitively with the purpose of extracting a profit.

Not describing completely all the assumptions on the economic relations included in the model, we will only discuss some of them:

1) subsystems are distinguished, experiencing the competitiveness of im-

port, and the export branches;

2) labor collectives and administration are interested in increasing salaries and, despite reduction of demand, achieve it with the help of mutual non-payment and soft loans from the Central Bank (CB); pure investments are absent, the productive capacities are decreasing;

3) variation of the production conditions influences the salary, but not the employment level; no bankruptcy of firms occurs, the nominal unemployment is insignificant;

4) only raw materials are exported, and only the goods are imported;

5) markets are controlled by an industrial – financial oligarchy, with exporters being at the top of it;

6) the policy of the state is reduced to the definition of rates of taxes, volumes the soft loans from the CB, state purchases, payments to the population from the budget and grants to companies, etc.

The formulated scenario is developed to a general model, so that a concrete model for the transition phase is obtained. From a mathematical viewpoint it represents a comprehensive and complex system of nonlinear ordinary differential equations (supplemented by a large number of algebraic equations) relative to several tens of basic economic variables (for example, production of various types of goods) and contain a lot of characteristics and parameters defining the solution (for example, inflationary expectations of the population). This input data is obtained and updated along with the scenario, in accordance with the current condition of the system.

For example, in one of the variants of the model it was considered that CB will not perform operations in an internal currency market, then in late 1993 the exchange rate would equal, according to the model, 4 000 roubles to the dollar. However from mid 1993 CB began corresponding operations, and in reality the rate reached "only" 1 300 roubles to the dollar. The modifications taking into account this new policy were introduced into the model, so that the time schedule predicted by it, appears to be in reasonably good accordance with reality (see Table 5).

Computing experiments, both with this and with other models of transforming economy constructed in similar way, enabled us to draw a series of rather common and important conclusions. In particular, it was established that the transition from the disorganization of the Soviet economy in the late 1980s and early 1990s to the effective equilibrium conditions of a new market economic system even at best will take not less than ten years, and will be accompanied by high structural unemployment and bankruptcy of many companies.

Another no less significant result of experiments with models is the prediction of the getting of post-reconstruction Russian economy into a special type of quasistationary condition, different from the investigated in classical political economical models. It is rather ineffective: in this state for the economic agents it is senseless to keep the resources or to invest them in industry, but instead it is rather favorable to save mutual non-payments and other delays in circulation of finances. The soft loans of CB at correct and precisely addressed dosage slightly improve the efficiency of such an equilibrium, but cannot drastically change the common picture (for partial variants of the model the existence of such equilibrium is established, as in section 2, Chapter IV, by relatively simple analytical methods and is described by simple steady-state solutions). Both described conclusions are in agreement with the macroeconomical situation in Russia in recent years.

With the help of models we can also carry out more detailed studies of various concrete problems of current economic policy. The latter concerns the natural requirement of "safety", particularly to avoid sharp destruction of although non-efficient but usual and already existing economic relations and structures. It is by no means a secondary problem, since the point is not someone's aim to deliberately destroy, but the "unprofessional" use of economic tools in a rather complex and unstable situation.

The typical problem is the determination of the size of soft loans providing by the state to the producers in fact with negative interest. The model has shown that extremes are rather dangerous. The lack of soft loans leads not only to the rapid suppression of inflation (and even to deflation), but also to the destruction of industrial structures, the majority already being used to inflation. Their incomes are so reduced, that this results in an "outflow" of workers from firms and an inevitable collapse of production. In the opposite case of soft loans which are too large and resulting hyperinflation, the system of business banks collapses. They estimate the profit, proceeding from rates of inflation. While the growth is not too big, their operations based even on rough prognosis, ensure steady profit. At hyperinflation the inevitable inaccuracy of the prognosis leads to systematic losses of banks and to the actual "disappearance" (in a relative sense) of their own capital.

We shall also mention two other actual events which were significant for the Russian economy and which have been analyzed with the help of computing experiments with a model. The first of them – "black Tuesday" of October 11, 1994 – involved a disastrous fall of the rate of the rouble relative to the dollar, which in a few days returned to approximately the former level. The adequacy of the model enabled us not only (post factum) to describe the dynamics of basic economic macroindices after the Tuesday, but also to reliably determine the economic agents, which had involuntarily won (basic branches, the incomes of the federal budget) and lost (majority of population, importers) as a result of that event.

Second – the war in Chechnya started in late 1994, and required significant additional taxpayer expenses to carry it out and to recover the economic and social life of the republic (according to different evaluations – from several trillions up to tens of trillions of roubles). The basic conclusion from the results of modeling is: although "the Chechen crisis" cannot cause hyperinflation, but, even at solid anti-inflationary policy of the state, it made a noticeable contribution to inflation and supports the decrease of actual income of the majority of the population.

Table 5

Date	Exchange rate	Calculated rate	Date	Exchange rate	Calculated rate
03.05	1820	1779	05.07	1998	1984
05	1854	1814	07	2011	2009
10	1859	1809	12	2020	2018
12	1869	1837	14	2022	2029
17	1877	1827	19	2028	2038
19	1881	1858	26	2052	2065
24	1895	1852	28	2052	2076
26	1901	1868	02.08	2060	2087
31	1916	1881	04	2081	2104
02.06	1918	1896	09	2087	2104
07	1940	1891	II	2108	2128
09	1952	1915	16	2117	2137
14	1952	1927	18	2141	2148
16	1959	1927	23	2161	2157
21	1971	1947	25	2156	2171
23	1977	1957	30	2153	2183
28	1985	1969	01.09	2204	2193
30	1989	1983			

3. Totalitarian and anarchic evolution of power distribution in hierarchies. The results of the analysis of the mathematical model of the "power–society" system obtained in section 4, Chapter IV, concerning mainly stationary distributions of power between the levels of the hierarchy in the conditions of a legal society. Despite the phenomenon the hierarchy leaving the framework of power, appearing at mismatch of characteristics of the system, in legal case dynamic stability does exist. This means the following: any non-stationary power distribution transforms at some time into a stationary one and hence, returns to the legal field (it is considered that stationary state itself belongs to a legal area).

The computing experiment clearly demonstrates this property of a legal system (in some sense it is incorporated in its definition). In Fig. 118 an example of such a return of the authorities to a legal field is shown, for the case of a basic model (subsection 4, section 4, Chapter IV) with parameters $\kappa_0 = 5 \cdot 10^{-3}$, $b = 0.9$, $\alpha = 0.75$, $k_1 = H = l = 1$, $t_0 = 0$. The initial data $p_0(x)$ are taken as "stairs": at $0 \leq x \leq 0.3$ we have $p_0(x) = 2 > p_2$, at $x > 0.3$ their values correspond to a stationary solution. The function $p(x, t)$ for moment t_1, t_2, t_3, t_4 is indicated by solid lines, *1, 2, 3, 4* respectively,

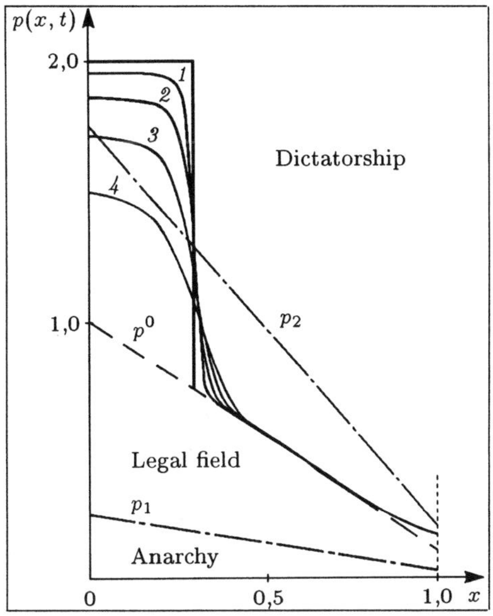

Fig.118. $t_1 = 0.05$, $t_2 = 0.15$, $t_3 = 0.35$, $t_4 = 0.75$

This scenario implies that the hierarchy (in this case its higher levels) had for some reasons noticeably exceeded its maximum imperious powers. However the response of the society has ensured the return of the solution to the legal framework and relaxation over a period of time to stationary power distribution (recall that for the sake of simplicity it is considered that the response of the hierarchy is equal to zero, i.e. "the bureaucrats" are indifferent to the level of their power, and they are quietly following the response of the society). The time of "relaxation" in good accuracy is equal to $t_{\text{set}} \sim 1/k_1$, i.e. at constant rest parameters of the system, the distribution of power reaches the legal area faster, the greater the intensity of the response of the society k_1.

If the partners in a system "power–society" aim to overcome the initial situation as soon as possible, the social response (elections, inquiries, etc.) have to be essentially strengthened and taken into account. This conclusion completely correlates with standard politological recipes.

From computing experiments (in combination with theoretical analysis) performed for a general model of a legal system follows its stability. In other words, in the legal case the hierarchical structures and civil society are in the condition of a stable dynamic equilibrium.

The situation at certain, negligible at first glance deviations of the public consciousness in relation to the legal one is completely different. In these cases it is impossible to speak even approximately about the existence of any dynamic equilibrium of the system "power-society". Below we consider a series of scenarios illustrating such an evolution of power distribution, obtained with the help of computing experiments.

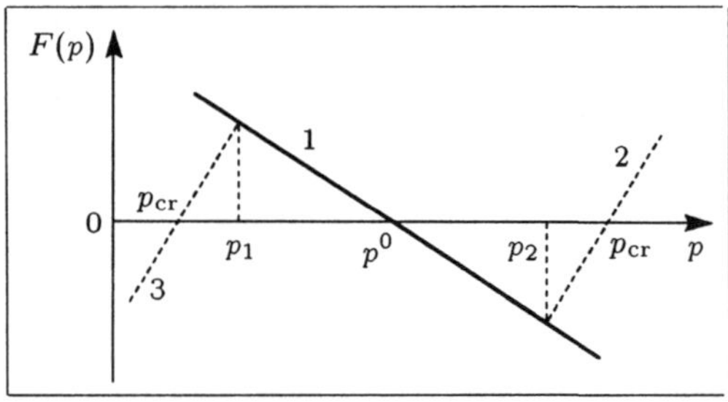

Fig.119.

I. *A totalitarian trap.* In relation to a legal system, the public consciousness (the response of society $F(p, x)$) is distorted as represented in Fig. 119 (the curve of $F(p, x)$ consists of a segment 1 from $p = 0$ up to $p = p_2$ and line 2). This means that at $p < p_2$ the reaction of society is legal, at $p_2 < p < p_{cr} = 2.5 \cdot p^0$ the response is "weakly legal", i.e. the resistance to the excess of power does exist, but decreases with growth of p. Finally, at $p > p_{cr}$ the society "requires" a realization of increasing power, which becomes stronger, as the value of p increases.

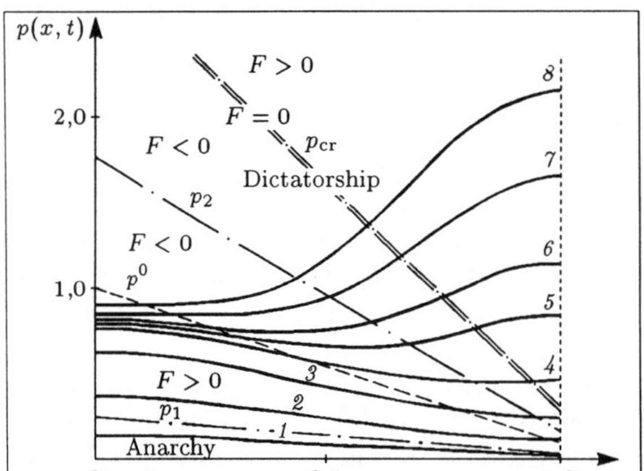

Fig.120. $t_1 = 0.155$, $t_2 = 0.635$, $t_3 = 1.915$, $t_4 = 6.395$, $t_5 = 15.995$, $t_6 = 17.915$, $t_7 = 19.195$, $t_8 = 19.835$

Fig. 120 represents the results of calculations of the basic model (but with $F(p)$ from Fig. 119) at $\kappa_0 = 7.5 \cdot 10^{-2}$ (the remaining parameters are the same, as in Fig. 118). The initial level of power in any link of the hierarchy is equal to zero ($p_0(x) = 0,\ 0 \le x \le 1$). We will now describe the stages of development of the scenario:

a) so far as at $t_0 = 0\quad p_0(x)) \equiv 0$, in an initial instant the reaction of society ensures the growth of power and its return to area $p_1 < p < p_2$ (curves *1*, *2*), so that the majority of hierarchical links is in area $p_1 < p < p^0$ (curve *3*) and as before receives "portions" of power from society. The solution grows everywhere, since the intensely working mechanism of transmitting power from higher to lower links ensures its magnification in the neighborhood of a point $x = 1$, where the response of society is already negative (for a part of the curve *3*, located in area $p^0 < p < p_{\text{cr}}$);

b) the situation represented by a curve *4*, is a key one for the understanding of the given scenario. Due to the "flatness" of the profile of power (the parameter κ_0 is big enough) a small number of the lowest levels appear in the area, where $p > p_{\text{cr}}$ and where they begin to receive "additional" power mainly from the society ($F(p) > 0$ at $p > p_{\text{cr}}$), and not from higher neighbors;

c) such replenishment leads to a slight raise of the power of the lowest instances in relation to a level of power of the nearest higher links (which are still in area $F(p) < 0,\ p^0 < p < p_{\text{cr}}$). In accordance with the postulate (subsection 2, section 4, Chapter IV) the higher links respond to this raise, "taking" a part of the power from the lower ones and hence, coming due to this "source" of power to the area $p > p_{\text{cr}},\ F(p) > 0$ (curves *5*, *6*);

d) "the wave" of excess power and exit to the area $p > p_{\text{cr}}$ is spread over the hierarchy from right to left (from below to above), and the function $p(x,\ t)$ irreversibly increases as much as possible (curves *7*, *8*, etc.) at all $0 \le x \le 1$ (this is the actual sense of the term "totalitarian").

The described scenario is rather non-trivial (even from a purely mathematical point of view). Indeed, the hierarchy was completely in the area of small (zero) values $p(x,\ t)$, but completely moves in time to the area of infinite values of $p(x,\ t)$, "overcoming" the peculiar "resistance band" defined by the legislation and the response of civil society (area $p^0 < p < p_{\text{cr}}$). Two phenomena are acting – distortion of the public consciousness and intense mechanisms of redistribution of power in a hierarchy. Just the latter have pushed the whole hierarchy link by link into the area $p > p_{\text{cr}}$, where initially only a minor part of lowest instances had entered (trap). When the quantity κ_0 diminishes the situation is normalized – the solution becomes stationary.

II. *An anarchic trap.* In the previous scenario (as well as in the model generally) no conscious aims of hierarchy towards the "dictatorship" have been included. At opposite distortion of public consciousness an inverse scenario is realized. The function $F(p)$ is represented in Fig. 119 via the part of a line *1* at $p > p_1$ and line *3*. At $p > p_1 = 0.85\,p^0$ the reaction is the same as in the legal case, at $p_{\text{cr}} = 0.7\,p^0 < p < p_1$ the response is "weakly legal", i.e. is positive, but decreases with the decrease of p. Finally, at $p < p_{cr}$ the response of society is directed to a decrease of power, which is greater, the larger is the decrease.

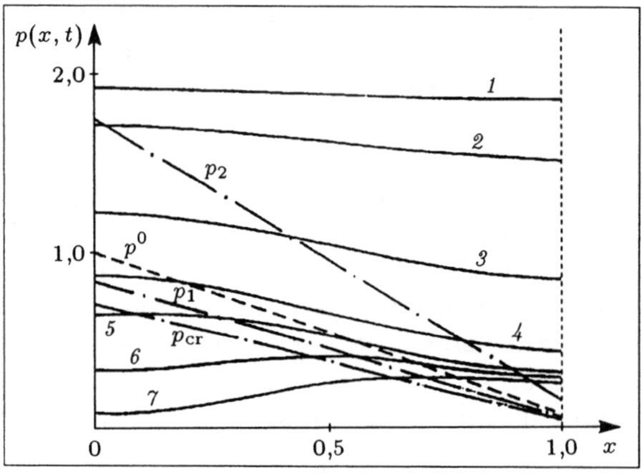

Fig.121. $t_1 = 0.075$, $t_2 = 0.315$, $t_3 = 1.275$, $t_4 = 3.195$, $t_5 = 8.315$, $t_6 = 10.875$, $t_7 = 11.515$

The results of the calculations are shown in Fig. 121. The parameters of a model are the same as in Fig. 120 (except, obviously, the form of $F(p)$). Initial data: $p_0(x))2$, $0 \leq x \leq 1$, i.e. at a moment $t_0 = 0$ the distribution of power is completely in the area $p > p_0$ (where $F(p) < 0$).

The evolution of power distribution proceeds via a scenario inverse to the case 1: the solution for a rather large value of t turns to zero for all $0 \leq x \leq 1$, passing the band $p_{cr} < p < p^0$, where $F(p) >$). Initially the level of power decreases, and the majority of the power profile is in the legal area $p_1 < p < p_2$. Then the part of higher links slightly falls into the area $p < p_{cr}$ (line *5*), however it is enough for other links of the hierarchy to move by time into this area and for there to be an irreversible decrease of a solution $p(x, t)$ up to zero at all $0 \leq x \leq 1$; as in the case 1, the mechanisms of redistribution of power are too strong, their decrease leads to a normal evolution towards a stationary solution.

III. *Fatigue of society and a "sleeping" trap.* For the parameters of the system "power–society" given in subsections I and II, the evolution of power distribution is predetermined in the sense that for any initial $p_0(x)$ the above described scenario will be realized (besides quantitative differences, the qualitative behavior of function $p(x, t)$ is the same).

A more complicated and diverse evolution can be realized in cases where the characteristics of the system vary in time.

Fig. 122 represents the results of calculations of a model with the same function $F(p)$, as for calculations in Fig. 120, but with $\kappa_0 = 10^{-2}$ (in this case the stationary solution for a legal system slightly enters the area $p > p_2$, but in such a way, that there is a noticeable gap between it and area $p > p_{cr}$). The

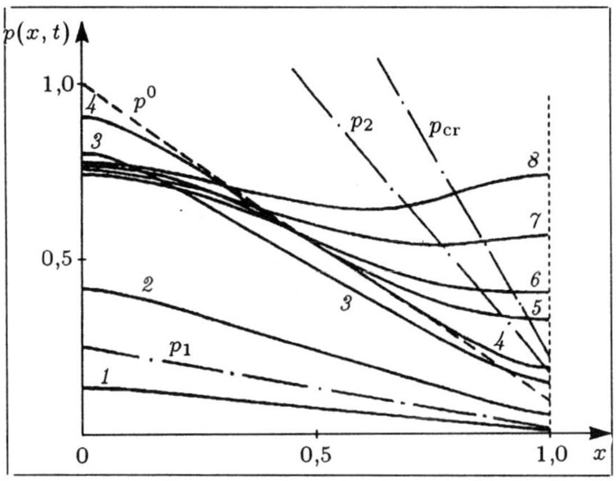

Fig.122. $t_1 = 0.155$, $t_2 = 0.63$, $t_3 = 2.555$, $t_4 = 7.035$, $t_5 = 21.115$, $t_6 = 55.675$, $t_7 = 99.835$

second difference is that in a moment $t = t_{cr} = 8$ both positive and negative parts of the social response decreases by 10 times, though qualitatively, as functions p, they have the previous form (line *1* at $p < p_2$ and line *2* in Fig. 119).

Such variation of $F(p)$ in time can be interpreted as a fatigue of society. The evolution of function $p(x, t)$ is as follows:

a) the solution grows and enters the area $p_1 < p < p_2$, as in Fig. 120 (curves *1*, *2*);

b) the solution during a certain time becomes close to the stationary one (curves *3*, *4*), which as distinct from the scenario I, exists at given κ_0;

c) the decrease of the response of society at a moment $t = t_{cr}$ leads to a spatial "smoothing" of power distribution and entering part of the lowest levels into the area $p > p_{cr}$ – curve *5* (the solution would remain stationary at former $F(p)$);

d) the trap connected with the form of $F(p, x, t)$, "awakes", and the further solution grows infinitely at all $0 \leq x \leq 1$ (curves *6*, etc.).

IV. *Activation of the society and the "reviving" of the stationary state.* As distinct from the previous calculations, in instants $6 < t < 10$ the amplitude of reaction will increase by 10 times (so that p_{cr} is increased as well). Such behavior of $F(p)$ can be interpreted as an activation of society.

The solution (Fig. 123) in the beginning behaves as in Fig. 122, coming to the area $p > p_{cr}$ (curves *3*, *4*). At a given value of $\kappa_0 = 10^{-1}$ and not varying in time function $F(p)$ the function $p(x, t)$ would grow infinitely (as in Fig. 122). However, the growth of the amplitude of $F(p)$ at $t > 6$ leads to the appearance of a stationary solution (curve *7*), to which the distribution of power by time (curves *5*, *6*) tends, b by avoiding the trap.

Certainly, the scenarios I–IV and their political interpretation are rather conditional (already because of the fact that at noticeable increase or de-

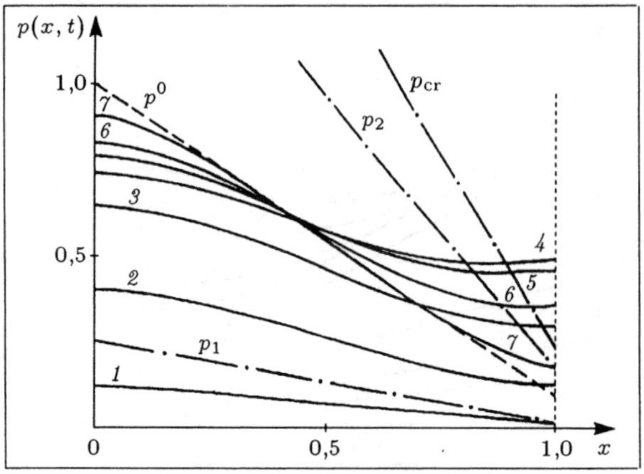

Fig.123. $t_1 = 0.15$, $t_2 = 0.75$, $t_3 = 2.35$, $t_4 = 5.55$, $t_5 = 7.15$, $t_6 = 7.55$, $t_7 = 19.90$

crease of power, the model, strictly speaking, is inapplicable, since the assumption about the legality of the system is violated).

However they demonstrate the potential diversity of possible behaviors of power distribution in a hierarchy included in a model, which contains various direct connections and feedbacks, nonlinearity and spatial-temporal variations of characteristics of the object, and also indicate the possibility of meaningful explanation of evolution of the investigated system.

For example, the anarchic trap can be attributed to Russian events of 1917 and in the of late 1980s – early 1990s (to a lesser degree). Indeed, this scenario describes well "the fall" of the supreme level of power (culminating in the resignation of the Tzar and the general secretary of the CPSU) under strongest pressure from the civil society and at "irresponsible" actions of the hierarchy with a consequent decrease and even "disappearance" of power influence of the rest of the hierarchy. The interpretation is only meaningful in the initial stage of the process (the general tendency is to "anarchy"). It cannot be continued further, since the "old" imperious structure has merely ceased to exist.

One more example is the political development of Russia in the last third of 1993. It corresponds to the scenario whereby power returns to its legal framework. Having formally left the boundaries (not precisely determined) of a legal field the hierarchy has at the same time ensured the civil society a possibility of amplifying the expression of the reaction by means of elections to the Parliament. This enabled the society to restore the power structure into the legal area, with an updated framework being established by its direct participation via a constitutional referendum.

Let us stress that the mathematical models of hardly formalizable objects always contain poorly or incompletely known behavioral characteristics of living beings. Therefore they cannot fulfill the conditions of adequacy and accuracy which are characteristic to the modeling of technological or natural scientific problems. However also in this highly complex area, the prognosis and decisions are essentially based on the use (almost without realizing it) of certain models and methods of modeling, often primitive ones. Therefore the broad application of the methodology of mathematical modeling and computing experiments seems to be inevitable even in this area of human activity.

Bibliography for Chapter VI: [4, 10, 13, 14, 19, 23, 27, 28, 32, 35, 36, 39, 41, 48, 49, 52–54, 58, 66, 81, 82].

REFERENCES

1. *Abramov A.P., Ivanilov Ju.P.* Physics and Mathematical Economy. Moscow: Znanie (1991) p. 32. (in Russian).
2. *Adjutov M.M., Klokov Ju.A., Mikhailov A.P.* Self-similar Thermal Structures with shortening half-width / Differential Equations. (1983) V. 19, No. 7. p. 1107–1114. (in Russian).
3. *Ajzerman M.A.* Classical Mechanics. Moscow: Nauka (1980) p. 368. (in Russian).
4. *Akhromeeva T.S., Kurdjumov S.P., Malinetsky G.G., Samarskii A.A.* Non-Stationary Structures and Diffusion Chaos. Moscow: Nauka (1992) p. 542. (in Russian).
5. *Aleksandrov V.V., Arhipov P.L., Parkhomenko V.P., Stenchikov G.L.* Global Model of a System Ocean-Atmosphere and the Study of its Sensitivity for Modification of Concentration of CO_2 / Izvestija AN USSR, Physics of Atmosphere and Ocean. (1983) V. 19, No. 5. p. 451–458. (in Russian).
6. *Amel'kin V.V.* Differential Equations in Applications. Moscow: Nauka (1987) p. 160. (in Russian).
7. *Anufrieva I.A., Mikhailov A.P.* Infinite Solutions of the Quasilinear Equations of Transfer. Preprint / IPM AN USSR. Moscow (1985) No. 34. p. 29. (in Russian).
8. *Arsen'ev A.A., Samarskii A.A.* What means Mathematical Physics? Moscow: Znanie (1983) p. 64. (in Russian).
9. *Ashmanov S.A.* Introduction to Mathematical Economy. Moscow: Nauka (1984) p. 296. (in Russian).
10. *Avula X.J.R.* Mathematical Modeling / Encyclopedia of Physical Science. (1987) V.7. p. 719–728.
11. *Baranov V.B., Krasnobaev N.V.* The Hydrodynamic Theory of Space Plasma. Moscow: Nauka (1977) p. 336. (in Russian).
12. *Barenblatt G.I.* Similarity, Self-similarity, Intermediate Asymptotics. Leningrad: Gidrometizdat (1982) p. 208. (in Russian).
13. *Bazykin A.D.* Mathematical Biophysics of Interacting Populations. Moscow: Nauka (1985) p. 182. (in Russian).
14. *Belotserkovsky O.M.* Numerical Modeling in Mechanics of Continuous Media. Moscow: Nauka (1994) p. 442 (in Russian).
15. *Bender P.* An Introduction to Mathematical Modeling. N.Y.: Wiley (1978).

16. *Bochkov M.V., Lovachev L.A., Chetverushkin B.N.* Chemical Kinetics of Formation of NO at Combustion of Methane in an Air / Mathematical Modeling. (1992) V. 4, No. 9. p. 3–36. (in Russian).
17. *Budak B.M., Samarskii A.A., Tikhonov A.N.* Collection of Problems in Mathematical Physics. Moscow: Nauka (1980) p. 686. (in Russian).
18. *Cross M., Moscardini A.O.* Learning the Art of Mathematical Modeling. N.Y.: Wiley (1985) p. 154.
19. *Demidov M.A., Mikhailov A.P.* Effects of Localization and Formation of Structures at Compression of a Finite Mass of Gas with Blow-up / PMM. (1986) V. 50, No. 1. p. 119–127. (in Russian).
20. *Dorodnitsyn A.A.* Computer Science: Subject and Problems / Cybernetics. Origins of Informatics. Moscow: Nauka (1996). (in Russian).
21. *Dorodnitsyn V.A., Elenin G.G.* Symmetry in Solution of Equations of Mathematical Physics. Moscow: Znanie (1984) p. 64. (in Russian).
22. *Dym C.L., lvey E.S.* Principles of Mathematical Modeling. N.Y.: Academic Press (1980) p. 256.
23. *Elizarova T.G., Pavlov A.N., Chetverushkin B.N.* Application of Kinetic Algorithm for Calculation of Gas Dynamical Flows / Differential equation. (1985) V. 21, No. 7. p. 1179–1185. (in Russian).
24. *Elizarova T.G., Pavlov A.N., Chetverushkin B.N.* Use of Quasihydrodynamic System for Calculation of a Gas Flow around a Body with a Needle / DAN USSR. (1987) V. 297, No. 2. p. 327–331. (in Russian).
25. *Elizarova T.G., Chetverushkin B.N.* Kinetic Algorithms for Calculation of Hydrodynamic Flow / ZHVM and MF. (1985) V. 25, No. 10. p. 1526–1533. (in Russian).
26. *Fedorenko R.P.* Introduction to Computing Physics. Moscow; MFTI Press (1994). p. 526. (in Russian).
27. *Gamaly E.G., Demchenko N.N., Lebo I.G. et al.* Theoretical Study of Stability of Compression of Targets with Thin Shells Irradiated by Lasers with Energy of Pulse of the Order 1 kJ / Quantum Electronics. (1988). V. 15, No. 8. p. 1622–1632. (in Russian).
28. *Gantmaher F.R.* Lectures on Analytical Mechanics. Moscow: Nauka (1968) p. 300. (in Russian).
29. *Godunov S.K., Rjaben'ky V.S.* Difference Schemes (Introduction to the Theory). Moscow: Nauka (1977) p. 440. (in Russian).
30. *Goloviznin V.M., Samarskii A.A., Favorsky A.P.* The Variational Approach to the Construction of Finite-Difference Mathematical Models in Hydrodynamics / DAN USSR. (1977) V. 235, No. 6. p. 1285–1288. (in Russian).
31. *Gol'din V.Ja., Anistratov D.Ju.* A Reactor on Fast Neutrons in a Self–Regulated Neutron– Nuclear Regime / Mathematical Modeling. (1995) V. 7, No. 10. p. 21–32. (in Russian).
32. *Hoerner S., von.* Population Explosion and Interstellar Expansion / Journal of British Interplanetary Society. (1975) V. 28, No. II. p. 691–712.
33. *Ibragimov N.H.* Groups of Transformations in Mathematical Physics. Moscow: Nauka, (1983) p. 280. (in Russian).
34. *Ivanova T.S., Ruzmajkin A.A.* Method of Solution of the Problem of Magnetohydrodynamic Dynamo of the Sun / ZHVM and MF. (1976) V. 16, No. 4. p. 956–968. (in Russian).

35. *Ivanova T.S., Ruzmajkin A.A.* Nonlinear Magnetohydrodynamic Model of Dynamo of the Sun / Astronomical Journal. (1977) V. 54. p. 846–858. (in Russian).
36. *Jacoby S.L.S, Kowalik J.S.* Mathematical Modeling with Computers, Englewood Cliffs, N.J.: Prentice Hall, Inc., (1980) p. 292.
37. *Kalitkin N.N., Mikhailov A.P.* Ideal Solution of a Problem of Offset of Mutual Debts / Mathematical Modeling. (1995) V. 7, No. 6. p. 111–117. (in Russian).
38. *Klokov Ju.A., Mikhailov A.P.* About one Boundary Problem of Neumann for the Integro–Differential equation / The Differential Equations. (1996) V. 32, No. 8, p. 1110–1113. (in Russian).
39. *Korobejnikov V.P.* Mathematical Modeling of Disastrous Natural Phenomena. Moscow: Znanie (1986) p.48. (in Russian).
40. *Krasnoshchekov P.S., Petrov A.A.* Principles of Construction of Models. Moscow: Moscow Univ. Press (1983) p. 264. (in Russian).
41. *Kriksin Ju.A., Samarskaja E.A., Tishkin V.F.* Balance Model of Propagation of an Admixture in Plane Filtering Flow / Mathematical Modeling. (1993) V. 5, No. 6. p. 69–84. (in Russian).
42. *Landau L.D., Lifshits E.M.* Mechanics. Moscow: Nauka (1973) p. 207. (in Russian).
43. *Landau L.D., Lifshits E.M.* Mechanics of Continuous Media. Moscow: Gostekhizdat (1953) p. 788. (in Russian).
44. *Lebo I.G., Rozanov V.B., Tishkin V.F. et al.* Numerical Simulation of Ruchtmyer-Meshkov Instability / Eds. R. Young, J. Glinun, B. Boston / Proc. of the Fifth Int. Workshop on Compressible Turbulent Mixing. N.Y.: Word Scientific (1995) p. 346–356.
45. *Lehman R.S.* Computer, Simulation and Modeling: An Introduction. N.Y.: Wiley (1977).
46. *Lejbenzon L.S.* Motion of Natural Fluids and Gases in a Porous Medium. Moscow–Leningrad: Gostekhizdat (1947) p. 244. (in Russian).
47. *Lojtsjansky L.G.* Mechanics of Fluids and Gas. Moscow: Gostekhizdat (1950) p. 676. (in Russian).
48. *Marchuk G.I.* Methods in Numerical Mathematics. Moscow: Nauka (1989) p. 608. (in Russian).
49. Mathematical Modeling /Eds. J.G.Andrews, R.R.McLone. London: Butterworths (1976).
50. Mathematical Modeling / Eds.J.Andrews, R.McLone; translation from English Moscow: Mir (1979) p. 278. (in Russian).
51. *Mikhailov A.P.* Mathematical Modeling of Distribution of Power in Hierarchic Structures / Mathematical Modeling. (1994) V. 6. No. 6. p. 108–138. (in Russian).
52. *Mikhailov A.P.* Modeling of Evolution of Distribution of Power in State Hierarchies / Comm. of Fund "Russian Political Center"/ (1996) No. 2. p. 26–39. (in Russian).
53. *Moiseev N.N.* Mathematical Problems of System Analysis. Moscow: Nauka (1981) p. 488. (in Russian).

54. *Ovsjannikov L.V.* Group Analysis of Differential Equations. Moscow: Nauka (1978) p. 400. (in Russian).
55. *Parkhomenko V.P., Stenchikov G.L.* Mathematical Modeling of Climate. Moscow: Znanie (1986) p. 32. (in Russian).
56. *Petrov A.A.* Economy. Models. Computing Experiment. Moscow: Nauka (1996). (in Russian).
57. *Petrov A.A., Pospelov I.G., Shananin A.A.* Experience of Mathematical Modeling of Economy. Moscow: Energoizdat, (1996) p. 544. (in Russian).
58. *Polubarinova-Kochina P.Ja.* The Theory of Motion of Ground Waters. Moscow: Nauka (1977). (in Russian).
59. *Pontrjagin L.S., Boltjansky V.G., Gamkrelidze R.V., Mishchenko E.F.* The Mathematical Theory of Optimal Processes. Moscow: Gostekhizdat (1961) p. 392. (in Russian).
60. *Popov Ju.P., Samarskii A.A.* Computing Experiment. Moscow: Znanie (1983) p. 64. (in Russian).
61. *Rapoport A.* Mathematical Models in the Social and Behavioral Sciences. N.Y.: Wiley (1983).
62. *Riznichenko G.Ju., Rubin A.B.* Mathematical Models of Biological Reproduction Processes. Moscow: Moscow Univ. Press (1993) p. 300. (in Russian).
63. *Saaty T.L., Alexander J.M.* Thinking with Models: Mathematical Models in the Physical, Biological and Social Sciences. N.Y.; Pergamon Press (1981).
64. *Samarskii A.A.* Introduction to Numerical Methods. Moscow: Nauka (1982) p. 272. (in Russian).
65. *Samarskii A.A.* Mathematical Modeling and Computing Experiment / Vestnik AN USSR. (1979) No. 5. p. 38–49. (in Russian).
66. *Samarskii A.A.* Theory of Difference Schemes. Moscow: Nauka (1989) p. 616. (in Russian).
67. *Samarskii A.A., Andreev V.B.* Difference Methods for Elliptical Equations. Moscow: Nauka (1976) p. 352. (in Russian).
68. *Samarskii A.A., Galaktionov V.A., Kurdjumov S.P., Mikhailov A.P.* Blow-up in Problems for Quasilinear Parabolic Equations. Moscow: Nauka (1987) p. 478. (in Russian).
69. *Samarskii A.A., Galaktionov V.A., Kurdyumov S.P., Mikhailov A.P.* Blow-up in Quasilinear Parabolic Equations. Berlin: Walter de Gruyter (1995) p. 534.
70. *Samarskii A.A., Gulin A.V.* Stability of Difference Schemes. Moscow: Nauka (1973) p. 416. (in Russian).
71. *Samarskii A.A., Koldoba A.V., Poveshchenko Ju.A. et al.* Difference Schemes on Irregular Grids. Minsk: ZAO "Kritery" (1996) p. 274. (in Russian).
72. *Samarskii A.A., Mikhailov A.P.* Computers and Life (Mathematical Modeling). Moscow: Pedagogika (1987) p. 128. (in Russian).
73. *Samarskii A.A., Nikolaev E.S.* Methods of Solution of Grid Equations. Moscow: Nauka (1978) p. 592. (in Russian).
74. *Samarskii A.A., Nikolaev E.S.* Numerical Methods for Grid Equations. V. 1, 2. Basel: Birkhauser Verlag (1981) p. 242, p. 502.

75. *Samarskii A.A., Popov Ju.P.* Difference Schemes of Gas Dynamics. Moscow: Nauka, 1980, p. 352. (in Russian).
76. *Samarskii A.A., Vabishchevich P.N.* Computational Heat Transfer. V. 1, 2. N.Y.: Wiley (1995) p. 406, p. 422.
77. *Sedov L.I.* Mechanics of Continuous Medium. v.1 Moscow: Nauka (1973) p. 536. (in Russian).
78. *Sedov L.I.* Methods of Similarity and Dimensionality in Mechanics. Moscow: Nauka (1981) p. 448. (in Russian).
79. *Sidorov A.F., Shapeev V.P., Janenko N.N.* Method of Differential Connections and its Application in Gas Dynamics. Novosibirsk: Nauka (1984) p. 272. (in Russian).
80. *Silin V.P.* Introduction to the Kinetic Theory of Gases. Moscow: Nauka (1971) p. 332. (in Russian).
81. *Tikhonov A.N., Kostomarov D.P.* The Introductory Lectures on Applied Mathematics. Moscow: Nauka (1984) p. 190. (in Russian).
82. *Tikhonov A.N., Samarskii A.A.* Equations of Mathematical Physics. Moscow: Nauka (1972) p.736. (in Russian).
83. *Vabishchevich P.N.* Numerical Modeling. Moscow: Moscow Univ. Press (1993) p. 152. (in Russian).
84. *Volosevich P.P., Dar'in N.A., Levanov E.I., Shirtladze N.M.* Problem of the Piston in Gas with Sources and Sinks (Self–Similar Solutions). Tbilisi: Tbilisi Univ. Press (1986) p. 238. (in Russian).
85. *Volosevich P.P., Levanov E.I.* Self–Similar Solutions of Problems of Gas Dynamics with Thermal Conduction. Moscow: MFTI Press (1996) p. 212. (in Russian).
86. *Zel'dovich Ja.B., Rajzer Ju.P.* Physics of Shock Waves and High–Temperature Hydrodynamic Phenomena. Moscow: Nauka (1966) p. 688. (in Russian).
87. *Zmitrenko N.V., Kurdjumov S.P., Mikhailov A.P.* The Theory of Blowing-up Regimes in Compressing Media / Reviews in Science and Technology. Modern Problems of Mathematics. Advanced Achievements. V. 28. Moscow: VINITI (1986) p. 3–94. (in Russian).
88. *Zmitrenko N.V., Kurdjumov S.P., Mikhailov A.P., Samarskii A.A.* Localization of Thermonuclear Combustion in Plasma with Electronic Thermal Conduction / JETP Lett. (1977) V. 26, No. 9. p. 620–624. (in Russian).
89. *Zmitrenko N.V., Mikhailov A.P.* Inertia of Heat. Moscow: Znanie (1982) p. 64. (in Russian).

Index